HANDBOOK
OF
ENERGY
ENGINEERING

Seventh Edition

HANDBOOK
OF
ENERGY
ENGINEERING

Seventh Edition

Albert Thumann, P.E., C.E.M.
D. Paul Mehta, Ph.D.

THE FAIRMONT PRESS, INC.

CRC Press
Taylor & Francis Group

Library of Congress Cataloging-in-Publication Data

Thumann, Albert
    Handbook of energy engineering/Albert Thumann, D. Paul Mehta--7th ed.
    p. cm.
    Includes bibliographical references and index.
    ISBN 0-88173-695-3 (alk. paper) -- ISBN 0-88173-696-1 (electronic) -- ISBN 978-1-4665-6161-8 (alk. paper) 1. Power resources--Handbooks, manuals, etc. 2. Power (Mechanics)--Handbooks, manuals, etc. I. Mehta, D. Paul, 1940- II. Title.

    TJ163.235.T48 2012
    621.042--dc23

                                                                    2012015773

1006988333

Published by The Fairmont Press, Inc.
700 Indian Trail
Lilburn, GA 30047
tel: 770-925-9388; fax: 770-381-9865
http://www.fairmontpress.com

Distributed by Taylor & Francis Ltd.
6000 Broken Sound Parkway NW, Suite 300
Boca Raton, FL 33487, USA
E-mail: orders@crcpress.com

Distributed by Taylor & Francis Ltd.
23-25 Blades Court
Deodar Road
London SW15 2NU, UK
E-mail: uk.tandf@thomsonpublishingservices.co.uk

Printed in the United States of America
10 9 8 7 6 5 4 3 2 1

ISBN 0-88173-695-3 (The Fairmont Press, Inc.)
ISBN 978-1-4665-6161-8 (Taylor & Francis Ltd.)

While every effort is made to provide dependable information, the publisher, authors, and editors cannot be held responsible for any errors or omissions.

Contents

Preface

The *Handbook of Energy Engineering*, now in its 7th edition, has served as an invaluable text for more than 22 years. Few books have shaped the energy engineering industry as much as this remarkable reference book. It is used by colleges and universities around the world and has served as an indispensable reference for professionals studying for the Certified Energy Manager (CEM) examination.

When the *Handbook of Energy Engineering* was first published in October of 1989, the field of energy engineering was in its infancy. Today, the energy engineering profession is poised for tremendous growth. Energy Engineers are at the forefront, helping their organizations gain the competitive edge by reducing operating costs. Energy engineers are essential as we enter the new world revolution to be "green" and play a vital role in reducing the risk of climate change.

The *Handbook of Energy Engineering* is more important today than ever before.

Wayne C. Turner
Energy Management Consultant
Editor-In-Chief, *Energy Engineering* Journal

1
CODES, STANDARDS, & LEGISLATION

INTRODUCTION

This chapter presents a historical perspective on key codes, standards, and regulations that have impacted energy policy and are still playing a major role in shaping energy usage. The context of past standards and legislation must be understood in order to properly implement the proper systems and be able to impact future codes. The Energy Policy Act, for example, has created an environment for retail competition. Electric utilities will drastically change the way they operate in order to provide power and lowest cost. This in turn will drastically reduce utility-sponsored incentive and rebate programs, which have influenced energy conservation adoption. This chapter attempts to cover a majority of the material that currently impacts the energy related industries, with relationship to their respective initial writing.

The main difference between standards, codes, and regulations is an increasing level of enforceability of the various design parameters. A group of interested parties (vendors, trade organizations, engineers, designers, citizens, etc.) may develop a standard in order to assure minimum levels of performance. The standard acts as a suggestion to those parties involved, but it is not enforceable until codified by a governing body (local or state agency), which makes the standard a code. Not meeting this code may prevent continuance of a building permit or result in the ultimate stoppage of work. Once the federal government makes the code part of the federal code, it becomes a regulation. Often this progression involves equipment development and commercialization prior to codification in order to assure that the standards are attainable.

THE AMERICAN RECOVERY AND REINVESTMENT
ACT OF 2009
The American Recovery and Reinvestment Act of 2009 funds:

Energy Efficiency
- $3.1 billion for state energy programs, which will encourage states to improve building energy improvement and retrofits.
- $5 billion for weatherization of low-income housing.
- $4.5 billion for green federal buildings.
- $3.6 billion for the Department of Defense for energy efficient projects and facilities upgrades
- $3.2 billion for energy efficiency and conservation block grants to help state and local governments implement energy efficiency
- $500 million for job training and programs to prepare individuals for careers in energy efficiency and renewable energy

Solar Energy—ARRA Contains Provisions for:
- *Renewable Energy Grant Program*: Offers Department of Energy grants equal to 30% of the cost of solar projects started in the next two years, including large-scale utility projects. This is a critical alternative to solar tax credits that are not functioning as Congress intended in the current economic climate.

- *Loan Guarantee Program*: Creates a new, streamlined loan guarantee program to support financing of renewable energy systems, including solar energy technology.

- *Manufacturing Investment Credit*: Creates a 30% investment tax credit for facilities engaged in the manufacture of renewable energy property or equipment.

Smart Grid/Advanced Battery/Energy Efficiency
- $34 billion will be provided for such initiatives as a new, smart power grid, advanced battery technology, and energy efficiency measures. According to the fact sheet, 500,000 energy jobs will be created.

THE ENERGY INDEPENDENCE
AND SECURITY ACT OF 2007 (H.R.6)

The Energy Independence and Security Act of 2007 (H.R.6) was enacted into law December 19, 2007. Key provisions of EISA 2007 are summarized below.

Title I Energy Security through Improved
Vehicle Fuel Economy
- Corporate average fuel economy (CAFE). The law sets a target of 35 miles per gallon for the combined fleet of cars and light trucks by 2020.
- The law establishes a loan guarantee program for advanced battery development, a grant program for plug-in hybrid vehicles, incentives for purchasing heavy-duty hybrid vehicles for fleets, and credits for various electric vehicles.

Title II Energy Security Through Increased
Production of Biofuels
- The law increases the renewable fuels standard (RFS), which sets annual requirements for the quantity of renewable fuels produced and used in motor vehicles. RFS requires 9 billion gallons of renewable fuels in 2008, increasing to 36 billion gallons in 2022.

Title III Energy Savings Through Improved
Standards for Appliances and Lighting
- The law establishes new efficiency standards for motors, external power supplies, residential clothes washers, dishwashers, dehumidifiers, refrigerators, refrigerator freezers, and residential boilers.
- The law contains a set of national standards for light bulbs. The first part of the standard would increase energy efficiency of light bulbs 30% and phase out most common types of incandescent light bulb by 2012-2014.
- Requires the federal government to substitute energy efficient lighting for incandescent bulbs.

Title IV Energy Savings in Buildings and Industry
- The law increases funding for the Department of Energy's weatherization program, providing $3.75 billion over five years.
- The law encourages the development of more energy efficient "green" commercial buildings. The law creates an Office of Commercial High Performance Green Buildings at the Department of Energy.
- A national goal is set to achieve zero-net energy use for new com-

mercial buildings built after 2025. A further goal is to retrofit all pre-construction 2025 buildings to zero-net energy by 2050.

- Requires that total energy use in federal buildings relative to the 2005 level be reduced 30% by 2015.
- Requires federal facilities to conduct a comprehensive energy and water evaluation for each facility at least once every four years.
- Requires new federal buildings and major renovations to reduce fossil fuel energy use 55% relative to 2003 level by 2010 and be eliminated (100 percent reduction) by 2030.
- Requires that each federal agency ensure that major replacements of installed equipment (such as heating and cooling systems) or renovation or expansion of existing space employ the most energy efficient designs, systems, equipment, and controls that are life cycle cost effective. For the purpose of calculating life cycle cost calculations, the time period will increase from 25 years in the prior law to 40 years.
- Directs the Department of Energy to conduct research to develop and demonstrate new process technologies and operating practices to significantly improve the energy efficiency of equipment and processes used by energy-intensive industries.
- Directs the Environmental Protection Agency to establish a recoverable waste energy inventory program. The program must include an ongoing survey of all major industry and large commercial combustion services in the United States.
- Includes new incentives to promote new industrial energy efficiency through the conversion of waste heat into electricity.
- Creates a grant program for healthy high performance schools that aims to encourage states, local governments, and school systems to build green schools.
- Creates a program of grants and loans to support energy efficiency and energy sustainability projects at public institutions.

Title V Energy Savings in Government and Public Institutions

- Promotes energy savings performance contracting in the federal government and provides flexible financing and training of federal contract officers.
- Promotes the purchase of energy efficient products and procurement of alternative fuels with lower carbon emissions for the fed-

eral government.

- Reauthorizes state energy grants for renewable energy and energy efficiency technologies through 2012.
- Establishes an energy and environmental block grant program to be used for seed money for innovative local best practices.

Title VI Alternative Research and Development
- Authorizes research and development to expand the use of geothermal energy.
- Improves the cost and effectiveness of thermal energy storage technologies that could improve the operation of concentrating solar power electric generation plants.

- Promotes research and development of technologies that produce electricity from waves, tides, currents, and ocean thermal differences.

- Authorizes a development program on energy storage systems for electric drive vehicles, stationary applications, and electricity transmission and distribution.

Title VII Carbon Capture and Carbon Sequestration
- Provides grants to demonstrate technologies to capture carbon dioxide from industrial sources.
- Authorizes a nationwide assessment of geological formations capable of sequestering carbon dioxide underground.

Title VIII Improved Management of Energy Policy
- Creates a 50% matching grants program for constructing small renewable energy projects that will have an electrical generation capacity less than 15 megawatts.
- Prohibits crude oil and petroleum product wholesalers from using any technique to manipulate the market or provide false information.

Title IX International Energy Programs
- Promotes US exports in clean, efficient technologies to India, China, and other developing countries.
- Authorizes US Agency for International Development (USAID) to increase funding to promote clean energy technologies in developing countries.

Title X Green Jobs
- Creates an energy efficiency and renewable energy worker training program for "green collar" jobs.
- Provides training opportunities for individuals in the energy field who need to update their skills.

Title XI Energy Transportation and Infrastructure
- Establishes an office of climate change and environment to coordinate and implement strategies to reduce transportation-related energy use.

Title XII Small Business Energy Programs
- Loans, grants, and debentures are established to help small businesses develop, invest in, and purchase energy efficient equipment and technologies.

Title XIII Smart Grid
- Promotes a "smart electric grid" to modernize and strengthen the reliability and energy efficiency of the electricity supply. The term "smart grid" refers to a distribution system that allows for flow of information from a customer's meter in two directions: both inside the house to thermostats, appliances, and other devices, and from the house back to the utility.

THE ENERGY POLICY ACT OF 2005

The first major piece of national energy legislation since the Energy Policy Act of 1992, EPAct 2005 was signed by President George W. Bush on August 8, 2005 and became effective January 1, 2006. The major thrust of EPAct 2005 is energy production. However, there are many important sections of EPAct 2005 that do help promote energy efficiency and energy conservation, as well as provide tax incentives to encourage participation in the private sector. There are also some significant impacts on federal energy management. Highlights are described below:

EPACT 2005 Highlights
Federal Energy Reduction – Existing Buildings
- Baseline changed to year 2003.
- An annual energy reduction goal of 2% is in place from fiscal year

2006 to fiscal year 2015, for a total energy reduction of 20% by year 2015.

Federal Facility Metering
- Electric metering is required in all federal building by the year 2012.

Energy Efficient Products
- Energy efficient specifications are required in procurement bids and evaluations.
- Energy efficient products to be listed in federal catalogs include Energy Star and FEMP products recommended by GSA and Defense Logistics Agency.

Federal—Energy Savings Performance Contracting (ESPC)
- ESPC authority extended through September 30, 2016.
- No caps, limitations or restrictions.
- Impacts all agencies.

Federal—Energy Efficient New Buildings
- New federal buildings will incorporate life cycle costing.
- New federal buildings are required to be designed 30% below ASHRAE standard or the International Energy Code (if life-cycle cost effective). Agencies must identify those that meet or exceed the standard.
- Incorporate sustainable design principles.

Federal Building Renewable Energy: Section 203
- Renewable electricity consumption by the federal government cannot be less than: 3% from fiscal year 2007-2009, 5% from fiscal year 2010-2012, and 7.5% from fiscal year 2013 and beyond.
- The goal for photovoltaic energy is to have 20,000 solar energy systems installed in federal buildings by the year 2012.
- Double credits are earned for renewables produced on the site, on federal lands and used at a federal facility, or produced on Native American lands.
- The goals are based on technical and economic feasibility.

Commercial Buildings
- A tax deduction of up to $1.80 per square foot for energy efficient upgrades to HVAC, lighting, hot water systems, and the building

envelope—and 60 cents per square foot for building subsystems that reduce annual power consumption 50 percent compared to the ASHRAE standard.

Residential Buildings

- 30% tax credit for purchase of qualifying residential solar water heating, photovoltaic equipment and fuel cell property. The maximum credit is $2000 (for solar equipment) and $500 for each kilowatt of capacity(fuel cell). The credit applies for property placed in service after 2005 and before 2008.
- Provides a 10% investment tax credit for expenditures with respect to improvements to building envelope.
- Allows tax credits for purchases of high efficiency HVAC systems; advanced main air circulating fans; natural gas, propane or fuel oil furnaces, or hot water boilers; and other energy efficient property. Credit applies to property placed in service after December 31, 2005 and prior to January 1, 2008. The lifetime maximum credit per tax payer is $500.
- Provides $1000 tax credit to eligible contractor for construction of a qualified new energy-efficient home. Tax credit applies to manufactured homes meeting Energy Star standards.

Appliances

- Energy Star Dishwashers (2007)—up to $100 tax credit.
- $75 tax credit for refrigerators that save 15 percent energy; $125 tax credit for refrigerators that save 20 percent or $175 that save 25 percent, based on 2001 standards.
- $100 tax credit for Energy Star Clothes washers (2007).

Fuel Cells, Microturbine Power Plants and Solar Energy

- Provides a 30% tax credit for purchase of qualified fuel cell power plants for businesses.
- Provides a 10 %tax credit for purchase of qualifying stationary micro turbines.
- Provides a 30% tax credit for purchase of qualifying of solar energy property. Tax credits apply to property placed in service after December 31, 2005 and before January 1, 2008.

Transportation

- Provides tax credits up to $3400 for purchase of hybrid and lean diesel vehicles (capped at 60,000 vehicles per manufacturer for 2006-2010).

Electricity
* Repeals Public Utility Holding Company Act (PUHCA).
* Requires mandatory reliability standards to make the electric power grid more reliable against blackouts.
* Tax incentives to expand investments in electric transmission and generation.

Domestic Production
* Reforms to clarify oil and gas permitting process.
* Authorizes full funding for clean coal research initiative.
* Establishes a new renewable fuel standard that requires the annual use of 7.5 billion gallons of ethanol and biodiesel in the nation's fuel supply by 2012.

THE ENERGY POLICY ACT OF 1992

The Energy Policy Act of 1992 is substantial and its implementation is impacting electric power deregulation, building codes, and new energy efficient products. Sometimes policy makers do not see the extensive impact of their legislation. This comprehensive legislation is far-reaching and impacts energy conservation, power generation, and alternative fuel vehicles, as well as energy production. The federal as well as private sectors are impacted by this comprehensive energy act. Highlights of EPACT 1992 are described below.

Energy Efficiency Provisions
Buildings
* Requires states to establish minimum commercial building energy codes and to consider minimum residential codes based on current voluntary codes.

Utilities
* Requires states to consider new regulatory standards that would require utilities to undertake integrated resource planning, allow efficiency programs to be at least as profitable as new supply options, and encourage improvements in supply system efficiency.

Equipment Standards
- Establishes efficiency standards for commercial heating and air-conditioning equipment, electric motors, and lamps.
- Gives the private sector an opportunity to establish voluntary efficiency information/labeling programs for windows, office equipment and luminaries (or the Department of Energy will establish such programs).

Renewable Energy
- Establishes a program for providing federal support on a competitive basis for renewable energy technologies. Expands program to promote export of these renewable energy technologies to emerging markets in developing countries.

Alternative Fuels
- Gives Department of Energy authority to require a private and municipal alternative fuel fleet program, starting in 1998. Provides a federal alternative fuel fleet program with phased-in acquisition schedule; also provides state fleet program for large fleets in large cities.

Electric Vehicles
- Establishes comprehensive program for the research and development, infrastructure promotion, and vehicle demonstration for electric motor vehicles.

Electricity
- Removes obstacles to wholesale power competition in the Public Utilities Holding Company Act by allowing both utilities and non-utilities to form exempt wholesale generators without triggering the PUHCA restrictions.

NATIONAL ENERGY CONSERVATION POLICY ACT 1978

Public Utilities Regulatory Policies Act (PURPA)

This legislation was part of the 1978 National Energy Act and has had perhaps the most significant effect on the development of cogeneration and other forms of alternative energy production in the past decade. Certain provisions of PURPA also apply to the exchange of electric power between utilities and cogenerators. PURPA provides a number of benefits to those cogenerators who can become qualifying facilities (QFs) under this act. Specifically, PURPA:

- Requires utilities to purchase the power made available by cogenerations at reasonable buy-back rates. These rates are typically based on the utilities' cost.

- Guarantees the cogeneration or small power producer interconnection with the electric grid and the availability of backup service from the utility.

- Dictates that supplemental power requirements of cogeneration must be provided at a reasonable cost.

- Exempts cogenerators and small power producers from federal and state utility regulations and associated reporting requirements of these bodies.

To assure a facility the benefits of PURPA, a cogenerator must become a qualifying facility. To achieve qualifying status, a cogenerator must generate electricity and useful thermal energy from a single fuel source. In addition, a cogeneration facility must be less than 50% owned by an electric utility or an electric utility holding company. Finally, the plant must meet the minimum annual operating efficiency standard established by the Federal Energy Regulatory Commission (FERC) when using oil or natural gas as the principal fuel source. The standard is that the useful electric power output, plus one half of the useful thermal output, of the facility must be no less than 42.5% of the total oil or natural gas energy input. The minimum efficiency standard increases to 45% if the useful thermal energy is less than 15% of the total energy output of the plant.

Natural Gas Policy Act

The Natural Gas Policy Act created a deregulated natural gas market for natural gas, which was the major objective of this regulation. It provides for incremental pricing of higher cost natural gas to fluctuate with the cost of fuel oil. Cogenerators classified as qualifying facilities under PURPA are exempt from the incremental pricing schedule established for industrial customers.

Public Utility Holding Company Act (PUHCA) of 1935

The Public Utility Holding Company Act of 1935 authorized the Securities and Exchange Commission (SEC) to regulate certain utility "holding companies" and their subsidiaries in a wide range of corporate transactions.

The utility industry and would-be owners of utilities lobbied Congress heavily to repeal PUHCA, claiming that it was outdated. On August 8, 2005, the Energy Policy Act of 2005 passed both houses of Congress and was signed into law, repealing PUHCA—despite consumer, environmental, union, and credit-rating agency objections. The repeal became effective on February 8, 2006.

ISO 50001 ENERGY MANAGEMENT STANDARD

This new standard for energy management systems could impact up to 60 percent of global energy demand.

The International Organization for Standardization's (ISO) new standard, ISO 50001, provides a framework for industrial facilities seeking to manage their energy use, drawing on the success of established environmental and quality management standards. Based in Geneva, ISO is responsible for creating many thousands of standards around the globe. It aims to bridge the efforts of both private and public organizations to create a common language as well as standards.

Certification under the standard demonstrates to customers and suppliers how a company is tracking and reducing its energy use.

The international standard provides guidance on how to measure, document and report energy use; define roles and responsibilities within an organization; build energy awareness within a company; and assist in goal-setting for energy reduction. The standard provides insight about where new energy-saving technologies could be most effectively implemented.

CODES AND STANDARDS

Energy codes specify how buildings must be constructed or perform, and are written in a mandatory, enforceable language. State and local governments adopt and enforce energy codes for their jurisdictions. Energy standards describe how buildings should be constructed to save energy cost effectively. They are published by national organizations such as the American Society of Heating, Refrigerating, and Air

Conditioning Engineers (ASHRAE). They are not mandatory but serve as national recommendations, with some variation for regional climate. State and local governments frequently use energy standards as the technical basis for developing their energy codes. Some energy standards are written in a mandatory, enforceable language, making it easy for jurisdictions to incorporate the provisions of the energy standards directly into their laws or regulations. The requirement for the federal sector to use ASHRAE 90.1 and 90.2 as mandatory standards for all new federal buildings is specified in the Code of Federal Regulations—10 CFR 435.

Most states use the ASHRAE 90 standard as their basis for the energy component of their building codes. ASHRAE 90.1 is used for commercial buildings and ASHRAE 90.2 is used for residential buildings. Some states have quite comprehensive building codes (for example: California Title 24).

ASHRAE Standard 90.1
Energy efficient design for new buildings
• Sets minimum requirements for the energy-efficient design of new buildings so they will be constructed, operated, and maintained in a manner that minimizes the use of energy without constraining the building function and productivity of the occupants.
• ASHRAE 90.1 addresses building components and systems that affect energy usage.
• Sections 5-10 are the technical sections that specifically address components of the building envelope, HVAC systems and equipment, service water heating, power, lighting, and motors. Each technical section contains general requirements and mandatory provisions. Some sections also include prescriptive and performance requirements.

ASHRAE Standard 90.2
Energy efficient design for new low-rise residential buildings
When the Department of Energy determines that a revision would improve energy efficiency, each state has two years to review the energy provisions of its residential or commercial building code. For residential buildings, a state has the option of revising its residential code to meet or exceed the residential portion of ASHRAE 90.2. For commercial buildings, a state is required to update its commercial code to meet or exceed the provision of ASHRAE 90.1.

ASHRAE standards 90.1 and 90.2 are developed and revised through voluntary consensus and public hearing processes that are critical to widespread support for their adoption. Both standards are continually maintained by separate, standing, standards projects committees. Committee membership varies from 10 to 60 voting members. Committee membership includes representatives from many groups to ensure balance among all interest categories. After the committee proposes revisions to the standard, it undergoes public review and comment. When a majority of the parties substantially agree, the revised standard is submitted to the ASHRAE board of directors. This entire process can take anywhere from two to ten years to complete. ASHRAE Standards 90.1 and 90.2 are automatically revised and published every three years. Approved interim revisions are posted on the ASHRAE website (www.ashrae.org) and are included in the next published version.

The energy-cost budget method permits trade-offs between building systems (lighting and fenestration, for example) if the annual energy cost estimated for the proposed design does not exceed the annual energy cost of a base design that fulfills the prescriptive requirements. Using the energy-cost budget method approach requires simulation software that can analyze energy consumption in buildings and model the energy features in the proposed design. ASHRAE 90.1 sets minimum requirements for the simulation software; suitable programs include BLAST, eQUEST, and TRACE.

ASHRAE Standard 62

Indoor air quality (IAQ) is an emerging issue of concern to building managers, operators, and designers. Recent research has shown that indoor air is often less clean than outdoor air, and federal legislation has been proposed to establish programs to deal with this issue on a national level. This, like the asbestos issue, will have an impact on building design and operations. Americans today spend long hours inside buildings, and building operators, managers, and designers must be aware of potential IAQ problems and how they can be avoided.

IAQ problems, sometimes termed "sick building syndrome," have become an acknowledged health and comfort problem. Buildings are characterized as sick when a significant number of occupants complain of acute symptoms such as headache, eye, nose, and throat irritations, dizziness, nausea, sensitivity odors, and difficulty in concentrating. The complaints may become more clinically defined such that an occupant may develop an actual building-related illness that is believed to be related to IAQ problems.

The most effective means to deal with an IAQ problem is to remove or minimize the pollutant source, when feasible. If not, dilution and filtration may be effective.

The purpose of ASHRAE Standard 62 is to specify minimum ventilation rates and indoor air quality that will be acceptable to human occupants with the intention of minimizing the potential for health effects. ASHRAE defines acceptable indoor air quality as the air in which there are no known contaminants at harmful concentrations, as determined by cognizant authorities and with which a substantial majority (80% or more) of those exposed do not express dissatisfaction.

ASHRAE Standard 55

ASHRAE Standard 55 for thermal environmental conditions for human occupancy covers several environmental parameters, including temperature, radiation, humidity, and air movement. The standard specifies conditions in which 80% of the occupants will find the environment thermally acceptable. This applies to healthy people in normal indoor environments, for winter and summer conditions. Adjustment factors are described for various activity levels and clothing levels.

ASHRAE Standard 189.1

The purpose of this standard is to provide minimum requirements for the siting, design, construction, and plan for high performance, green buildings to:

a) Balance environmental responsibility, resource efficiency, occupant comfort and wellbeing, and community sensitivity.

b) Support the goal of development that meets the needs of the present without compromising the ability of future generations to meet their own needs.

The scope of this Standard provides minimum criteria that apply to the following elements of building projects:

• New buildings or new building portions and their systems
• New systems and equipment in existing buildings

The Standard addresses site sustainability, water use efficiency, energy efficiency, IEQ, and the building's impact on the atmosphere, materials, and resources.

This Standard does not apply to single-family houses, multi-family structures of 3 stories or fewer above grade, mobile homes, or modular homes. It also does not apply to buildings that do not use electricity, fossil fuel, or water.

This Standard shall not be used to circumvent any safety, health, or environmental requirements.

MEASUREMENT AND VERIFICATION

The international performance measurement and verification protocol (IPMVP) is used for commercial and industrial facility operators. The IPMVP offers standards for measurement and verification of energy and water efficiency projects. The IPMVP volumes are used to: 1) develop a measurement and verification strategy and plan for quantifying energy and water savings in retrofits and new construction, 2) monitor indoor environmental quality, and 3) quantify emissions reduction (www.evo-world.org).

SUMMARY

The dynamic process of revisions to existing codes, plus the introduction of new legislation, will impact the energy industry and bring a dramatic change. Energy conservation and creating new power generation supply options will both be required to meet the energy demands of the twenty-first century.

2

Energy
Economic Analysis

To justify the energy investment cost, a knowledge of life-cycle costing is required.

The life-cycle cost analysis evaluates the total owning and operating cost. It takes into account the "time value" of money and can incorporate fuel cost escalation into the economic model. This approach is also used to evaluate competitive projects. In other words, the life-cycle cost analysis considers the cost over the life of the system rather than just the first cost.

THE TIME VALUE OF MONEY CONCEPT

To compare energy utilization alternatives, it is necessary to convert all cash flow for each measure to an equivalent base. The life-cycle cost analysis takes into account the "time value" of money; thus a dollar in hand today is more valuable than one received at some time in the future. This is why a time value must be placed on all cash flows into and out of the company.

DEVELOPING CASH FLOW MODELS

The cash flow model assumes that cash flows occur at discrete points in time as lump sums and that interest is computed and payable at discrete points in time.

To develop a cash flow model which illustrates the effect of "compounding" of interest payments, the cash flow model is developed as follows:

End of Year 1: $P + i(P) = (1 + i)P$

Year 2: $(1 + i) P + (1 + i) Pi = (1 + i)P [(1 + i)]$

   $= (1 + i)^2 P$

Year 3: $(1 + i)^3 P$

Year n $(1 + i)^n P$ or $F = (1 + i)^n P$

Where P = present sum

   I = interest rate earned at the end of each interest period

   n = number of interest periods

   F = future value

$(1 + i)^n$ is referred to as the "Single Payment Compound Amount" factor (F/P) and is tabulated for various values of i and n in Appendix A.

The cash flow model can also be used to find the present value of a future sum F.

$$P = \left(\frac{1}{(1 + i)^n}\right) \cdot F$$

Cash flow models can be developed for a variety of other types of cash now as illustrated in Figure 2-1.

To develop the cash flow model for the "Uniform Series Compound Amount" factor, the following cash flow diagram is drawn.

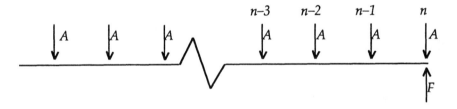

Where A is a uniform series of year-end payments and F is the future sum of A payments for n interest periods.

The A dollars deposited at the end of the nth period earn no interest and, therefore, contribute A dollars to the fund. The A dollars deposited at the end of the $(n - 1)$ period earn interest for 1 year and will, therefore, contribute $A (1 + i)$ dollars to the fund. The A dollars deposited at the end of the $(n - 2)$ period earn interest for 2 years and will, therefore, contribute

$A(1 + i)^2$. These years of earned interest in the contributions will continue to increase in this manner, and the A deposited at the end of the first period will have earned interest for $(n - 1)$ periods. The total in the fund F is, thus, equal to $A + A(1 + i) + A(1 + i)^2 + A(1 + i)^3 + A(1 + i)^4 + \dots + A(1 + i)^{n-2} + A(1 + i)^{n-1}$. Factoring out A,

(1) $F = A[1 + (1 + i) + (1 + i)^2 \dots + (1 + i)^{n-2} + (1 + i)^{n-1}$
Multiplying both sides of this equation by $(1 + i)$;

(2) $(1+i)F = A[(1+i)+(1+i)^2 + (1+i)^3 + \dots + (1+i)^{n-1} + (1+i)n]$
Subtracting equation (1) from (2):

$$(1+i)F - F = A[(1+i) + (1+i)^2 + (1+i)^3$$
$$+ (1+i)^{n-1} + (1+i)^n] - A[1 + (1+i)$$
$$+ (1+i)^2 + \dots + (1+i)^{n-2} + (1+i)^{n-1}]$$
$$iF = A[(1+i)^{n-1}]$$

$$F = A\left[\frac{(1+i)^{n-1}}{i}\right]$$

Interest factors are seldom calculated. They can be determined from computer programs, and interest tables included in the Appendix. Each factor is defined when the number of periods (n) and interest rate (i) are specified. In the case of the gradient present worth factor the escalation rate must also be stated.

The three most commonly used methods in life-cycle costing are the annual cost, present worth and rate-of-return analysis.

In the present worth method a minimum rate of return (i) is stipulated. All future expenditures are converted to present values using the interest factors. The alternative with lowest effective first cost is the most desirable.

A similar procedure is implemented in the annual cost method. The difference is that the first cost is converted to an annual expenditure. The alternative with lowest effective annual cost is the most desirable.

In the rate-of-return method, a trial-and-error procedure is usually required. Interpolation from the interest tables can determine what rate of return (i) will give an interest factor which will make the overall cash flow balance. The rate-of-return analysis gives a good indication of the overall ranking of independent alternates.

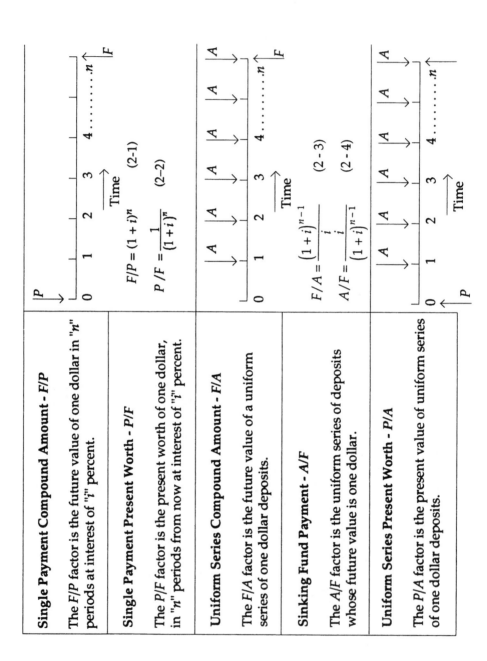

Single Payment Compound Amount - *F/P*

The *F/P* factor is the future value of one dollar in "*n*" periods at interest of "*i*" percent.

$$F/P = (1+i)^n \qquad (2\text{-}1)$$

Single Payment Present Worth - *P/F*

The *P/F* factor is the present worth of one dollar, in "*n*" periods from now at interest of "*i*" percent.

$$P/F = \frac{1}{(1+i)^n} \qquad (2\text{-}2)$$

Uniform Series Compound Amount - *F/A*

The *F/A* factor is the future value of a uniform series of one dollar deposits.

$$F/A = \frac{(1+i)^{n}-1}{i} \qquad (2\text{-}3)$$

Sinking Fund Payment - *A/F*

The *A/F* factor is the uniform series of deposits whose future value is one dollar.

$$A/F = \frac{i}{(1+i)^{n}-1} \qquad (2\text{-}4)$$

Uniform Series Present Worth - *P/A*

The *P/A* factor is the present value of uniform series of one dollar deposits.

Capital Recovery - A/P

The *A/P* factor is the uniform series of deposits whose present value is one dollar.

$$P/A = \frac{(1+i)^n - 1}{i(1+i)^n} \quad (2\text{-}5)$$

$$A/P = \frac{i\,(1+i)^n}{(1+i)^n - 1} \quad (2\text{-}6)$$

Gradient Present Worth - *GPW*

The GPW factor is the present value of a gradient series.

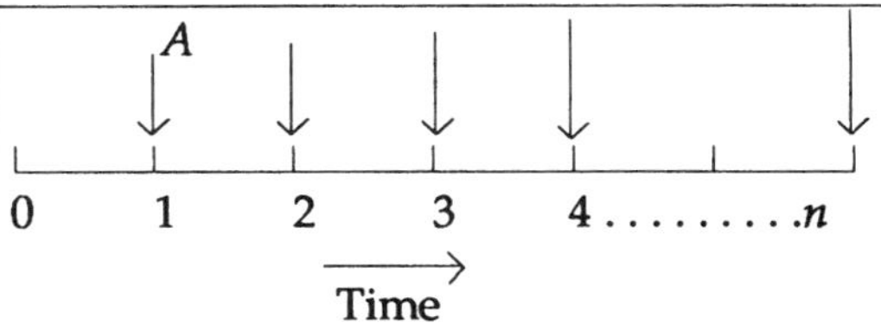

$$GPW = P/A = \frac{\dfrac{1+e}{1+i}\left[1-\left(\dfrac{1+e}{1+i}\right)^n\right]}{1-\dfrac{1+e}{1+i}} \quad (2\text{-}7)$$

NOTES

where

P is the present worth (occurs at the beginning of the interest period).
F is the future worth (occurs at the end).
n is the number of periods that the interest is compounded.
i is the interest rate or desired rate of return.
A is the uniform series of deposits (occurs at the end of the interest period).
e is the escalation rate

Figure 2-1. Interest Factors

The effect of escalation in fuel costs can influence greatly the final decision. When an annual cost grows at a steady rate it may be treated as a gradient and the gradient present worth factor can be used.

Special appreciation is given to Rudolph R. Vaneck and Dr. Robert Brown for use of their specially designed interest and escalation tables used in this text.

When life-cycle costing is used to compare several alternatives, the differences between costs are important. For example, if one alternate forces additional maintenance or an operating expense to occur, then these factors as well as energy costs need to be included. Remember, what was previously spent for the item to be replaced is irrelevant. The only factor to be considered is whether the new cost can be justified based on projected savings over its useful life.

PAYBACK ANALYSIS

The simple payback analysis is sometimes used instead of the methods previously outlined. The simple payback is defined as initial investment divided by annual savings after taxes. The simple payback method does not take into account the effect of interest or escalation rate.

Since the payback period is relatively simple to calculate, and due to the fact managers wish to recover their investment as rapidly as possible, the payback method is frequently used.

It should be used in conjunction with other decision-making tools. When used by itself as the principal criterion, it may result in choosing less profitable investments which yield high initial returns for short periods as compared with more profitable investments which provide profits over longer periods of time.

Example Problem 2-1

An electrical energy audit indicates electrical motor consumption is 4×10^6 kWh per year. By upgrading the motor spares with high efficiency motors, a 10% savings can be realized. The additional cost for these motors is estimated at $80,000. Assuming an 8¢ per kWh energy charge and 20-year life, is the expenditure justified based on a minimum rate of return of 20% before taxes? Solve the problem using the present worth, annual cost, and rate-of-return methods.

Analysis

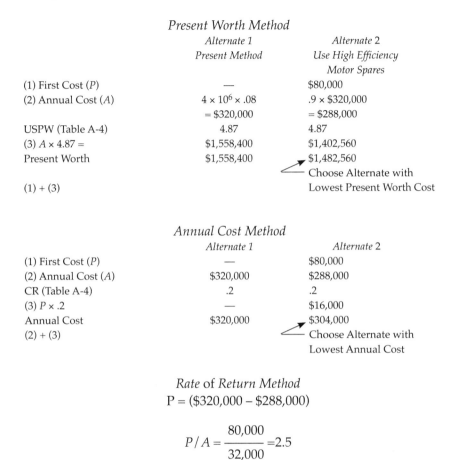

Present Worth Method

	Alternate 1 *Present Method*	Alternate 2 *Use High Efficiency Motor Spares*
(1) First Cost (*P*)	—	$80,000
(2) Annual Cost (*A*)	$4 \times 10^6 \times .08$	$.9 \times \$320,000$
	= $320,000	= $288,000
USPW (Table A-4)	4.87	4.87
(3) *A* × 4.87 =	$1,558,400	$1,402,560
Present Worth	$1,558,400	$1,482,560
(1) + (3)		Choose Alternate with Lowest Present Worth Cost

Annual Cost Method

	Alternate 1	Alternate 2
(1) First Cost (*P*)	—	$80,000
(2) Annual Cost (*A*)	$320,000	$288,000
CR (Table A-4)	.2	.2
(3) *P* × .2	—	$16,000
Annual Cost	$320,000	$304,000
(2) + (3)		Choose Alternate with Lowest Annual Cost

Rate of Return Method

$$P = (\$320,000 - \$288,000)$$

$$P/A = \frac{80,000}{32,000} = 2.5$$

What value of *i* will make P/A = 2.5? *i* = 40% (Table A-7).

Example Problem 2-2

Show the effect of 10% escalation on the rate-of-return analysis given the

Energy equipment investment	=	$20,000
After tax savings	=	$ 2,600
Equipment life (*n*)	=	15 years

Analysis

Without escalation

$$CR = \frac{A}{P} = \frac{2,600}{20,000} = .13$$

From Table A-1, the rate of return in 10%.

With 10% escalation assumed:

$$GPW = \frac{P}{A} = \frac{20,000}{2,600} = 7.69$$

From Table A-11, the rate of return is 21%.

Thus we see that taking into account a modest escalation rate can dramatically affect the justification of the project.

TAX CONSIDERATIONS

Depreciation

Depreciation affects the "accounting procedure" for determining profits and losses and the income tax of a company. In other words, for tax purposes the expenditure for an asset such as a pump or motor cannot be fully expensed in its first year. The original investment must be charged off for tax purposes over the useful life of the asset. A company wishes to expense an item as quickly as possible.

The Internal Revenue Service allows several methods for determining the annual depreciation rate.

Straight-line Depreciation: The simplest method is referred to as a straight-line depreciation and is defined as

$$D = \frac{P - L}{n} \qquad\qquad\qquad Formula\ (2\text{-}8)$$

Where

D is the annual depreciation rate

L is the value of equipment at the end of its useful life, commonly referred to as salvage value

n is the life of the equipment which is determined by Internal Revenue Service guidelines

P is the initial expenditure.

Sum-of-Years Digits: Another method is referred to as the sum-of-years digits. In this method the depreciation rate is determined by finding the sum of digits using the following formula:

$$N = n \;\frac{n+1}{2}$$

Formula (2-9)

Where n is the life of equipment.

Each year's depreciation rate is determined as follows:

First year
$$D = \frac{n}{N} \; (P-L)$$
Formula (2-10)

Second year
$$D = \frac{n-1}{N} \; (P-L)$$
Formula (2-11)

n year
$$D = \frac{1}{N} \; (P-L)$$
Formula (2-12)

Declining-Balance Depreciation: The declining-balance method allows for larger depreciation charges in the early years, which is sometimes referred to as fast write-off.

The rate is calculated by taking a constant percentage of the declining undepreciated balance. The most common method used to calculate the declining balance is to predetermine the depreciation rate. Under certain circumstances a rate equal to 200% of the straight-line depreciation rate may be used. Under other circumstances the rate is limited to 1-1/2 or 1-1/4 times as great as straight-line depreciation. In this method the salvage value or undepreciated book value is established once the depreciation rate is pre-established.

To calculate the undepreciated book value, Formula 2-13 is used:

$$D = 1 - \left(\frac{L}{P}\right)^{1/N}$$

Formula (2-13)

Where
  D is the annual depreciation rate
  L is the salvage value
  P is the first cost

Example Problem 2-3
  Calculate the depreciation rate using the straight-line, sum-of-years digit, and declining-balance methods.
  Salvage value is 0.
  $n = 5$ years
  $P = 150,000$
  For declining balance use a 200% rate.

Straight-line Method

$$D = \frac{P - L}{n} = \frac{150,000}{5} = \$30,000 \text{ per year}$$

Sum-of-years Digits

$$N = \frac{n(n + 1)}{2} = \frac{5(6)}{2} = 15$$

$$D_1 = \frac{n}{N}\ (P) = \frac{5}{15}\ (150,000) = 50,000$$

N	P
1 =	$54,000
2 =	40,000
3 =	30,000
4 =	20,000
5 =	10,000

Declining-balance Method

$D = 2 \times 20\% = 40\%$ (Straight-line Depreciation Rate = 20%)

Year	*Undepreciated Balance* *At Beginning of Year*	*Depreciation Charge*
1	150,000	60,000
2	90,000	36,000
3	54,000	21,600
4	32,400	12,960
5	19,440	<u>7,776</u>
	TOTAL	138,336

Undepreciated Book Value (150,000 – 138,336) = $11,664

Cogeneration Equipment Depreciation

Most cogeneration equipment is depreciated over a 15- or 20-year period, depending on the particular type of equipment involved, using the 150% declining balance method switching to straight-line to maximize deductions. Gas and combustion turbine equipment used to produce electricity for sale is depreciated over a 15-year period. Equipment used in the steam power production of electricity for sale (including combustion turbines operated in combined cycle with steam units), as well as assets used to produce steam for sale, are normally depreciated over a 20-year period.

However, most electric and steam generation equipment owned by a taxpayer and producing electric or thermal energy for use by the taxpayer in its industrial process and plant activity, and not ordinarily for sale to others, is depreciated over a 15-year period. Electrical and steam transmission and distribution equipment will be depreciated over a 20-year period at the same 150 percent declining balance rate.

Energy Efficiency Equipment and Real Property Depreciation

Energy conservation equipment, still classified as real property, is depreciated on a straight line basis over a recovery period. Equipment installed in connection with residential real property qualifies for a 27-1/2-year period, while equipment placed in nonresidential facilities is subject to a 31-1/2-year period. Other real property assets are depreciated over the above period, depending on their residential or nonresidential character.

After-tax Analysis

Tax-deductible expenses such as maintenance, energy, operating costs, insurance and property taxes reduce the income subject to taxes.

For the after-tax life-cycle cost analysis and payback analysis, the actual incurred annual savings is given as follows:

$$AS=(1 - I) E + ID \qquad \qquad \text{Formula (2-14)}$$

Where

AS = yearly annual after-tax savings (excluding effect of tax credit)

E = yearly annual energy savings (difference between original expenses and expenses after modification)

D = annual depreciation rate

I = income tax bracket

Formula 2-14 takes into account that the yearly annual energy savings is partially offset by additional taxes which must be paid due to reduced operating expenses. On the other hand, the depreciation allowance reduces taxes directly.

To compute a rate of return which accounts for taxes, depreciation, escalation and tax credits, a cash-flow analysis is usually required. This method analyzes all transactions including first and operating costs. To determine the after-tax rate of return, a trial and error or computer analysis is required.

The present worth factors tables in the Appendix, can be used for this analysis. All money is converted to the present assuming an interest rate. The summation of all present dollars should equal zero when the correct interest rate is selected, as illustrated in Figure 2-2.

This analysis can be made assuming a fuel escalation rate by using the gradient present worth interest of the present worth factor.

Example Problem 2-4

Comment on the after-tax rate of return for the installation of a heat-recovery system given the following:
- First Cost $100,000
- Year Savings   36,363
- Straight-line depreciation life and equipment life of 5 years
- Income tax bracket 34%

Year	1 Investment	2 Tax Credit	3 After Tax Savings (AS)	4 Single Payment Present Worth Factor	(2 + 3) x 4 Present Worth
0	−P				−P
1		+TC	AS_1	$SPPW_1$	$+P_1$
2			AS_2	$SPPW_2$	P_2
3			AS_3	$SPPW_3$	P_3
4			AS_4	$SPPW_4$	P_4
Total					ΣP

$$AS = (1 - I)\,E + ID$$
Trial & Error Solution:
Correct i when $\Sigma P = 0$

Figure 2-2. Cash Flow Rate of Return Analysis

Analysis

$D = 100,000/5 = 20,000$

$AS = (1 - I)\,E + ID = .66(36,363) + .34(20,000) = \$30,800$

First Trial $i = 20\%$

Investment	After Tax Savings	SPPW 20%	PW
0-100,000			−100,000
1	30,800	.833	25,656
2	30,800	.694	21,374
3	30,800	.578	17,802
4	30,800	.482	14,845
5	30,800	.401	12,350
			$\Sigma - 7,972$

Since summation is negative a higher present worth factor is required. Next try is 15%.

	After Tax	SPPW	
Investment	Savings	15%	PW
0-100,000			-100,000
1	30,800	.869	+ 26,765
2	30,800	.756	+ 23,284
3	30,800	.657	+ 20,235
4	30,800	.571	+ 17,586
5	30,800	.497	+ 15,307
			+ 3,177

Since rate of return is bracketed, linear interpolation will be used.

$$\frac{3177 + 7972}{-5} = \frac{3177 - 0}{15 - i\%}$$

$$i = \frac{3177}{2229.6} + 15 = 16.4\%$$

Impact of Fuel Inflation on Life-Cycle Costing

As illustrated by Problem 2-2, a modest estimate of fuel inflation has a major impact on improving the rate of return on investment of the project. The problem facing the energy engineer is how to forecast what the future of energy costs will be. All too often no fuel inflation is considered because of the difficulty of projecting the future. In making projections the following guidelines may be helpful:

- Is there a rate increase that can be forecast based on new nuclear generating capacity? In locations such as Georgia, California, and Arizona electric rates will rise at a faster rate due to commissioning of new nuclear plants and rate increases approved by the Public Service Commission of that state.
- What has been the historical rate increase for the facility? Even with fluctuations there are likely to be trends to follow.
- What events on a national or international level would impact on your costs? New state taxes, new production quotas by OPEC and other factors will affect your fuel prices.
- What do the experts say? Energy economists, forecasting services, and your local utility projections all should be taken into account.

The rate of return on investment becomes more attractive when life-cycle costs are taken into account. Tables A-9 through A-12 can be used to show the impact of fuel inflation on the decision-making process.

Example Problem 2-5
Develop a set of curves that indicate the capital that can be invested to give a rate of return of 15% after taxes for each $1000 saved for the following conditions:
1. The effect of escalation is not considered.
2. A 5% fuel escalation is considered.
3. A 10% fuel escalation is considered.
4. A 14% fuel escalation is considered.
5. A 20% fuel escalation is considered.

Calculate for 5-, 10-, 15-, 20-year life.
Assume straight-line depreciation over useful life, 34% income tax bracket, and no tax credit.

Answer

$$AS = (1 - I)E + ID$$
$$I = 0.34, E = \$1000$$
$$AS = 660 + \frac{0.34P}{N}$$

Thus, the after-tax savings (*AS*) is composed of two components. The first component is a uniform series of $660 escalating at *e* percent/year. The second component is a uniform series of 0.34P/N.

Each component is treated individually and convened to present day values using the GPW factor and the USPW factor, respectively. The sum of these two present worth factors must equal *P*. In the case of no escalation, the formula is

$$P = 660\ P/A + \frac{0.34P}{N}\ P/A$$

In the case of escalation

$$P = 660\ GPW + \frac{0.34P}{N}\ P/A$$

Since there is only one unknown, the formulas can be readily solved. The results are indicated below.

	N=5 $P	N=10 $P	N= 15 $P	N=20 $P
e = 0	2869	4000	4459	4648
e = 10%	3753	6292	8165	9618
e = 14%	4170	7598	10,676	13,567
e = 20%	4871	10,146	16,353	23,918

Figure 2-3 illustrates the effects of escalation. This figure can be used as a quick way to determine after-tax economics of energy utilization expenditures.

Example Problem 2-6
It is desired to have an after-tax savings of 15%. Comment on the investment that can be justified if it is assumed that the fuel rate escalation should not be considered and the annual energy savings is $2000 with an equipment economic life of 10 years.
Comment on the above, assuming a 10% fuel escalation.

Answer
From Figure 2-3, for each $1000 energy savings, an investment of $3600 is justified or $8000 for a $2000 savings for which no fuel increase is accounted.

With a 10% fuel escalation rate on investment of $6300 justified for each $1000 energy savings, $12,600 can be justified for $2000 savings. Thus, a 57% higher expenditure is economically justifiable and will yield the same after tax rate of return of 15% when a fuel escalation of 10% is considered.

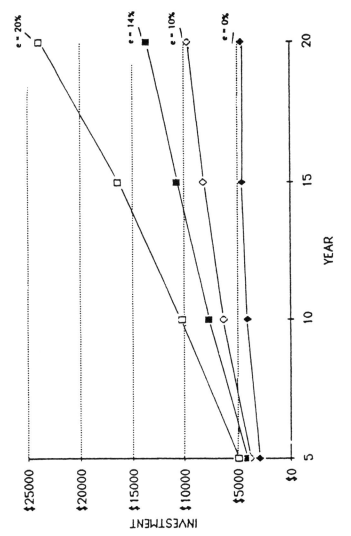

Figure 2-3. Effects of Escalation on Investment Requirements

Note: Maximum investment in order to attain a 15% after-tax rate of return on investment for annual savings of $1000.

3

Energy Auditing and Accounting

TYPES OF ENERGY AUDITS

The simplest definition for an energy audit is as follows: An energy audit serves the purpose of identifying where a building or plant facility uses energy and identifies energy conservation opportunities.

There is a direct relationship to the cost of the audit (amount of data collected and analyzed) and the number of energy conservation opportunities to be found. Thus, a first distinction is the cost of the audit which determines the type of audit to be performed.

The second distinction is the type of facility. For example, a building audit may emphasize the building envelope, lighting, heating, and ventilation requirements. On the other hand, an audit of an industrial plant emphasizes the process requirements.

Most energy audits fall into three categories or types, namely, walk-through, mini-audit, or maxi-audit.

Walk-through—This type of audit is the least costly and identifies preliminary energy savings. A visual inspection of the facility is made to determine maintenance and operation energy saving opportunities plus collection of information to determine the need for a more detailed analysis.

Mini-audit—This type of audit requires tests and measurements to quantify energy uses and losses and determine the economics for changes.

Maxi-audi—This type of audit goes one step further than the mini-audit. It contains an evaluation of how much energy is used for each function such as lighting, process, etc. It also requires a model analysis, such as a computer simulation, to determine energy use patterns and predictions on a year-round basis, taking into account such variables as weather data.

As noted in the audit definition, there are two essential parts, namely, data acquisition and data analysis.

Data Acquisition

This phase requires the accumulation of utility bills, establishing a baseline to provide historical documentation and a survey of the facility.

All energy flows should be accounted for; thus all "energy in" should equal "energy out." This is referred to as an energy balance.

All energy costs should be determined for each fuel type. The energy survey is essential. Instrumentation commonly used in conducting a survey is discussed at the conclusion of the chapter.

Data Analysis

As a result of knowing how energy is used, a complete list of "Energy Conservation Opportunities" (ECOs) will be generated. The life-cycle costing techniques presented in Chapter 2 will be used to determine which alternative should be given priority.

A very important phase of the overall program is to continuously monitor the facility even after the ECOs have been implemented. Documentation of the cost avoidance or savings is essential to the audit.

Remember, in order to have a continuous ongoing program, *individuals must be made accountable* for energy use. As part of the audit, recommendations should be made as to where to add "root" or submetering.

ENERGY USE PROFILES

The energy audit process for a building emphasizes building envelope, heating and ventilation, air conditioning, plus lighting functions. For an industrial facility the energy audit approach includes process consideration. Figures 3-1 through 3-3 illustrate how energy is used for a typical industrial plant. It is important to account for total consumption, cost, and how energy is used for each commodity such as steam, water, air and natural gas. This procedure is required to develop the appropriate energy conservation strategy.

The top portion of Figure 3-1 illustrates how much energy is used by fuel type and its relative percentage. The pie chart below shows how much is spent for each fuel type. Using a pie-chart presentation or nodal flow diagram can be very helpful in visualizing how energy is being used.

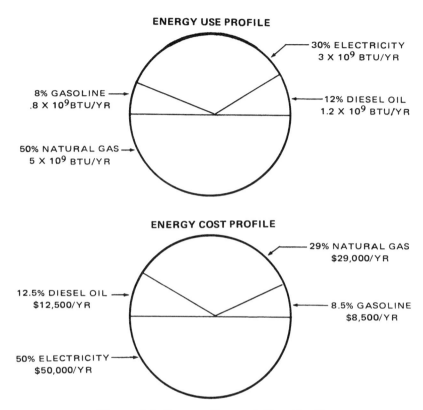

Figure 3-1. Energy Use and Cost Profile

Figure 3-2, on the other hand, shows how much of the energy is used for each function such as lighting, process, and building heating and ventilation. Pie charts similar to the right-hand side of the figure should be made for each category such as air, steam, electricity, water and natural gas.

Figure 3-3 illustrates an alternate representation for the steam distribution profile.

Several audits are required to construct the energy use profiles, such as:

Envelope Audit—This audit surveys the building envelope for losses or gains due to leaks, building construction, doors, glass, lack of insulation, etc.

Functional Audit—This audit determines the amount of energy required for a particular function and identifies energy conservation opportunities. Functional audits include:

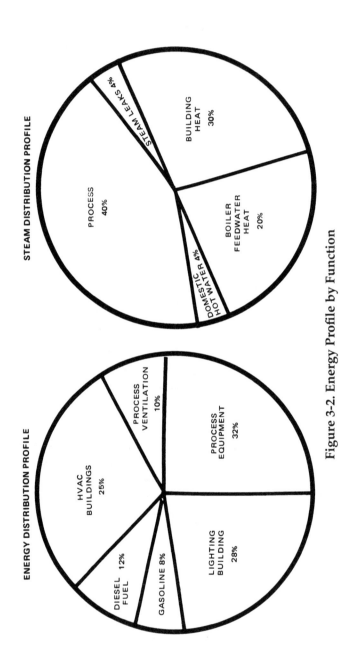

Figure 3-2. Energy Profile by Function

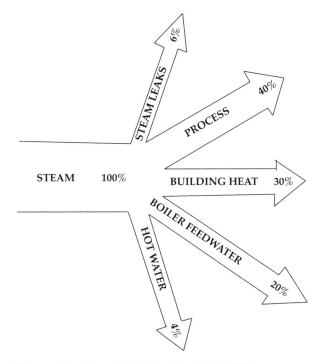

Figure 3-3. Steam Distribution Nodal Diagram

- Heating, ventilation and air conditioning
- Building
- Lighting
- Domestic hot water
- Air distribution

Process Audit—This audit determines the amount of energy required for each process function and identifies energy conservation opportunities. Process functional audits include:

- Process machinery
- Heating, ventilation and air-conditioning process
- Heat treatment
- Furnaces

Transportation Audit—This audit determines the amount of energy required for forklift trucks, cars, vehicles, trucks, etc.

Utility Audit—This audit analyzes the monthly, daily or yearly energy usage for each utility.

ENERGY USERS

Energy use profiles for several end-users are summarized in Tables 3-1 through 3-11.

Table 3-1. Energy Use in Apartment Buildings

	Range (%)	Norms (%)
Environmental Control	50 to 80	70
Lighting and Wall Receptacles	10 to 20	15
Hot Water	2 to 5	3
Special Functions		
Laundry, Swimming Pool, Restaurants,		
Parking, Elevators, Security Lighting	5 to 20	10

Table 3-2. Energy Use in Bakeries

Housekeeping Energy	Percent
Space Heating	21.5
Air Conditioning	1.6
Lighting	1.4
Domestic Hot Water	1.8
TOTAL	26.3
Process Energy	*Percent*
Baking Ovens	49.0
Pan Washing	10.6
Mixers	4.1
Freezers	3.3
Cooking	3.0
Fryers	1.8
Proof Boxes	1.8
Other Processes	1.1
TOTAL	73.7

Data are for a 27,000-square-foot bakery in Washington, D.C.

Table 3-3. Energy Use in Die Casting Plants

Housekeeping Energy	*Percent*
Space Heating	24
Air Conditioning	2
Lighting	2
Domestic Hot Water	2
TOTAL	30
Process Energy	*Percent*
Melting Hearth	30
Quiet Pool	20
Molding Machines	10
Air Compressors	5
Other Processes	5
TOTAL	70

Table 3-4. Energy Use in Hospital Buildings

	Range (%)	*Norms* (%)
Environmental Control	40 to 65	58
Lighting and Wall Receptacles	10 to 20	15
Laundry	8 to 15	12
Food Service, Kitchen Operations		
Medical Equipment, Sterilization,	5 to 10	7
Incinerator, Parking, Elevators,		
Security Lighting	5 to 15	8

Table 3-5. Energy Use in Hotels and Motels

	Range (%)	*Norms* (%)
Space Heating	45 to 70	60
Lighting	5 to 15	11
Air Conditioning	3 to 15	10
Refrigeration	0 to 10	4
Special Functions	5 to 20	15
Laundry, Kitchen, Restaurant,		
Swimming Pool, Garage,		
Security Lighting, Hot Water		

Table 3-6. Energy Use in Retail Stores

	Range (%)	Norms (%)
HVAC	20 to 50	30
Lighting	40 to 75	60
Special Functions	5 to 20	10
Elevators, General Power, Parking		
Security Lighting, Hot Water		

Table 3-7. Energy Use in Restaurants

	Table Restaurant Norms (%)	Fast Food Restaurant Norms (%)
HVAC	32	36
Lighting	8	26
Special Functions		
Food Preparation	45	27
Food Storage	2	6
Sanitation	12	1
Other	1	4

Table 3-8. Energy Use in Schools

	Range (%)	Norms (%)
Environmental Control	45 to 80	65
Lighting and Wall Receptacles	10 to 20	15
Food Service	5 to 10	7
Hot Water	2 to 5	3
Special Functions	0 to 20	10

Table 3-9. Energy Use in Transportation Terminals

	Range (%)	Norms (%)
Space Heating	50 to 75	60
Lighting	5 to 25	15
Air Conditioning	5 to 25	15
Special Functions	3 to 20	10
Elevators, General Power,		
Parking, Security Lighting,		
Hot Water		

Table 3-10. Energy Use in Warehouses and Storage Facilities

(Vehicles Not Included)	Range (%)	Norms (%) *
Space Heating	45 to 80	67
Air Conditioning	3 to 10	6
Lighting	4 to 12	7
Refrigeration	0 to 40	12
Special Functions	5 to 15	8
Elevators, General Power, Parking, Security Lighting, Hot Water		

*Norms for a warehouse or storage facility are strongly dependent on the products and their specific requirements for temperature and humidity control.

Table 3-11. Comparative Energy Use by System

		Heating & Ventilation	Cooling & Ventilation	Lighting	Power & Process	Domestic Hot Water
Schools	A	4	3	1	5	
	B	1	4	2	5	3
	C	1	4	2	5	3
Colleges	A	5	2	1	4	3
	B	1	3	2	5	4
	C	1	5	2	4	3
Office Bldg.	A	3	1	2	4	5
	B	1	3	2	4	5
	C	1	3	2	4	5
Commercial Stores	A	3	1	2	4	5
	B	2	3	1	4	5
	C	1	3	2	4	5
Religious Bldg.	A	3	2	1	4	5
	B	1	3	2	4	5
	C	1	3	2	4	5
Hospitals	A	4	1	2	5	3
	B	1	3	4	5	2
	C	1	5	3	4	2

Climatic Zone A: Fewer than 2500 degree days
Climate Zone B: 2500-5500 degree days
Climate Zone C: 5500-9500 degree days
Source: Guidelines For Saving Energy In Existing Buildings ECM-1
Note: Numbers indicate energy consumption relative to each other
    (1) greatest consumption
    (5) least consumption

ENERGY USE IN BUILDINGS[1]

More than 36% of the nation's primary energy was consumed in providing services to residential and commercial buildings in 1985 (26.9 quads). The residential sector consumed 15.3 quads; the commercial sector (including public buildings) consumed 11.6 quads.

The largest end use is space heating in both residential (40%), and commercial (34%) buildings. Water heating is second in residences (17%), followed by refrigerators (9%), air conditioning (7%), lighting (7%), and kitchen ranges and ovens (6%). In commercial buildings, lighting is second (25%), followed by ventilation (12%), air conditioning (10%) and water heating (6%). (See Table 3-12.)

Table 3-12. Building Sector Energy Use

BUILDING SECTOR ENERGY USE
BY END USE
1985-2010 FORECAST

	1985	2010
Space Heating	38%	34%
Air Conditioning and Ventilation	15	16
Lighting	15	15
Water Heating	12	12
Refrigeration	7	5
Other	13	18

The forecast changes as the type of energy used becomes more significant. The analysis estimates a major increase in the use of electricity by the building sector, increasing from 17.0 quads in 1985 to 26.8 quads in 2010. These figures represent the source (primary) energy used to generate electricity.

During this period, natural gas use is forecast to increase slightly from 7.0 quads to 7.2 quads. The use of solar and renewable energy is ex-

---

[1]Source: Energy Conservation Goals for Buildings, A Report to the Congress of The United States, May 1988.

pected to more than double from 1.2 quads to 2.9 quads; the use of oil is expected to remain flat at 2.6 quads.

THE ENERGY SURVEY

As part of the data acquisition phase, a detailed survey should be conducted. The various types of instrumentation commonly used in the survey are discussed in this section.

Infrared Equipment

Some companies may have the wrong impression that infrared equipment can meet most of their instrumentation needs.

The primary use of infrared equipment in an energy utilization program is to detect building or equipment losses. Thus it is just one of the many options available.

Several energy managers find infrared in use in their plant prior to the energy utilization program. Infrared equipment, in many instances, was purchased by the electrical department and used to detect electrical hot spots.

Infrared energy is an invisible part of the electromagnetic spectrum. It exists naturally and can be measured by remote heat-sensing equipment. Within the last four years lightweight portable infrared systems became available to help determine energy losses. Differences in the infrared emissions from the surface of objects cause color variations to appear on the scanner. The hotter the object, the more infrared radiated. With the aid of an isotherm circuit, the intensity of these radiation levels can be accurately measured and quantified. In essence the infrared scanning device is a diagnostic tool which can be used to determine building heat losses. Equipment costs range from $400 to $25,000.

An overview energy scan of the plant can be made through an aerial survey using infrared equipment. Several companies offer aerial scan services starting at $1500. Aerial scans can determine underground stream pipe leaks, hot gas discharges, leaks, etc.

Since IR detection and measurement equipment have gained increased importance in the energy audit process, a summary of the fundamentals is reviewed in this section.

The visible portion of the spectrum runs from .4 to .75 micrometers (μm). The infrared or thermal radiation begins at this point and extends to approximately 1000 μm. Objects such as people, plants, or buildings will emit radiation with wavelengths around 10 p.m. (See Figure 3-4.)

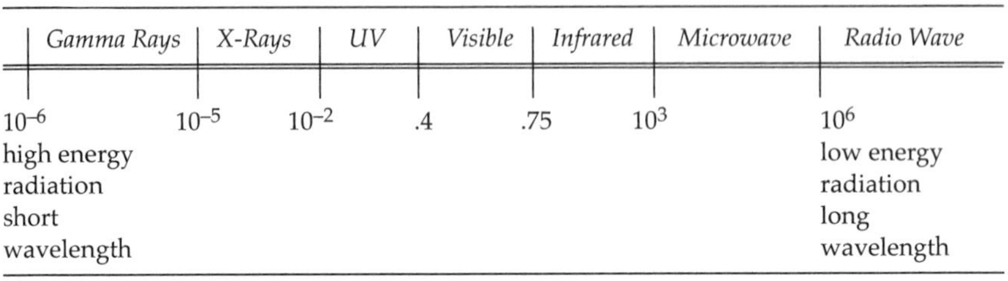

Gamma Rays	X-Rays	UV	Visible	Infrared	Microwave	Radio Wave

10^{-6} 10^{-5} 10^{-2} .4   .75 10^3 10^6

high energy radiation short wavelength / low energy radiation long wavelength

Figure 3-4. Electromagnetic Spectrum

Infrared instruments are required to detect and measure the thermal radiation. To calibrate the instrument, a special "black body" radiator is used. A black body radiator absorbs all the radiation that impinges on it and has an absorbing efficiency or emissivity of 1.

The accuracy of temperature measurements by infrared instruments depends on the three processes which are responsible for an object acting like a black body. These processes—absorbed, reflected, and transmitted radiation—are responsible for the total radiation reaching an infrared scanner.

The real temperature of the object is dependent only upon its emitted radiation.

Corrections to apparent temperatures are made by knowing the emissivity of an object at a specified temperature.

The heart of the infrared instrument is the infrared detector. The detector absorbs infrared energy and converts it into electrical voltage or current. The two principal types of detectors are the thermal and photo type. The thermal detector generally requires a given period of time to develop an image on photographic film. The photo detectors are more sensitive and have a higher response time. Television-like displays on a cathode ray tube permit studies of dynamic thermal events on moving objects in real time.

There are various ways of displaying signals produced by infrared detectors. One way is by use of an isotherm contour. The lightest areas of the picture represent the warmest areas of the subject, and the darkest areas represent the coolest portions. These instruments can show thermal variations of less than 0.1°C and can cover a range of −30° C to over 2000° C.

The isotherm can be calibrated by means of a black body radiator so that a specific temperature is known. The scanner can then be moved and the temperatures of the various parts of the subject can be made.

MEASURING ELECTRICAL SYSTEM PERFORMANCE

The ammeter, voltmeter, wattmeter, power factor meter, and foot-candle meter are usually required to do an electrical survey. These instruments are described below.

Ammeter and Voltmeter

To measure electrical currents, ammeters are used. For most audits, alternating currents are measured. Ammeters used in audits are portable and are designed to be easily attached and removed.

There are many brands and styles of snap-on ammeters commonly available that can read up to 1000 amperes continuously. This range can be extended to 4000 amperes continuously for some models with an accessory step-down current transformer.

The snap-on ammeters can be either indicating or recording with a printout. After attachment, the recording ammeter can keep recording current variations for as long as a full month on one roll of recording paper. This allows the study of current variations in a conductor for extended periods without constant operator attention.

The ammeter supplies a direct measurement of electrical current, which is one of the parameters needed to calculate electrical energy. The second parameter required to calculate energy is voltage, and it is measured by a voltmeter.

Several types of electrical meters can read the voltage or current. A voltmeter measures the difference in electrical potential between two points in an electrical circuit.

In series with the probes are the galvanometer and a fixed resistance (which determine the voltage scale). The current through this fixed resistance circuit is then proportional to the voltage, and the galvanometer deflects in proportion to the voltage.

The voltage drops measured in many instances are fairly constant and need only be performed once. If there are appreciable fluctuations, additional readings or the use of a recording voltmeter may be indicated.

Most voltages measured in practice are under 600 volts and there are many portable voltmeter/ammeter clamp-ons available for this and lower ranges.

Wattmeter and Power Factor Meter

The portable wattmeter can be used to indicate by direct reading electrical energy in watts. It can also be calculated by measuring voltage, current and the angle between them (power factor angle).

The basic wattmeter consists of three voltage probes and a snap-on current coil which feeds the wattmeter movement.

The typical operating limits are 300 kilowatts, 650 volts, and 600 amperes. It can be used on both one- and three-phase circuits.

The portable power factor meter is primarily a three-phase instrument. One of its three voltage probes is attached to each conductor phase and a snap-on jaw is placed about one of the phases. By disconnecting the wattmeter circuitry, it will directly read the power factor of the circuit to which it is attached.

It can measure power factor over a range of 1.0 leading to 1.0 lagging with "ampacities" up to 1500 amperes at 600 volts. This range covers the large bulk of the applications found in light industry and commerce.

The power factor is a basic parameter whose value must be known to calculate electric energy usage. Diagnostically, it is a useful instrument to determine the sources of poor power factor in a facility.

Portable digital kWh and kW demand units are now available.

Digital read-outs of energy usage in both kWh and kW demand or in dollars and cents, including instantaneous usage, accumulated usage, projected usage for a particular billing period, alarms when over-target levels are desired for usage, and control-outputs for load shedding and cycling are possible.

Continuous displays or intermittent alternating displays are available at the touch of a button for any information needed such as the cost of operating a production machine for one shift, one hour or one week.

Footcandle Meter

Footcandle meters measure illumination in units of footcandles through a light-sensitive barrier layer of cells contained within them. They are usually pocket-size and portable and are meant to be used as field instruments to survey levels of illumination. These meters differ from conventional photographic lightmeters in that they are color and cosine corrected.

TEMPERATURE MEASUREMENTS

To maximize system performance, knowledge of the temperature of a fluid, surface, etc. is essential. Several types of temperature devices are described in this section.

Thermometer

There are many types of thermometers that can be used in an energy audit. The choice of what to use is usually dictated by cost, durability, and application.

For air-conditioning, ventilation and hot-water service applications (temperature ranges 50°F to 250°F), a multipurpose portable battery-operated thermometer is used. Three separate probes are usually provided to measure liquid, air or surface temperatures.

For boiler and oven stacks (1000°F) a dial thermometer is used. Thermocouples are used for measurements above 1000°F.

Surface Pyrometer

Surface pyrometers are instruments which measure the temperature of surfaces. They are somewhat more complex than other temperature instruments because their probe must make intimate contact with the surface being measured.

Surface pyrometers are of immense help in assessing heat losses through walls and also for testing steam traps.

They may be divided into two classes: low-temperature (up to 250°F) and high-temperature (up to 600°F to 700°F). The low-temperature unit is usually part of the multipurpose thermometer kit. The high-temperature unit is more specialized but needed for evaluating fired units and general steam service.

There are also noncontact surface pyrometers which measure infrared radiation from surfaces in terms of temperature. These are suitable for general work and also for measuring surfaces which are visually but not physically accessible.

A more specialized instrument is the optical pyrometer. This is for high-temperature work (above 1500°F) because it measures the temperature of bodies which are incandescent because of their temperature.

Psychrometer

A psychrometer is an instrument which measures relative humidity based on the relation of the dry-bulb temperature and the wetbulb temperature.

Relative humidity is of prime importance in HVAC and drying operations. Recording psychrometers are also available. Above 200°F humidity studies constitute a specialized field of endeavor.

Portable Electronic Thermometer

The portable electronic thermometer is an adaptable temperature measurement tool. The battery-powered basic instrument, when housed in a carrying case, is suitable for laboratory or industrial use.

A pocket-size digital, battery-operated thermometer is especially convenient for spot checks or where a number of rapid readings of process temperatures need to be taken.

Thermocouple Probe

No matter what sort of indicating instrument is employed, the thermocouple used should be carefully selected to match the application and properly positioned if a representative temperature is to be measured. The same care is needed for all sensing devices-thermocouple, bimetals, resistance elements, fluid expansion, and vapor pressure bulbs.

Suction Pyrometer

Errors arise if a normal sheathed thermocouple is used to measure gas temperatures, especially high ones. The suction pyrometer overcomes these by shielding the thermocouple from wall radiation and drawing gases over it at high velocity to ensure good convective heat transfer. The thermocouple thus produces a reading which approaches the true temperature at the sampling point rather than a temperature between that of the walls and the gases.

MEASURING COMBUSTION SYSTEMS

To maximize combustion efficiency, it is necessary to know the composition of the flue gas. By obtaining a good air-fuel ratio, substantial energy will be saved.

Combustion Tester

Combustion testing consists of determining the concentrations of the products of combustion in a stack gas. The products of combustion usually considered are carbon dioxide and carbon monoxide. Oxygen is tested to assure proper excess air levels.

The definitive test for these constituents is an Orsat apparatus. This test consists of taking a measured volume of stack gas and measuring successive volumes after intimate contact with selective absorbing solutions. The reduction in volume after each absorption is the measure of each constituent.

The Orsat has a number of disadvantages. The main ones are that it requires considerable time to set up and use and that its operator must have a good degree of dexterity and be in constant practice.

Instead of an Orsat, there are portable and easy to use absorbing instruments which can easily determine the concentrations of the constituents of interest on an individual basis. Setup and operating times are minimal and just about anyone can learn to use them.

The typical range of concentrations are CO_2, 0-20%; O_2, 0-21 %; and CO, 0-0.5%. The CO_2 or O_2 content, along with knowledge of flue gas temperature and fuel type, allows the flue gas loss to be determined off standard charts.

Boiler Test Kit

The boiler test kit contains the following:

CO_2 Gas analyzer
O_2 Gas analyzer
      Inclined manometer
CO   Gas analyzer.

The purpose of the components of the kit is to help evaluate fireside boiler operation. Good combustion usually means high carbon dioxide (CO_2), low oxygen (O_2), and little or no trace of carbon monoxide (CO).

Gas Analyzers

The gas analyzers are usually of the Fyrite type. The Fyrite type differs from the Orsat apparatus in that it is more limited in application and less accurate. The chief advantages of the Fyrite are that it is simple and easy to use and is inexpensive. This device is used many times in an energy audit. Three readings using the Fyrite analyzer should be made and the results averaged.

Draft Gauge

The draft gauge is used to measure pressure. It can be the pocket type or the inclined manometer type.

Smoke Tester

To measure combustion completeness the smoke detector is used. Smoke is unburned carbon, which wastes fuel, causes air pollution, and fouls heat-exchanger surfaces. To use the instrument, a measured volume of flue gas is drawn through filter paper with the probe. The smoke spot is compared visually with a standard scale and a measure of smoke density is determined.

Combustion Analyzer

The combustion electronic analyzer permits fast, close adjustments. The unit contains digital displays. A standard sampler assembly with probe allows for stack measurements through a single stack or breaching hole.

MEASURING HEATING, VENTILATION
AND AIR-CONDITIONING (HVAC)
SYSTEM PERFORMANCE

Air Velocity Measurement

The following suggests the preference, suitability, and approximate costs of particular equipment:

- *Smoke pellets*—limited use but very low cost. Considered to be useful if engineering staff has experience in handling.
- *Anemometer* (deflecting vane)—good indication of air movement with acceptable order of accuracy. Considered useful (approximately $50).
- *Anemometer* (revolving vane) —good indicator of air movement with acceptable accuracy. However, easily subject to damage. Considered useful (approximately $100).
- *Pitot tube*—a standard air measurement device with good levels of accuracy. Considered essential. Can be purchased in various lengths— 12" about $20, 48" about $35. Must be used with a monometer. These vary considerably in cost, but could be on the order of $20 to $60.
- *Impact tube*—usually packaged air flow meter kits, complete with various jets for testing ducts, grills, open areas, etc. These units are convenient to use and of sufficient accuracy. The costs vary around $150 to $300, and therefore this order of cost could only be justified for a large system.

- *Heated thermocouple*—these units are sensitive and accurate but costly. A typical cost would be about $500 and can only be justified for regular use in a large plant.
- *Hot wire anemometer*—not recommended. Too costly and too complex.

Temperature Measurement

The temperature devices most commonly used are as follows:

- *Glass thermometers*—considered to be the most useful to temperature measuring instruments-accurate and convenient but fragile. Cost runs from $5 each for 12" long mercury in glass. Engineers should have a selection of various ranges.
- *Resistance thermometers*—considered to be very useful for A/C testing. Accuracy is good and they are reliable and convenient to use. Suitable units can be purchased from $150 up, some with a selection of several temperature ranges.
- *Thermocouples*—similar to resistance thermocouple but do not require battery power source. Chrome-Alum or iron types are the most useful and have satisfactory accuracy and repeatability. Costs start from $50 up.
- *Bimetallic thermometers*—considered unsuitable.
- *Pressure bulb thermometers*—more suitable for permanent installation. Accurate and reasonable in cost—$40 up.
- *Optical pyrometers*—only suitable for furnace settings and therefore limited in use. Cost from $300 up.
- *Radiation pyrometers*—limited in use for A/C work and costs from $500 up.
- *Indicating crayons*—limited in use and not considered suitable for A/C testing—costs around $2/crayon.
- *Thermographs*—use for recording room or space temperature; gives a chart indicating variations over a 12- or 168-hour period. Reasonably accurate. Low cost at around $30 to $60. (Spring-wound drive.)

Pressure Measurement (Absolute and Differential)

Common devices used for measuring pressure in HVAC applications (accuracy, range, application, and limitations are discussed in relation to HVAC work) are as follows:

- *Absolute pressure manometer*—not really suited to HVAC test work.
- *Diaphragm*—not really suited to HVAC test work.

- *Barometer (Hg manometer)*—not really suited to HVAC test work.
- *Micromanometer*—not usually portable, but suitable for fixed measurement of pressure differentials across filter, coils, etc. Cost around $30 up.
- *Draft gauges*—can be portable and used for either direct pressure or pressure differential. From $30 up.
- *Manometers*—can be portable. Used for direct pressure reading and with Pitot tubes for air flows. Very useful. Costs from $20 up.
- *Swing vane gauges*—can be portable. Usually used for air flow. Costs about $30.
- *Bourdon tube gauges*—very useful for measuring all forms of system fluid pressures from 5 psi up. Costs vary greatly, from $10 up. Special types for refrigeration plants.

Humidity Measurement

The data given below indicate the type of instruments available for humidity measurement. The following indicates equipment suitable for HVAC applications:

- *Psychrometers*—basically these are wet and dry bulb thermometers. They can be fixed on a portable stand or mounted in a frame with a handle for revolving in air. Costs are low ($10 to $30) and they are convenient to use.
- *Dewpoint hygrometers*—not considered suitable for HVAC test work.
- *Dimensional change*—device usually consists of a "hair," which changes in length proportionally with humidity changes. Not usually portable, fragile, and only suitable for limited temperature and humidity ranges.
- *Electrical conductivity*—can be compact and portable but of a higher cost (from $200 up). Very convenient to use.
- *Electrolytic*—as above, but for very low temperature ranges. Therefore unsuitable for HVAC test work.
- *Gravimeter*—not suitable.

IDENTIFYING STEAM AND UTILITY COSTS

Steam and utility costs are significant for most plants. It is important to quantify usage, fuel costs as a function of production. Figure 3-5 illustrates a typical steam and utility cost report. This report enables the plant manager to evaluate the total Btus of fuel consumed, the total fuel cost, and the total steam generation cost as a function of production. This report is issued monthly.

Since each plant has the same report, plant to plant comparisons are made and the effectiveness of the energy use is measured.

STEAM AND UTILITY COST REPORT

Preparatory Production [ ] **Lbs C & F** **Month** [ ] **19** [ ]

Tire Production ______ **Operating Days** [ ]

A. STEAM — Quantities

1. Actual Steam Generated [000 lbs]
   a) Press and Temp ___ psi ___ °F ___ 000 lbs
   b) Press and Temp ___ psi ___ °F ___ 000 lbs
   c) Press and Temp ___ psi ___ °F ___ 000 lbs
2. Factor of Evaporation (ASME Standard) a) ___ b) ___ c) ___
3. Equivalent Steam (No. 1 x No. 2) [000 lbs]
4. Blowdown [ ] lbs
   Heat Loss in BTU from Blowdown ______ BTU
   BTU/Lb-Gal-Cu Ft

	Total $	Total	
5. Coal [ ]	[ ]	[ ]	tons
6. Oil [ ]	[ ]	[ ]	gals
7. Gas [ ]	[ ]	[ ]	MCF
8. Misc [ ]	[ ]	[ ]	Units ______ Fuel ______

9. Total BTU of Fuel Consumed [000,000 BTU]
10. Boiler Efficiency $\frac{(970.3 \times \text{No. } 3) + \text{No. } 4}{\text{No. } 9}$ = [ ] %

B. OPERATING CONDITIONS AND DATA IN BOILER PLANT

Quantity of Feedwater Makeup [000] (Gal.)

Water Treating Chemicals
a) Phosphate ______ lbs
b) Salt ______ lbs
c) ______ ______ lbs
d) ______ ______ lbs
e) ______ ______ lbs
f) ______ ______ lbs
g) ______ ______ lbs

Ave. Feedwater Temp. ______ °F

Max Steam Demand ______ lbs/hr

Equivalent Steam Consumption Per 100 lbs C&F ______ lbs

	Cost	Unit Cost $/1000 Lbs Equiv. Steam
11. Total Fuel Cost (No. 5 + No. 6 + No. 7 + No. 8)	$ [ ]	$ ______
12. Operating Labor (include wage benefits)	$ [ ]	$ ______
13. Operating Supplies (water, chemicals, etc.)	$ ______	
14. Electricity for Power Plant Auxiliaries ______ KWH	$ ______	
15. Maintenance Charges	$ ______	
16. Other Miscellaneous Charges	$ ______	
17. Total Operating Cost (No. 11 thru No. 16)	$ [ ]	$ ______
18. Total Steam Generation Cost (No. 17 + fixed Cost)	$ [ ]	$ ______

Figure 3-5. Steam and Utility Cost Report

CALCULATING THE ENERGY CONTENT OF THE PROCESS

Knowing the energy content of the plant's process is the first step in understanding how to reduce its cost. Using energy more efficiently reduces the product cost, thus increasing profits. In order to account for the process energy content, all energy that enters and leaves a plant during a given period must be measured. Figure 3-6 illustrates energy content of a process report. The report applies to any manufacturing operation, whether it is a pulp mill, steel mill, or assembly line. This report enables one to quickly identify energy inefficient operations.

Attention can then be focused on which equipment should be replaced and what maintenance programs should be initiated. This report also focuses attention on the choice of raw materials. By using Btus per unit of production, measurable goals can be set. This report will also identify opportunities where energy usage can be reduced. The energy content of raw materials can be estimated by using the heating values indicated in Table 3-13.

Example Problem 3-1

Comment on energy content by modifying process No.1 as follows:

	Monthly Usage Rate	
	Process No.1	*Modified Process No.1*
Ethane	30,000 lb	50,000 lb
Steam	250×10^6 lb	200×10^6 lb
Electricity	0.5×10^6 kWh	0.8×10^6 kWh
Natural gas	350×10^6 ft^3	300×10^6 ft^3

Assuming a Btu content of steam of 1077 Btu/lb, compute the net energy content per process.

Answer
Process No.1
Energy of Raw Materials

	Total Usage	*Btus Per Unit*	*Total Btus*
Ethane	30,000 lb	From Table 3-13	0.6×10^9
		22,323 Btu/lb	

ENERGY CONTENT OF PROCESS REPORT

Process_____ 〔1〕 Total Units Produced_____

ENERGY OF RAW MATERIALS AND MISCELLANEOUS REQUIREMENTS

	Total Usage		Btu's Per Unit		Total Btu's
a.	_____	x	_____	=	_____
b.	_____		_____		_____
c.	_____		_____		_____
d.	_____		_____		_____

〔2〕 Total Raw Material _____ Btu's

ENERGY OF MAJOR UTILITIES
(ELECTRICAL, STEAM, COOLING WATER, NATURAL GAS, ETC.)

	Total Usage		Btu's Per Unit		Total Btu's
a.	_____	x	_____	=	_____
b.	_____		_____		_____
c.	_____		_____		_____

〔3〕 Total Energy Re-
quired from Utilities_____ Btu's

ENERGY CREDIT AS A RESULT OF BY-PRODUCTS

	Total Output		Btu's Per Unit		Total Btu's
a.	_____	x	_____	=	_____
b.	_____		_____		_____
c.	_____		_____		_____

〔4〕 Energy Credit _____ Btu's

〔5〕 NET ENERGY CONTENT OF PROCESS 〔2〕 + 〔3〕 - 〔4〕 _____ Btu's

〔6〕 ENERGY CONTENT PER UNIT OF
PRODUCTION 〔5〕 ÷ 〔1〕 _____ Btu's
Per
Unit

Figure 3-6. Energy Content of Process Report

Energy of Major Utilities

Steam	250×10^6	lb	1077 Btu/lb	269.2×10^9
Electricity	0.5×10^6	kWh	From Table 1-2	5×10^9
			10,000 Btu/kWh	
Natural gas	350×10^9	ft^3	From Table 1-1	350×10^9
			1000 Btu/ft^3	_____
		Net energy content for Process No.1		624×10^9

Table 3-13. Heat of Combustion for Raw Materials

	Formula	Gross Heat of Combustion Btu/lb
Raw Material		
Carbon	C	14,093
Hydrogen	H_2	61,095
Carbon monoxide	CO	4,347
Paraffin Series		
Methane	CH_4	23,875
Ethane	C_2H_4	22,323
Propane	C_3H_8	21,669
n-Butane	C_4H_{10}	21,321
Isobutane	C_4H_{10}	21,271
n-Pentane	C_5H_{12}	21,095
Isopentane	C_5H_{12}	21,047
Neopentane	C_5H_{12}	20,978
n-Hexane	C_6H_{14}	20,966
Olefin Series		
Ethylene	C_2H_4	21,636
Propylene	C_3H_6	21,048
n -Butene	C_4H_8	20,854
Isobutene	C_4H_8	20,737
n -Pentene	C_5H_{10}	20,720
Aromatic Series		
Benzene	C_6H_6	18,184
Toluene	C_7H_8	18,501
Xylene	C_8H_{10}	18,651
Miscellaneous Gases		
Acetylene	C_2H_2	21,502
Naphthalene	$C_{10}H_8$	17,303
Methyl alcohol	CH_3OH	10,258
Ethyl alcohol	C_2H_5OH	13,161
Ammonia	NH_3	9,667

Source: NBS Handbook 115.

Answer

Process No. 1
Energy of Raw Materials

	Total Usage		Btus Per Unit	Total Btus
Ethane	30,000	lb	From Table 3-13	0.6×10^9
	22,323	Btu/lb		

Energy of Major Utilities

Steam	250×10^6	lb	1077 Btu/lb	269.2×10^9
Electricity	0.5×10^6	kWh	From Table 1-2	5×10^9
	10,000	Btu/kWh		
Natural gas	350×10^6	ft^3	From Table 1-1	350×10^9
			1000 Btu/ft^3	_____
		Net energy content for Process No. 1		624×10^9

Modified Process No.1
Energy of Raw Materials

Ethane	50,000	lb	From Table 3-13	1.1×10^9
			22,323 Btu/lb	

Energy of Major Utilities

Steam	200×10^6	lb	1077 Btu/lb	215.4×10^9
Electricity	0.8×10^6	lb	From Table 1-2	8×10^9
			10,000 Btu/kWh	
Natural gas	300×10^6	ft^3	From Table 1-1	300×10^9
			1000 Btu/ft^3	_____
		Net energy content for Process No. 1		524×10^9

Modified Process No. 1 saves 100×10^9 Btus per month.

4

Electrical System Optimization

This chapter is divided into three main sections: *Power Distribution, Lighting Efficiency,* and *Energy Management Systems.*

By understanding the basic concepts of electrical systems it is possible to reduce energy consumption 25% or more. The first step is to analyze the billing structure. It may be possible to negotiate a better tariff rate with the local utility or modify the facility operation to qualify for a lower rate. In addition, specified charges or discounts for power factor, time of day or demand will determine if certain electrical efficiency measures are economically justified. This chapter reviews the basic parameters required to make sound energy engineering decisions.

I—Power Distribution

ELECTRICAL RATE TARIFF

The basic electrical rate charges contain the following elements:

Billing Demand—The maximum kilowatt requirement over a 15-, 30-, or 60-minute interval.

Load Factor—The ratio of the average load over a designated period to the peak demand load occurring in that period.

Power Factor—The ratio of resistive power to apparent power. Traditionally electrical rate tariffs have a decreasing kilowatt hour (kWh) charge with usage. This practice is likely to gradually phase out. New tariffs are containing the following elements:

Time of Day—Discounts are allowed for electrical usage during off-peak hours.

Ratchet Rate—The billing demand is based on 80-90% of peak demand for any one month. The billing demand will remain at that ratchet for 12 months even though the actual demand for the succeeding months may be less.

THE POWER TRIANGLE

The total power requirement of a load is made up of two components, namely the resistive part and the reactive part. The resistive portion of a load cannot be added directly to the reactive component since it is essentially 90 degrees out of phase with the other. The pure resistive power is known as the watt, while the reactive power is referred to as the reactive volt amperes. To compute the total volt ampere load, it is necessary to analyze the power triangle indicated below.

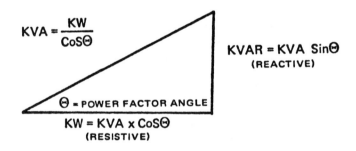

K =	1000
W=	Watts
VA =	Volt Amperes
VAR =	Volt Amperes Reactive
Θ =	Angle Between kVA and kW
$CoS\Theta$ =	Power Factor

$$\text{Tan } \Theta = \frac{\text{kVAR}}{\text{kW}}$$

The windings of transformers and motors are usually connected in a wye or delta configuration. The relationships for line and phase voltages and currents are illustrated by Figure 4-1.

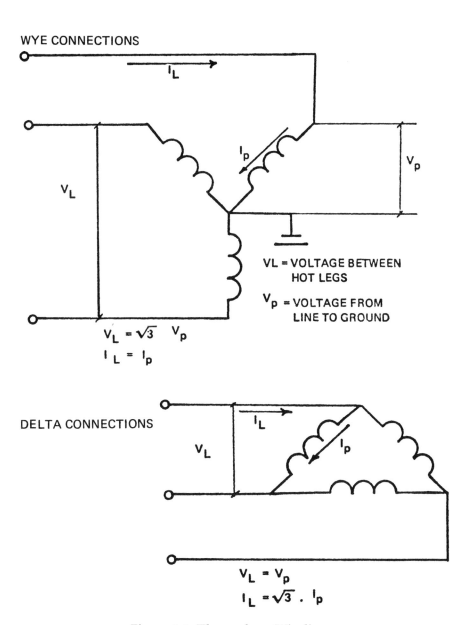

Figure 4-1. Three-phase Windings

For a balanced 3-phase load

$$\text{Power} = \underbrace{\sqrt{3} \quad V_L \quad I_L} \quad \text{CoS}\Theta \qquad \qquad \textit{Formula (4-1)}$$

Watts =	Volt	Power
	Amperes	Factor

For a balanced 1-phase load

$$P = V_L\, I_L \, \text{CoS}\Theta \qquad \qquad \textit{Formula (4-2)}$$

The primary windings of 13.8 kV – 480-volt unit substations are usually delta-connected with the secondary wye-connected.

MOTOR HORSEPOWER

The standard power rating of a motor is referred to as a horsepower. In order to relate the motor horsepower to a kilowatt (kW), multiply the horsepower by .746 (conversion factor) and divide by the motor efficiency.

$$\text{kVA} = \frac{\text{HP} \times .746}{\eta \times \text{P.F.}} \qquad \qquad \textit{Formula (4-3)}$$

$$\begin{aligned}
\text{HP} &= \text{Motor Horsepower} \\
\eta &= \text{Efficiency of Motor} \\
\text{P.F.} &= \text{Power Factor of Motor}
\end{aligned}$$

Motor efficiencies and power factors vary with load. Typical values are shown in Table 4-1. Values are based on totally enclosed fan-cooled motors (TEFC) running at 1800 rpm "T" frame.

POWER FLOW CONCEPT

Power flowing is analogous to water flowing in a pipe. To supply several small water users, a large pipe services the plant at a high pressure. Several branches from the main pipe service various loads. Pressure re-

Table 4.1.

HP RANGE	3-30	40-100
η% at		
1/2 Load	83.3	89.2
3/4 Load	85.8	90.7
Full Load	86.2	90.9
P.F. at		
1/2 Load	70.1	79.2
3/4 Load	79.2	85.4
Full Load	83.5	87.4

ducing stations lower the main pressure to meet the requirements of each user. Similarly, a large feeder at a high voltage services a plant. Through switchgear breakers, the main feeder is distributed into smaller feeders. The switchgear breakers serve as a protector for each of the smaller feeders. Transformers are used to lower the voltage to the nominal value needed by the user.

ELECTRICAL EQUIPMENT

Electrical equipment commonly specified is as follows:
- *Switchgear—Breakers*—used to distribute power.
- *Unit Substation*—used to step down voltage. Consists of a high voltage disconnect switch, transformer and low-voltage breakers. Typical 480-volt transformer sizes are 300 kVA, 500 kVA, 750 kVA, 1000 kVA, 2500 kVA and 3000 kVA.
- *Motor Control Center* (MCC)—a structure which houses starters and circuit breakers or fuses for motor control. It consists of the following:
  (1) Thermal overload relays which guard against motor over loads,
  (2) Fuse disconnect switches or breakers which protect the cable and motor and can be used as a disconnecting means,
  (3) Contactors (relays) whose contacts are capable of opening and closing the power source to the motor.

MOTORS

- *Squirrel Cage Induction Motors* are commonly used. These motors require three power loads. For two-speed applications several different types of motors are available. Depending on the process requirements such as constant horsepower or constant torque, the windings of the motor are connected differently. The theory of two-speed operation is based on Formula 4-4.

$$\text{Frequency} = \frac{\text{No. of poles} \times \text{speed}}{120} \qquad \textit{Formula (4-4)}$$

Thus, if the frequency is fixed, the effective number of motor poles should be changed to change the speed. This can be accomplished by the manner in which the windings are connected. Two-speed motors require six power leads.

- *D.C. Motors* are used where speed control is essential. The speed of a D.C. motor is changed by varying the field voltage through a rheostat. A D.C. motor requires two power wires to the armature and two smaller cables for the field.

- *Synchronous Motors* are used when constant speed operation is essential. Synchronous motors are sometimes cheaper in the large horsepower categories when slow speed operation is required. Synchronous motors also are considered for power factor correction. A .8 P.F. synchronous motor will supply corrective kVARs to the system. A synchronous motor requires A.C. for power and D.C. for the field. Since many synchronous motors are self-excited, only the power cables are required to the motor.

IMPORTANCE OF POWER FACTOR

Transformer size is based on kVA The closer Θ equals $0°$ or power factor approaches unity, the smaller the kVA. Many times utility companies have a power factor clause in their contract with the customer. The statement usually causes the customer to pay an additional power rate if the power factor of the plant deviates substantially from unity. The utility company wishes to maximize the efficiency of its transformers and associated equipment.

POWER FACTOR CORRECTION

One problem facing the energy engineer is to estimate the power factor of a new plant and to install equipment such as capacitor banks or synchronous motors so that the overall power factor will meet the utility company's objectives.

Capacitor banks lower the total reactive kVAR by the value of the capacitors installed.

A second problem is to retrofit an existing plant so that the overall power factor desired is obtained.

Example Problem 4-1

A total motor horsepower load of 854 is made up of motors ranging from 40-100 horsepower. Calculate the connected kVA.

It is desired to operate the plant at a power factor of .95. What approximate capacitor bank is required?

Answer

$$kVA = \frac{HP \times .746}{Motor\ Eff. \times Motor\ P.F.}$$

From Table 4-1 at full load

$$Eff. = .909\ and\ P.F. = .87$$

$$kVA = \frac{854 \times .746 = 806}{.909 \times .87}$$

The plant is operating at a power factor of .87. The power factor of .87 corresponds to an angle of 29°.

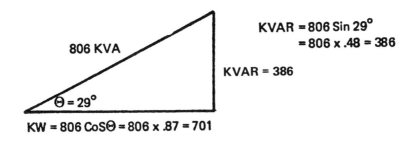

806 KVA

Θ = 29°

KW = 806 CoSΘ = 806 x .87 = 701

KVAR = 806 Sin 29°
= 806 x .48 = 386

KVAR = 386

A power factor of .95 is required.

$$CoS\Theta = .95$$
$$\Theta = 18°$$

The kVAR of 386 needs to be reduced by adding capacitors.

Remember kW does not change with different power factors, but kVA does.

Thus, the desired power triangle would look as follows:

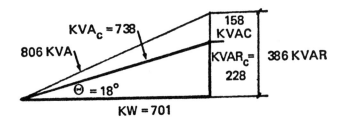

$$CoS\Theta = .95$$
$$\Theta = 18° \quad Sin\ 18° = .31$$

$$kVA_C = \frac{701}{CoS\Theta} = \frac{701}{.95} = 738$$

Note: Power factor correction reduces total kVA.

$$kVAR_C = 738\ Sin\ 18° = \quad 738 \times .31 = 228$$
Capacitance Bank $\quad = \quad 386\text{-}228$
$$158\ kVAC$$

Example Problem 4-2

The client wishes to know the expected power factor for a new plant. The lighting load is 40 kW. The plant is composed of two identical modules (2 motors for each equipment number listed). Remember that kVAs at different power factors cannot be added directly.

Motor List — Module 1

Motor No.	Description	HP	Voltage	Phase
AG-1	Agitator Motor	60	460	3
CF-3	Centrifuge Motor	100	460	3
FP-4	Feed Pump Motor	30	460	3
TP-5	Transfer Pump Motor	10	460	3
CTP-6	Cooling Tower	25	460	3
CT-9	Cooling Tower Motor	20	460	3
HF-I0	H&V Supply Fan Motor	40	460	3
HF-11	H&V Exhaust Fan Motor	20	460	3
BC-13	Brine Compressor Motor	50	460	3
C-16	Conveyor Motor	20	460	3
H-17	Hoist Motor	5	460	3

Answer

Based on the motor list, the plant power factor is estimated as follows:

Module 1

Lighting $kW_3 = 40$ Total

Motors 3-30	Motors 40-100
30	60
10	100
25	40
20	50
20	250
20	
5	
130	

At Full Load:

P.F. = 83.5	P.F. = 87.4
η = 86.2	η = 90.9

$$kVA_1 = \frac{130 \times .746}{.83 \times .86} = 135 \qquad kVA_2 = \frac{250 \times .746}{.90 \times .87} = 238$$

$$kW_1 = kVA\ CoS\Theta$$
$$= kVA\ .83$$
$$= 112$$

$$kW_2 = kVA_2\ CoS\Theta$$
$$= kVA\ .87$$
$$= 207$$

$$kVAR_1 = kVA_1\ Sin\Theta$$
$$\Theta = 33°$$
$$kVAR_1 = kVA_1 \times .54$$
$$= 135 \times .54$$
$$= 73$$

$$kVAR_2 = kVA_2\ Sin\Theta$$
$$\Theta = 29°$$
$$kVAR_2 = kVA_2 \times .48$$
$$kVAR_2 = 115$$

$$kW_{total} = kW_1 + kW_2 + kW_1 + kW_2 + kW_3 = 678\ kW$$
$$\qquad\qquad \text{Module 1} \qquad \text{Module 2}$$

$$kVAR_{total} = kVAR_1 + kVAR_2 + kVAR_1 + kVAR_2$$
$$\qquad\qquad\quad \text{Module 1} \qquad\qquad \text{Module 2}$$
$$= 73 + 115 + 73 + 115 = 376$$

$$KVA_{total} = \sqrt{(687)^2 + (376)^2} = 775\ KVA$$

$$CoS\Theta = \frac{kW_{total}}{kVA_{total}} = \frac{678}{775} = .87$$

Shortcut Methods

A handy shortcut table which can be used to find the value of the capacitor required to improve the plant power factor is illustrated by Table 4-2.

WHERE TO LOCATE CAPACITORS

As indicated, the primary purpose of capacitors is to reduce the power consumption. Additional benefits are derived by capacitor location. Figure 4-2 indicates typical capacitor locations. Maximum benefit of capacitors is derived by locating them as close as possible to the load. At this location, its kilovars are confined to the smallest possible segment, decreasing the load current. This, in turn, will reduce power losses of the system substantially. Power losses are proportional to the square of the current. When power losses are reduced, voltage at the motor increases; thus, motor performance also increases.

Locations C1A, C1B and C1C of Figure 4-2 indicate three different ar-

Table 4-2. Shortcut Method — Power Factor Correction

KW MULTIPLIERS FOR DETERMINING CAPACITOR KILOVARS

DESIRED POWER-FACTOR IN PERCENTAGE

ORIGINAL POWER FACTOR IN PERCENTAGE

	80	81	82	83	84	85	86	87	88	89	90	91	92	93	94	95	96	97	98	99	100
50	.982	1.008	1.034	1.060	1.086	1.112	1.139	1.165	1.192	1.220	1.248	1.276	1.303	1.337	1.369	1.403	1.441	1.481	1.529	1.590	1.732
51	.936	.962	.988	1.014	1.040	1.066	1.093	1.119	1.146	1.174	1.202	1.230	1.257	1.291	1.323	1.357	1.395	1.435	1.483	1.544	1.688
52	.894	.920	.946	.972	.998	1.024	1.051	1.077	1.104	1.132	1.160	1.188	1.215	1.249	1.281	1.315	1.353	1.393	1.441	1.502	1.644
53	.850	.876	.902	.928	.954	.980	1.007	1.033	1.060	1.088	1.116	1.144	1.171	1.205	1.237	1.271	1.309	1.349	1.397	1.458	1.600
54	.809	.835	.861	.887	.913	.939	.966	.992	1.019	1.047	1.075	1.103	1.130	1.164	1.196	1.230	1.268	1.308	1.356	1.417	1.559
55	.769	.795	.821	.847	.873	.899	.926	.952	.979	1.007	1.035	1.063	1.090	1.124	1.156	1.190	1.228	1.268	1.316	1.377	1.519
56	.730	.756	.782	.808	.834	.860	.887	.913	.940	.968	.996	1.024	1.051	1.085	1.117	1.151	1.189	1.229	1.277	1.338	1.480
57	.692	.718	.744	.770	.796	.822	.849	.875	.902	.930	.958	.986	1.013	1.047	1.079	1.113	1.151	1.191	1.239	1.300	1.442
58	.655	.681	.707	.733	.759	.785	.812	.838	.865	.893	.921	.949	.976	1.010	1.042	1.076	1.114	1.154	1.202	1.263	1.405
59	.618	.644	.670	.696	.722	.748	.775	.801	.828	.856	.884	.912	.939	.973	1.005	1.039	1.077	1.117	1.165	1.226	1.368
60	.584	.610	.636	.662	.688	.714	.741	.767	.794	.822	.849	.878	.905	.939	.971	1.005	1.043	1.083	1.131	1.192	1.334
61	.549	.575	.601	.627	.653	.679	.706	.732	.759	.787	.815	.843	.870	.904	.936	.970	1.008	1.048	1.096	1.157	1.299
62	.515	.541	.567	.593	.619	.645	.672	.698	.725	.753	.781	.809	.836	.870	.902	.936	.974	1.014	1.062	1.123	1.265
63	.483	.509	.535	.561	.587	.613	.640	.666	.693	.721	.749	.777	.804	.838	.870	.904	.942	.982	1.030	1.091	1.233
64	.450	.476	.502	.528	.554	.580	.607	.633	.660	.688	.716	.744	.771	.805	.837	.871	.909	.949	.997	1.058	1.200
65	.419	.445	.471	.497	.523	.549	.576	.602	.629	.657	.685	.713	.740	.774	.806	.840	.878	.918	.966	1.027	1.169
66	.388	.414	.440	.466	.492	.518	.545	.571	.598	.626	.654	.682	.709	.743	.775	.809	.847	.887	.935	.996	1.138
67	.358	.384	.410	.436	.462	.488	.515	.541	.568	.596	.624	.652	.679	.713	.745	.779	.817	.857	.905	.966	1.108
68	.329	.355	.381	.407	.433	.459	.486	.512	.539	.567	.595	.623	.650	.684	.716	.750	.788	.828	.876	.937	1.079
69	.299	.325	.351	.377	.403	.429	.456	.482	.509	.537	.565	.593	.620	.654	.686	.720	.758	.798	.840	.907	1.049
70	.270	.296	.322	.348	.374	.400	.427	.453	.480	.508	.536	.564	.591	.625	.657	.691	.729	.769	.811	.878	1.020
71	.242	.268	.294	.320	.346	.372	.399	.425	.452	.480	.508	.536	.563	.597	.629	.663	.701	.741	.783	.850	.992
72	.213	.239	.265	.291	.317	.343	.370	.396	.423	.451	.479	.507	.534	.568	.600	.634	.672	.712	.754	.821	.963
73	.186	.212	.238	.264	.290	.316	.343	.369	.396	.424	.452	.480	.507	.541	.573	.607	.645	.685	.727	.794	.936
74	.159	.185	.211	.237	.263	.289	.316	.342	.369	.397	.425	.453	.480	.514	.546	.580	.618	.658	.700	.767	.909
75	.132	.158	.184	.210	.236	.262	.289	.315	.342	.370	.398	.426	.453	.487	.519	.553	.591	.631	.673	.740	.882
76	.105	.131	.157	.183	.209	.235	.262	.288	.315	.343	.371	.399	.426	.460	.492	.526	.564	.604	.652	.713	.855
77	.079	.105	.131	.157	.183	.209	.236	.262	.289	.317	.345	.373	.400	.434	.466	.500	.538	.578	.620	.687	.829
78	.053	.079	.105	.131	.157	.183	.210	.236	.263	.291	.319	.347	.374	.408	.440	.474	.512	.552	.594	.661	.803
79	.026	.052	.078	.104	.130	.156	.183	.209	.236	.264	.292	.320	.347	.381	.413	.447	.485	.525	.567	.634	.776
80	.000	.026	.052	.078	.104	.130	.157	.183	.210	.238	.266	.294	.321	.355	.387	.421	.450	.499	.541	.608	.750
81	—	.000	.026	.052	.078	.104	.131	.157	.184	.212	.240	.268	.295	.329	.361	.395	.433	.473	.515	.582	.724
82	—	—	.000	.026	.052	.078	.105	.131	.158	.186	.214	.242	.269	.303	.335	.369	.407	.447	.489	.556	.698
83	—	—	—	.000	.026	.052	.079	.105	.132	.160	.188	.216	.243	.277	.309	.343	.381	.421	.463	.530	.672
84	—	—	—	—	.000	.026	.053	.079	.106	.134	.162	.190	.217	.251	.283	.317	.355	.395	.437	.504	.645
85	—	—	—	—	—	.000	.027	.053	.080	.108	.136	.164	.191	.225	.257	.291	.329	.369	.417	.478	.620

Example: Total kw input of load from wattmeter reading 100 kw at a power factor of 60%. The leading reactive kvar necessary to raise the power factor to 90% is found by multiplying the 100 kw by the factor found in the table, which is .849. Then 100 kw × 0.849 = 84.9 kvar. Use 85 kvar.

Reprinted by permission of Federal Pacific Electric Company.

rangements at the load. Note that in all three locations extra switches are not required, since the capacitor is either switched with the motor starter or the breaker before the starter. Case C1A is recommended for new installation, since the maximum benefit is derived and the size of the motor thermal protector is reduced. In Case C1B, as in Case C1A, the capacitor is energized only when the motor is in operation. Case C1B is recommended in cases where the installation is existing and the thermal protector does not need to be re-sized. In position C1C, the capacitor is permanently connected to the circuit but does not require a separate switch, since it can be disconnected by the breaker before the starter.

It should be noted that the rating of the capacitor should *not* be greater than the no-load magnetizing kVAR of the motor. If this condition exists, damaging overvoltage or transient torques can occur. This is why most motor manufacturers specify maximum capacitor ratings to be applied to specific motors.

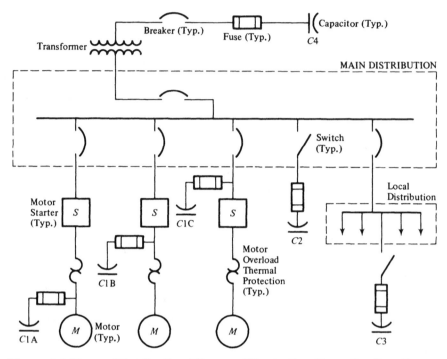

Figure 4-2. Power Distribution Diagram Illustrating Capacitor Locations

The next preference for capacitor locations as illustrated by Figure 4-2 is at locations C2 and C3. In these locations, a breaker or switch will be required. Location C4 requires a high voltage breaker. The advantage of locating capacitors at power centers or feeders is that they can be grouped together. When several motors are running intermittently, the capacitors are permitted to be on-line all the time, reducing the total power regardless of load.

ENERGY EFFICIENT MOTORS

Energy efficient motors are now available. These motors are approximately 30% more expensive than their standard counterpart. Based on the energy cost, it can be determined if the added investment is justified. With the emphasis on energy conservation, new lines of energy efficient motors are being introduced. Figures 4-3 and 4-4 illustrate a typical comparison between energy efficient and standard motors.

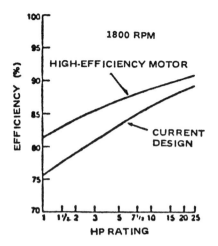

Figure 4-3. Efficiency vs. Horsepower Rating (Dripproof Motors)

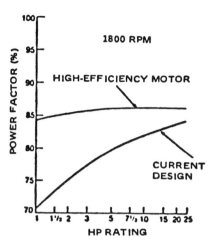

Figure 4-4. Power Factor vs. Horsepower Rating (Dripproof Motors)

II—LIGHTING EFFICIENCY

LIGHTING BASICS

About 20 percent of all electricity generated in the United States today is used for lighting.

By understanding the basics of lighting design, several ways to improve the efficiency of lighting systems will become apparent.

There are two common lighting methods used. One is called the lumen method, while the other is the point-by-point method. The lumen method assumes an equal footcandle level throughout the area. This method is used frequently by lighting designers since it is simplest; however, it wastes energy, since it is the light "at the task" which must be maintained and not the light in the surrounding areas. The point-by-point method calculates the lighting requirements for the task in question.

The point-by-point method makes use of the inverse-square law, which states that the illuminance at a point on a surface perpendicular to the light ray is equal to the luminous intensity of the source at that point divided by the square of the distance between the source and the point of calculation, as illustrated in Formula 4-5.

$$E = \frac{I}{D^2}$$ *Formula (4-5)*

Where
> $E =$ Illuminance in footcandles
> $I =$ Luminous intensity in candles
> $D =$ Distance in feet between the source and the point of calcu-
> lation.

If the surface is not perpendicular to the light ray, the appropriate trigonometrical functions must be applied to account for the deviation.

Lumen Method

A footcandle is the illuminance on a surface of one square foot in area having a uniformly distributed flux of one lumen. From this definition, the lumen method is developed and illustrated by Formula 4-6.

$$N = \frac{F_1 \times A}{Lu \times L_1 \times L_2 \times Cu}$$ *Formula (4-6)*

Where
> N is the number of lamps required.
> F_1 is the required footcandle level at the task. A footcandle is a measure of illumination; one standard candle power measured one foot away.
> A is the area of the room in square feet.
> Lu is the lumen output per lamp. A lumen is a measure of lamp intensity: its value is found in the manufacturer's catalogue.
> Cu is the coefficient of utilization. It represents the ratio of the lumens reaching the working plane to the total lumens generated by the lamp. The coefficient of utilization makes allowances for light absorbed or reflected by walls, ceilings, and the fixture itself. Its values are found in the manufacturer's catalogue.
> L_1 is the lamp depreciation factor. It takes into account that the lamp lumen depreciates with time. Its value is found in the manufacturer's catalogue.

L_2 is the luminaire (fixture) dirt depreciation factor. It takes into account the effect of dirt on a luminaire and varies with type of luminaire and the atmosphere in which it is operated.

The lumen method formula illustrates several ways lighting efficiency can be improved.

Faced with the desire to reduce their energy use,[1] lighting consumers have four options: i) reduce light levels, ii) purchase more efficient equipment, iii) provide light when needed at the task at the required level, and iv) add control and reduce lighting loads automatically. The multitude of equipment options to meet one or more of the above needs permits the consumer and the lighting designer-engineer to consider the trade-offs between the initial and operating costs based upon product performance (life, efficacy, color, glare, and color rendering).

Some definitions and terms used in the field of lighting will be presented to help consumers evaluate and select lighting products best suiting their needs. Then, some state-of-the-art advances will be characterized so that their benefits and limitations are explicit.

Lighting Terminology

Efficacy—Is the amount of visible light (lumens) produced for the amount of power (watts) expended. It is a measure of the efficiency of a process but is a term used in place of efficiency when the input (W) has different units than the output (lm) and expressed in lm/W.

Color Temperature—A measure of the color of a light source relative to a black body at a particular temperature expressed in degrees Kelvin (°K). Incandescents have a low color temperature (~2800°K) and have a red-yellowish tone; daylight has a high color temperature (~6000°K) and appears bluish. Today, the phosphors used in fluorescent lamps can be blended to provide any desired color temperature in the range from 2800°K to 6000°K.

Color Rendering—A parameter that describes how a light source renders a set of colored surfaces with respect to a black body light source at the same color temperature. The color rendering index (CRI) runs from 0 to 100. It depends upon the specific wavelengths of which the light is composed. A black body has a continuous spectrum and contains all of the colors in the visible spectrum. Fluorescent lamps and high intensity

---

[1]Source: *Lighting Systems Research*. R.R. Verderber

discharge lamps (HID) have a spectrum rich in certain colors and devoid in others. For example, a light source that is rich in blues and low in reds could appear white, but when reflected from a substance, it would make red materials appear faded. The same material would appear different when viewed with an incandescent lamp, which has a spectrum that is rich in red.

LIGHT SOURCES[2]

Figure 4-5 indicates the general lamp efficiency ranges for the generic families of lamps most commonly used for both general and supplementary lighting systems. Each of these sources is discussed briefly here. It is important to realize that in the case of fluorescent and high intensity discharge lamps, the figures quoted for "lamp efficacy" are for the lamp only and do not include the associated ballast losses. To obtain the total system efficiency, ballast input watts must be used rather than lamp watts to obtain an overall system lumen per watt figure. This will be discussed in more detail in a later section.

Incandescent lamps have the lowest lamp efficacies of the commonly used lamps. This would lead to the accepted conclusion that incandescent lamps should generally not be used for large-area, general lighting systems where a more efficient source could serve satisfactorily. However, this does not mean that incandescent lamps should never be used. There are many applications where the size, convenience, easy control, color rendering, and relatively low cost of incandescent lamps are suitable for a specific application.

General service incandescent lamps do not have good lumen maintenance throughout their lifetime. This is the result of the tungsten's evaporation off the filament during heating as it deposits on the bulb wall, thus darkening the bulb and reducing the lamp lumen output.

Efficient Types of Incandescents for Limited Use

Attempts to increase the efficiency of incandescent lighting while maintaining good color rendition have led to the manufacture of a number of energy-saving incandescent lamps for limited residential use.

Tungsten Halogen—These lamps vary from the standard incandescent

---

[2]Source: *Selection Criteria for Lighting Energy Management.* Roger L. Knott

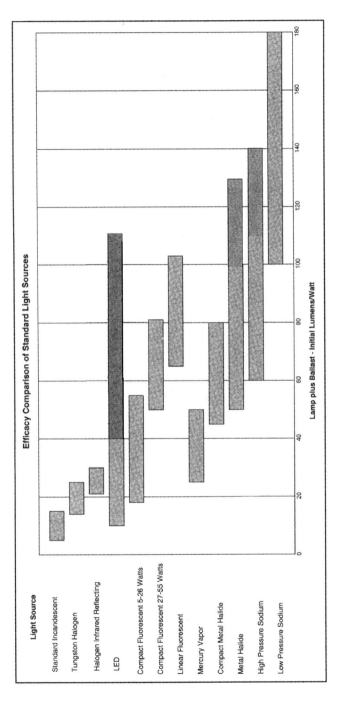

Figure 4-5. Efficiency of Various Light Sources

by the addition of halogen gases to the bulb. Halogen gases keep the glass bulb from darkening by preventing the filament's evaporation, thereby increasing lifetime up to four times that of a standard bulb. The lumen-per-watt rating is approximately the same for both types of incandescents, but tungsten halogen lamps average 94% efficiency throughout their extended lifetime, offering significant energy and operating cost savings. However, tungsten halogen lamps require special fixtures, and during operation the surface of the bulb reaches very high temperatures, so they are not commonly used in the home.

Reflector or R-Lamps—Reflector lamps are incandescents with an interior coating of aluminum that directs the light to the front of the bulb. Certain incandescent light fixtures, such as recessed or directional fixtures, trap light inside. Reflector lamps project a cone of light out of the fixture and into the room, so that more light is delivered where it is needed. In these fixtures, a 50-watt reflector bulb will provide better lighting and use less energy when substituted for a 100-watt standard incandescent bulb.

Reflector lamps are an appropriate choice for task lighting (because they directly illuminate a work area) and for accent lighting. Reflector lamps are available in 25, 30, 50, 75, and 150 watts. While they have a lower initial efficiency (lumens per watt) than regular incandescents, they direct light more effectively, so that more light is actually delivered than with regular incandescents. (See Figure 4-6.)

PAR Lamps—Parabolic aluminized reflector (PAR) lamps are reflector lamps with a lens of heavy, durable glass, which makes them an appropriate choice for outdoor flood and spot lighting. They are available in 75, 150, and 250 watts. They have longer lifetimes with less depreciation than standard incandescents.

ER Lamps—Ellipsoidal reflector (ER) lamps are ideally suited for recessed fixtures, because the beam of light produced is focused two inches ahead of the lamp to reduce the amount of light trapped in the fixture. In a directional fixture, a 75-watt ellipsoidal reflector lamp delivers more light than a 150-watt R-lamp. (See Figure 4-6.)

Mercury vapor lamps find limited use in today's lighting systems because fluorescent and other high intensity discharge (HID) sources have surpassed them in both lamp efficacy and system efficiency. Typical ratings for mercury vapor lamps range from about 30 to 70 lumens per watt. The primary advantages of mercury lamps are a good range of color, availability in sizes as low as 30 watts, long life and relatively low cost. However, fluorescent systems are available today which can do many of

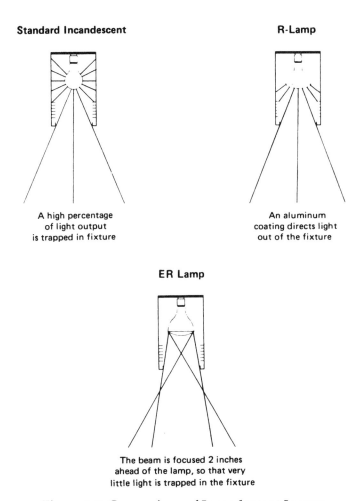

Standard Incandescent

A high percentage
of light output
is trapped in fixture

R-Lamp

An aluminum
coating directs light
out of the fixture

ER Lamp

The beam is focused 2 inches
ahead of the lamp, so that very
little light is trapped in the fixture

Figure 4-6. Comparison of Incandescent Lamps

the jobs mercury used to do and they do it more efficiently. There are still
places for mercury vapor lamps in lighting system design, but they are becoming fewer as technology advances in fluorescent and higher efficacy
HID sources.

Fluorescent lamps have made dramatic advances in the last 10
years. From the introduction of reduced wattage lamps in the mid-1970s,
to the marketing of several styles of low wattage, compact lamps recently, there has been a steady parade of new products. Lamp efficacy now
ranges from about 30 lumens per watt to near 90 lumens per watt. The

range of colors is more complete than mercury vapor, and lamp manufacturers have recently made significant progress in developing fluorescent and metal halide lamps which have much more consistent color rendering properties allowing greater flexibility in mixing these two sources without creating disturbing color mismatches. The recent compact fluorescent lamps open up a whole new market for fluorescent sources. These lamps permit design of much smaller luminaries which can compete with incandescent and mercury vapor in the low cost, square or round fixture market which the incandescent and mercury sources have dominated for so long. While generally good, lumen maintenance throughout the lamp lifetime is a problem for some fluorescent lamp types.

Energy Efficient "Plus" Fluorescents[3]

The energy efficient "plus" fluorescents represent the second generation of improved fluorescent lighting. These bulbs are available for replacement of standard 4-foot, 40-watt bulbs and require only 32 watts of electricity to produce essentially the same light levels. To the authors' knowledge, they are not available for 8-foot fluorescent bulb retrofit. The energy efficient plus fluorescents require a ballast change. The light output is similar to the energy efficient bulbs, and the two types may be mixed in the same area if desired.

Examples of energy efficient plus tubes include the Super-Saver Plus by Sylvania and General Electric's Watt Mizer Plus.

Energy Efficient Fluorescents System Change

The third generation of energy efficient fluorescents requires both a ballast and a fixture replacement. The standard 2-foot by 4-foot fluorescent fixture, containing four bulbs and two ballasts, requires approximately 180 watts (40 watts per tube and 20 watts per ballast). The new-generation fluorescent manufacturers claim the following:

- General Electric—"Optimizer" requires only 116 watts with a slight reduction in light output.
- Sylvania—"Octron" requires only 132 watts with little reduction in light level.
- General Electric—"Maximizer" requires 169 watts but supplies 22 percent more light output.

---

[3]Source: *Fluorescent Lighting—An Expanding Technology,* R.E. Webb, M.G. Lewis, W.C. Turner.

The fixtures and ballasts designed for the third-generation fluorescents are not interchangeable with earlier generations.

Metal halide lamps fall into a lamp efficacy range of approximately 75-125 lumens per watt. This makes them more energy efficient than mercury vapor but somewhat less so than high pressure sodium. Metal halide lamps generally have fairly good color rendering qualities. While this lamp displays some very desirable qualities, it also has some distinct drawbacks including relatively short life for an HID lamp, long restrike time to restart after the lamp has been shut off (about 15-20 minutes at 70°F) and a pronounced tendency to shift colors as the lamp ages. In spite of the drawbacks, this source deserves serious consideration and is used very successfully in many applications.

High pressure sodium lamps introduced a new era of extremely high efficacy (60-140 lumens/watt) in a lamp which operates in fixtures having construction very similar to those used for mercury vapor and metal halide. When first introduced, this lamp suffered from ballast problems. These have now been resolved and luminaries employing high quality lamps and ballasts provide very satisfactory service. The 24,000-hour lamp life, good lumen maintenance and high efficacy of these lamps make them ideal sources for industrial and outdoor applications where discrimination of a range of colors is not critical.

The lamp's primary drawback is the rendering of some colors. The lamp produces a high percentage of light in the yellow range of the spectrum. This tends to accentuate colors in the yellow region. Rendering of reds and greens shows a pronounced color shift. This can be compensated for in the selection of the finishes for the surrounding areas, and, if properly done, the results can be very pleasing. In areas where color selection, matching and discrimination are necessary, high pressure sodium should not be used as the only source of light. It is possible to gain quite satisfactory color rendering by mixing high pressure sodium and metal halide in the proper proportions. Since both sources have relatively high efficacies, there is not a significant loss in energy efficiency by making this compromise.

High pressure sodium has been used quite extensively in outdoor applications for roadway, parking and façade or security lighting. This source will yield a high efficiency system; however, it should be used only with the knowledge that foliage and landscaping colors will be severely distorted where high pressure sodium is the only, or predominant, illumi-

nant. Used as a parking lot source, there may be some difficulty in identification of vehicle colors in the lot. It is necessary for the designer or owner
to determine the extent of this problem and what steps might be taken to
alleviate it.

Recently lamp manufacturers have introduced high pressure sodium lamps with improved color rendering qualities. However, the improvement in color rendering was not gained without cost—the efficacy
of the color-improved lamps is somewhat lower approximately 90 lumens
per watt.

Low pressure sodium lamps provide the highest efficacy of any of
the sources for general lighting with values ranging up to 180 lumens per
watt. Low pressure sodium produces an almost pure yellow light with
very high efficacy, and renders all colors gray except yellow or near yellow. This effect results in no color discrimination under low pressure sodium lighting; it is suitable for use in a very limited number of applications.
It is an acceptable source for warehouse lighting where it is only necessary
to read labels but not to choose items by color. This source has application
for either indoor or outdoor safety or security lighting as long as color rendering is not important.

In addition to these primary sources, there are a number of retrofit
lamps which allow use of higher efficacy sources in the sockets of existing
fixtures. Therefore, metal halide or high pressure sodium lamps can be ret-
rofitted into mercury vapor fixtures, or self-ballasted mercury lamps can
replace incandescent lamps. These lamps all make some compromises in
operating characteristics, life and/ or efficacy.

Figure 4-7 presents data on the efficacy of each of the major lamp
types in relation to the wattage rating of the lamps. Without exception, the
efficacy of the lamp increases as the lamp wattage rating increases.

The lamp efficacies discussed here have been based on the lumen
output of a new lamp after 100 hours of operation or the "initial lumens."
Not all lamps age in the same way. Some lamp types, such as lightly loaded fluorescent and high pressure sodium, hold up well and maintain their
lumen output at a relatively high level until they are into, or past, middle
age. Others, as represented by heavily loaded fluorescent, mercury vapor
and metal halide, decay rapidly during their early years and then coast
along at a relatively lower lumen output throughout most of their useful
life. These factors must be considered when evaluating the various sources for overall energy efficiency.

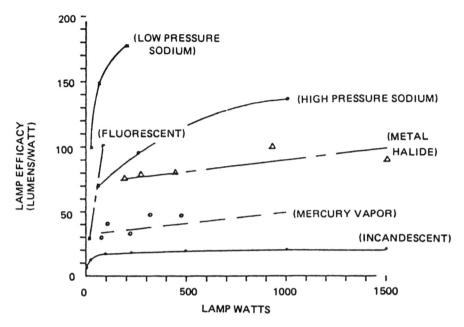

Figure 4-7. Lamp Efficacy (Does Not Include Ballast Losses)

Incandescent Replacement

The most efficacious lamps that can be used in incandescent sockets are the compact fluorescent lamps. The most popular systems are the twin tubes and double twin tubes. These are closer to the size and weight of the incandescent lamp than the earlier type of fluorescent (circline) replacements.

Twin tubes with lamp wattages from 5 to 13 watts provide amounts of light ranging from 240 to 850 lumens. Table 4-3 lists the characteristics of various types of incandescent and compact fluorescent lamps that can be used in the same type sockets.

The advantages of the compact fluorescent lamps are larger and increased efficacy, longer life and reduced total cost. The cost per 10^6 lumen hours of operating the 75-watt incandescent and the 18-watt fluorescent is $5.47 and $3.29, respectively. This is based upon an energy cost of $0.075 per kWh and lamp costs of $0.70 and $17 and the 75-watt incandescent and 18-watt fluorescent lamps, respectively. The circline lamps were much larger and heavier than the incandescents and would fit in a limited number of fixtures. The twin tubes are only slightly heavier and larger than the

Table 4-3. Lamp Characteristics

Lamp Type (Total Input Power)*	Lamp Power (W)	Light Output (lumens)	Lamp Life (hour)	Efficacy (lm/W)
100 W (Incandescent)	100	1750	750	18
75 W (Incandescent)	75	1200	750	16
60 W (Incandescent)	60	890	1000	15
40 W (Incandescent)	40	480	1500	12
25 W (Incandescent)	25	238	2500	10
22 W (Fl. Circline)	18	870	9000	40
44 W (Fl. Circline)	36	1750	9000	40
7 W (Twin)	5	240	10000	34
10 W (Twin)	7	370	10000	38
13 W (Twin)	9	560	10000	43
19 W (Twin)	13	850	10000	45
18 W (Solid-State)**	—	1100	7500	61

*Includes ballast losses.
**Operated at high frequency.

equivalent incandescent lamp. However, there are some fixtures that are too small for them to be employed.

The narrow tube diameter compact fluorescent lamps are now possible because of the recently developed rare earth phosphors. These phosphors have an improved lumen depreciation at high lamp power loadings. The second important characteristic of these narrow band phosphors is their high efficiency in converting the ultraviolet light generated in the plasma into visible light. By proper mixing of these phosphors, the color characteristics (color temperature and color rendering) are similar to the incandescent lamp.

There are two types of compact fluorescent lamps. In one type of lamp system, the ballast and lamp are integrated into a single package; in the second type, the lamp and ballast are separate, and when a lamp burns out it can be replaced. In the integrated system, both the lamp and the ballast are discarded when the lamp burns out.

It is important to recognize when purchasing these compact fluorescent lamps that they provide the equivalent light output of the lamps being replaced. The initial lumen output for the various lamps is shown in Table 4-3.

Lighting Efficiency Options

Several lighting efficiency options are illustrated below: (Refer to Formula 4-6.)

Footcandle Level—The footcandle level required is that level at the task. Footcandle levels can be lowered to one third of the levels for surrounding areas such as aisles. (A minimum 20-footcandle level should be maintained.)

The placement of the lamp is also important. If the luminaire can be lowered or placed at a better location, the lamp wattage may be reduced.

Coefficient of Utilization (Cu)—The color of the walls, ceiling, and floors, the type of luminaire, and the characteristics of the room determine the *Cu*. This value is determined based on manufacturer's literature. The *Cu* can be improved by analyzing components such as lighter colored walls and more efficient luminaires for the space.

Lamp Depreciation Factor and *Dirt Depreciation Factor*—These two factors are involved in the maintenance program. Choosing a luminaire which resists dirt build-up, group relamping and cleaning the luminaire will keep the system in optimum performance. Taking these factors into account can reduce the number of lamps initially required.

The *Light Loss Factor (LLF)* takes into account that the lamp lumen depreciates with time (L_1), that the lumen output depreciates due to dirt build-up (L_2), and that lamps burn out (L_3). Formula 4-7 illustrates the relationship of these factors.

$$LLF = L_1 \times L_2 \times L_3 \qquad\qquad Formula\ (4\text{-}7)$$

To reduce the number of lamps required which in turn reduces energy consumption, it is necessary to increase the overall light loss factor. This is accomplished in several ways. One is to choose the luminaire which minimizes dust build-up. The second is to improve the maintenance program to replace lamps prior to burn-out. Thus if it is known that a group relamping program will be used at a given percentage of rated life, the appropriate lumen depreciation factor can be found from manufacturer's data. It may be decided to use a shorter relamping period in order to increase (L_1) even further. If a group relamping program is *used*, (L_3) is assumed to be unity.

Figure 4-8 illustrates the effect of dirt build-up on (L_2) for a dustproof luminaire. Every luminaire has a tendency for dirt buildup. Manufacturer's data should be consulted when estimating (L_2) for the luminaire in question.

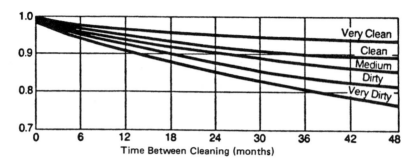

Figure 4-8. Effect of Dirt Build-Up on Dustproof Luminaires for Various Atmospheric Conditions

EMERGING LIGHT TECHNOLOGIES

When the U.S. Congress passed The Energy Independence and Security Act of 2007, one component of this bill was to phase out the use of incandescent bulbs. The phasing out of incandescent bulbs will start in 2012 and continue until 2014. This phasing out of the incandescent bulbs, which are the least efficient lighting source, is because of the development of some emerging technologies for more efficient lighting. For example, the U.S. Department of Energy (DOE) had sponsored a competition called the L-price competition. This competition was to encourage manufacturers of lighting sources to develop a replacement for the 60 W incandescent lamp and the PAR3 halogen lamp. The condition was that the new bulb must use 17% of the energy used by the above-mentioned bulbs and produce the same lumen output. The DOE had authorized up to $20 million in prizes. In August of 2011, CREE delivered a Light Emitting Diode (LED) lamp that could deliver more than 1,300 lumens for just 8.7 Watts, which translates to 152 lumens/W. It will take a good amount of time for a product like this to hit the market, but it just goes to show that these types of efficiencies in lighting are possible to obtain.

Earlier the government had come up with a new program called Energy Star. This was to promote energy efficient lights as well as appliances and HVAC systems. In order to qualify as an Energy Star light there are three qualifications that the light needs to meet. First, the light must be able to save over $40 when compared to a conventional incandescent light. Second, the light must last 6 times longer and be 75% more efficient. And lastly, the light must generate 75% less heat than an incandescent light, which is safer to operate and can cut energy costs associated with house cooling. The two bulbs that meet these criteria are CFLs, which were previously mentioned, and light emitting diodes (LED) bulbs. Additional comments on some of the emerging lighting technologies are summarized below.

Fluorescent Lights
T5 tubes

In the 1990s, T5 tubes were designed. The lengths of these tubes are designed to fit within 300 mm modular units (such as modular ceilings, modular cupboards, etc.), each being a multiple of 300 mm, less a constant fixed amount for end-caps and the construction of the unit end. The use of T5 tubes (rather than T8 or T12) allows the tubes to be fitted into smaller spaces, and the smaller light source also enables more accurate control of beam direction by means of optics (reflectors and lenses in the luminaire). Each tube length is available in both a lower power high efficiency (HE) version, and a higher power (but lower efficiency) high output (HO) version. The loading (watts per unit length) of the T5 HE tubes is similar to the original 4/6/8/13 W T5 tubes, and some manufacturers produce a range of fittings spanning both these ranges of tubes. When originally developed in Europe, operation from both switch start series ballasts and electronic ballasts were rapidly taking over at the time, particularly in the commercial lighting space where these tubes are most commonly used, and switch start series ballast operation is no longer specified by manufacturers, only electronic ballasts. Table 4-4 shows specifications of currently available T5 lamps.

Table 4-4. Specifications of T5 Lamps

Tube Diameter in 1/8 in (3.175 mm)	Length	Watts HE	Watts HO	Notes
T5	563 mm (22.2 in)	14 W	24W	Fits within a 600mm modular unit
T5	863 mm (34.0 in)	21 W	39 W	Fits within a 900mm modular unit
T5	1,163mm (45.8 in)	28 W	54 W	Fits within a 1200mm modular unit
T5	1,463 mm (57.6 in)	35 W, 49 W	80 W	Fits within a 1500mm modular unit

T5 fluorescent is the first linear lamp type to be served only by electronic ballasts. It is smaller than T8 and T12 lamps, with a miniature bi-pin base. It is notable for its lumens-per-watt efficiency, due to its peak light output occurring at 35°C (95°F) air temperature. There are three types of ballasts available for T5 lamps: instant start, rapid start, and programmed start electronic ballasts. Built-in microprocessors enable the ballast to automatically recognize and operate whichever lamp type is connected.

General Electric's line of T5 ballasts incorporates benefits of programmed start ballasts with new energy saving, fast starting, and parallel lamp operation of instant start ballasts. The GE UltraStart T5 ballast has low energy loss and high efficiency components along with continuous cathode cutout (CCC) technology. This technology results in fewer watts than standard 4-lamp 54W T5 ballasts. The UltraStart T5 ballasts meet 95% efficiency, a 44% improvement over standard T5 ballasts and a new industry threshold for high efficiency ballasts. Such ballasts can be replaced in most any industrial light settings such as warehouses and manufacturing facilities.

T5 luminaries that utilize a sleep mode or motion sensors to operate can generate even larger cost savings. In a 2008 field test in a warehouse scenario, using a standard T5 lighting system as a replacement to a metal halide system had potential cost savings of 23%. However, when using a T5 system with a sleep mode to replace a metal halide system, building owners had potential cost savings of 34-75% depending on sleep and wake control modes used. T5 lamps generally last for 20,000 hours, as compared to T8 lamps, which last for 15,000 hours.

HID Lighting

New developments in high intensity discharge (HID) lights now help reduce energy cost and potentially fixture count with great enhancements for color. PulseArc HID lighting provides better than average lumen maintenance with a long life and great lumen output. Special ballasts are necessary for such HID lighting, however can be direct replacements for most 250W, 320W, 350W and 400W pulse start lamps. These newer HID lamps have the ability to reach full light output much faster than standard metal halide lamps. They offer great energy savings potential with 20% lower initial lumens. Longevity of new HID provides a longer life compared to most 1000W metal halides currently used in industrial locations.

LED Lighting

Solid-State lighting (SSL) uses a semiconductor light emitting-diode (LED), an organic light-emitting diode (OLED), or a polymer light-emitting diode (PLED) to generate light, instead of electrical filaments, plasma, or gas. The U.S. Department of Energy (DOE) estimates that by switching to LEDs the country could save $120 billion in energy costs over the next twenty years. This equates to reducing the country's electricity consumption by 25 percent, and would prevent 246 million metric tons of carbon emission. LEDs have a long life span of 30,000-50,000 hours, compared to 1,000 hours for an incandescent and 10,000 hours for a compact fluorescent (CFL). LEDs have many other good characteristics including: compact size, resistance to breakage and vibration, good performance in cold temperatures, do not produce infrared or ultraviolet emissions, may be used with dimmers, and do not require any warm up time when turned on. LEDs are naturally suited for colored lighting applications such as stoplights and exit signs. White light can be made with LEDs two ways; phosphor conversion, a phosphor is used on or near the LED to emit white light, RGB systems, where light from multiple LEDs is mixed to create white light.

LEDs have recently become commercially available as a lighting source. They have extremely long life spans, are energy efficient, and come in a variety of colors. As research continues, LEDs continue to improve and be used in new applications. An increasingly popular residential use is decorative light strings, or holiday lights. Colored LEDs are now commonly used commercially in exit signs and traffic signals, which can significantly reduce maintenance costs. While LED technology is being explored for common residential use, technical and cost barriers remain viz: Initial cost: medium-high, color rendering ability: low-medium, and energy consumption: low-medium.

Why Choose LED Lighting?

LED bulbs switch on instantly, generate very little heat and use less energy than traditional incandescent and fluorescent bulbs. LED lights have a higher color rendering, so everything under them looks more vibrant. LED lights deliver optimal light distribution, softly washing walls and illuminating work surfaces without glare or hotspots. And LEDs are environmentally friendly—containing no mercury or lead—reducing your carbon footprint. They also last significantly longer (years and years) than incandescent and fluorescent light bulbs. So LEDs are the most cost-effective replacement bulbs.

Some of the advantages of LED lamps over traditional light bulbs:

• Significantly lower energy consumption
• Less heat
• Longer lifetime
• No UV & IR radiation
• No environmentally hazardous substances
• Small size
• Faster switching (perfectly bright in an instant)
• No bugs! (LEDs don't attract insects)

Most LED bulbs are rated to last 100,000 hours or more. That's about 34 years when used for an average of eight hours a day. So you change them less frequently, saving time and labor. Which, again, means saving money. And LED light bulbs burn cool, so they don't waste energy. A traditional incandescent light bulb dissipates about 92% of its energy as heat; an LED bulb, about 10%. Which again means LEDs are cost effective in the long run by cutting down the air-conditioning load of the occupied spaces.

SEL Lighting

Spectrally enhanced lighting (SEL) uses existing products and technologies to significantly reduce energy usage. SEL can produce 20-40% energy savings for very little cost, as the lamps cost approximately the same as traditional lamps. Recent findings indicate that when ambient lighting is whiter, or more like the color of daylight, our eyes respond the same way as if the lighting level was increased. By using lights with a higher correlated color temperature (CT) and a lower light output the perceived brightness will be unchanged, but the electricity used will be reduced. SEL lights can last 100,000 hours with minimal decrease in light output, and do not produce any humming or stroboscopic effects. SEL bulbs are available for existing fixtures, but can also be used with their own dedicated fixtures. Additional controls are not required for SEL.

Plasma Lighting

While LEP is a category of solid-state lighting, it is not an LED. The fundamental difference is that LEDs use the solid-state device itself for light generation whereas LEP light sources use a solid-state device to generate RF (radio frequency) energy to power a *plasma light source*. LEP is able to combine the reliability of the solid-state technology with the high brightness and full spectrum of HID (high-intensity discharge) sources. This is the reason LEP is called a solid-state high intensity light source.

The LEP technology shares many of the same benefits with the LED technology but also has some basic differences in performance. The similarities that it shares with LEDs are the reliability of the solid-state electronics, the directionality of the light output, and the ability to dim instantaneously. The differences in performance are that LEP has an order of magnitude higher lumen density (amount of light from one device), a full color spectrum with a CRIU (color-rendering index) up to 94, and better source efficacy. LED luminaries work best in low and mid illuminance applications, LEP™ has superior performance in higher illuminance applications such as streetlights, parking lots, big box retail, distribution centers and factories.

LUXIM's LEP® brand of Light Emitting Plasma™ is a new class of solid state high intensity light sources brining clean lighting solutions in general and specialty lighting. With energy efficiency, long useful lifetime, full spectrum color, and dimming, LEP lighting applications work better compared to conventional approaches such as HID or even newer sources such as LED in many applications. This technology brief describes the general construction of LEP lighting systems and the basic technology building blocks behind their function.

LEP Construction
The LEP product consists of three primary sub-assemblies:
• Emitter (including bulb)
• RF Driver
• Power supply

An RF (radio-frequency) signal is generated by the solid-state RF Driver and is guided into an electric field about the bulb. The high concentration of energy in the electric field vaporizes the contents of the bulb to a plasma state at the bulb's center; this controlled plasma generates an intense source of light.

Function of the Bulb Assembly
At the heart of LEP™ is the bulb sub-assembly where a sealed bulb is embedded in a dielectric material. This design is more reliable than conventional light sources that insert degradable electrodes into the bulb. The dielectric material serves two purposes; first as a waveguide for the RF

energy transmitted by the PA and second as an electric field concentrator that focuses energy in the bulb. The energy from the electric field rapidly heats the material in the bulb to a plasma state that emits light of high intensity and full spectrum.

Reasons to Use LEP

- *Superior Light Distribution*: Create optimal light distribution from a single, directional point source.
- *Natural Illumination*: Enhance visibility from a full spectrum plasma arc (up to 95 CRI).
- *Uncompromised Energy Savings*: Reduce energy usage by 50% without sacrificing brightness levels.
- *Worry-free Reliability*: Eliminate failure modes and lumen degradation found in most lighting.
- *Seamless Controls Integration*: Connect to any lighting controls via build in control gear.

INDUCTION LIGHTING

SYLVANIA ICETRON light bulbs are an electrodeless type of fluorescent, where the electrodes are replaced with magnetic induction technology. This results in a 100,000 hour service life. ICETRON lights produce up to 75 lumens per watt, and have a color-rendering index of 80. These lights do not require a warm up or cool down period, so they can be turned on and off at any time. The bulbs are available in 3500K and 4100K color temperatures, and color shift is minimal over time.

REMOTE CONTROLLER

One of the more recent advancements in lighting is wireless RF devices. These devices allow for control, via Internet, of a buildings' lighting, from anywhere around the globe. If an office is closed for the weekend and the owner remembers, while on vacation halfway around the world, that he/she had left a light on, wasting electricity, the wireless RF device would allow him/her to shut off the light, thus saving approximately forty-eight hours worth of electricity being supplied to that particular light bulb. This not only saves the electricity, but also prolongs the useful life of the light bulb, as those forty-eight hours would be completely wasted. Along with wireless RF devices, other apparatuses have become increasingly available for use to save on lighting.

To conclude, these recent innovations in the lighting industry are sure to become forerunners in the race for maximum efficiency. Not only are these lighting types highly efficient, but feature no drawbacks associated with previous high efficiency lighting systems. By having no warm up period and no restrike time they will all be able to be used in conjunction with other energy efficiency strategies.

CONTROL EQUIPMENT

Table 4-5 lists various types of equipment that can be components of a lighting control system, with a description of the predominant characteristic of each type of equipment. Static equipment can alter light levels semipermanently. Dynamic equipment can alter light levels automatically over short intervals to correspond to the activities in a space. Different sets of components can be used to form various lighting control systems in order to accomplish different combinations of control strategies.

FLUORESCENT LIGHTING CONTROL SYSTEMS

The control of fluorescent lighting systems is receiving increased attention. Two major categories of lighting control are available—personnel sensors and lighting compensators.

Personnel Sensors

There are three classifications of personnel sensors—ultrasonic, infrared and audio.

Ultrasonic sensors generate sound waves outside the human hearing range and monitor the return signals. Ultrasonic sensor systems are generally made up of a main sensor unit with a net work of satellite sensors providing coverage throughout the lighted area. Coverage per sensor is dependent upon the sensor type and ranges between 500 and 2,000 square feet. Sensors may be mounted above the ceiling, suspended below the ceiling or mounted on the wall. Energy savings are dependent upon the room size and occupancy. Advertised savings range from 20 to 40 percent.

Table 4-5. Lighting Control Equipment

System	Remarks
STATIC:	
Delamping	Method for reducing light level 50%.
Impedance Monitors	Method for reducing light level 30, 50%.
DYNAMIC:	
Light Controllers	
Switches/Relays	Method for on-off switching of large banks of lamps.
Voltage/Phase Control	Method for controlling light level continuously 100 to 50%.
Solid-State Dimming Ballasts	Ballasts that operate fluorescent lamps efficiently and can dim them continuously (100 to 10%) with low voltage.
SENSORS:	
Clocks	System to regulate the illumination distribution as a function of time.
Personnel	Sensor that detects whether a space is occupied by sensing the motion of an occupant.
Photocell	Sensor that measures the illumination level of a designated area.
COMMUNICATION:	
Computer /Microprocessor	Method for automatically communicating instructions and/or input from sensors to commands to the light controllers.
Power-Line Carrier	Method for carrying information over existing power lines rather than dedicated hard-wired communication lines.

Several companies manufacture ultrasonic sensors including Novita and Unenco.

Infrared sensor systems consist of a sensor and control unit. Coverage is limited to approximately 130 square feet per sensor. Sensors are mounted on the ceiling and usually directed towards specific work stations. They can be tied into the HVAC control and limit its operation also. Advertised savings range between 30 and 50 percent. (See Figure 4-9.)

Audio sensors monitor sound within a working area. The coverage of the sensor is dependent upon the room shape and the mounting

Figure 4-9. Transformer, relay and wide view infrared sensor to control lights. (*Photograph courtesy of Sensor switch***)**

height. Some models advertise coverage of up to 1,600 square feet. The first cost of the audio sensors is approximately one half that of the ultrasonic sensors. Advertised energy savings are approximately the same as the ultrasonic sensors. Several restrictions apply to the use of the audio sensors. First, normal background noise must be less than 60 dB. Second, the building should be at least 100 feet from the street and may not have a metal roof.

Lighting Compensators

Lighting compensators are divided into two major groups—switched and censored.

Switched compensators control the light level using a manually operated wall switch. These particular systems are used frequently in residential settings and are commonly known as "dimmer switches." Based on discussions with manufacturers, the switched controls are available for the 40-watt standard fluorescent bulbs only. The estimated savings are difficult to determine, as usually switched control systems are used to control room mood. The only restriction to their use is that the luminaire must have a dimming ballast.

Sensored compensators are available in three types. They may be very simple or very complex. They may be integrated with the building's energy management system or installed as a stand-alone system. The first type of system is the excess light turn-off (ELTO) system. This system senses daylight levels and automatically turns off lights as the sensed light level approaches a programmed upper limit. Advertised paybacks for these types of systems range from 1.8 to 3.8 years.

The second type of system is the daylight compensator (DAC) system. This system senses daylight levels and automatically dims lights to achieve a programmed room light level. Advertised savings range from 40 to 50 percent. The primary advantage of this system is it maintains a uniform light level across the controlled system area. The third system type is the daylight compensator + excess light turn-off system. As implied by the name, this system is a combination of the first two systems. It automatically dims light outputs to achieve a designated light level and, as necessary, automatically turns off lights to maintain the desired room conditions.

Specular reflectors: Fluorescent fixtures can be made more efficient by the insertion of a suitably shaped specular reflector. The specular reflector material types are aluminum, silver and multiple dielectric film mirrors. The latter two have the highest reflectivity while the aluminum reflectors are less expensive.

Measurements show the fixture efficiency with higher reflectance specular reflectors (silver or dielectric films) is improved by 15 percent compared to a new fixture with standard diffuse reflectors.

Specular reflectors tend to concentrate more light downward with reduced light at high exit angles. This increases the light modulation in the space, which is the reason several light readings at different sites around the fixture are required for determining the average illuminance. The increased downward component of candle power may increase the potential for reflected glare from horizontal surfaces.

When considering reflectors, information should be obtained on the new candle power characteristics. With this information a lighting designer or engineer can estimate the potential changes in modulation and reflected glare.

III—Energy Management

The availability of computers at moderate costs and the concern for reducing energy consumption have resulted in the application of computer-based controllers to more than just industrial process applications. These controllers, commonly called energy management systems (EMS), can be used to control virtually all non-process energy using pieces of equipment in buildings and industrial plants. Equipment controlled can include fans, pumps, boilers, chillers and lights. This section will investigate the various types of energy management systems which are available and illustrate some of the methods used to reduce energy consumption.

THE TIME CLOCK

One of the simplest and most effective methods of conserving energy in a building is to operate equipment only when it is needed. If, due to time, occupancy, temperature or other means, it can be determined that a piece of equipment does not need to operate, energy savings can be achieved without affecting occupant comfort by turning off the equipment.

One of the simplest devices to schedule equipment operation is the mechanical time clock. The time clock consists of a rotating disk which is divided into segments corresponding to the hour of the day and the day of the week. This disk makes one complete revolution in, depending on the type, a 24-hour or a 7-day period. (See Figure 4-10.)

On and off "lugs" are attached to the disk at appropriate positions corresponding to the schedule for the piece of equipment. As the disk rotates, the lugs cause a switch contact to open and close, thereby controlling equipment operation.

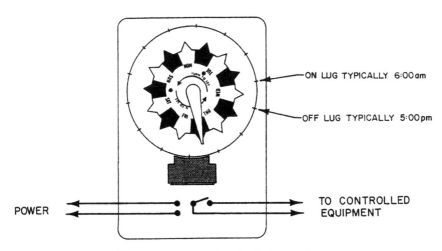

Figure 4-10. Mechanical Time Clock

A common application of time clocks is scheduling office building HVAC equipment to operate during business hours Monday through Friday and to be off all other times. As is shown in the following problem, significant savings can be achieved through the correct application of time clocks.

Example Problem 4-3
    An office building utilizes two 50 hp supply fans and two 15 hp return fans which operate continuously to condition the building. What are the annual savings that result from installing a time clock to operate these fans from 7:00 a.m. to 5:00 p.m., Monday through Friday? Assume an electrical rate of $0.08/kWh.

Answer
    Annual Operation Before Time clock =
        52 weeks × 7 days/week × 24 hours/day = 8736 hours

    Annual Operation After Time clock =
        52 × (5 days/week × 10 hours/day) = 2600 hours

    Savings = 130 hp × 0.746 kW/hp × (8736-2600) hours ×
        $0.08/kWh = $47,600

Although most buildings today utilize some version of a time clock, the magnitude of the savings value in this example illustrates the importance of correct time clock operation and the potential for additional costs if this device should malfunction or be adjusted inaccurately. Note that the above example also ignores heating and cooling savings which would result from the installation of a time clock.

PROBLEMS WITH MECHANICAL TIME CLOCKS

Although the use of mechanical time clocks in the past has resulted in significant energy savings, they are being replaced by energy management systems because of problems that include the following:

•   The on/off lugs sometimes loosen or falloff.

•   Holidays, when the building is unoccupied, cannot easily be taken into account.

- Power failures require the time clock to be reset or it is not synchronized with the building schedule.

- Inaccuracies in the mechanical movement of the time clock prevent scheduling any closer than ±15 minutes of the desired times.

- There are a limited number of on and off cycles possible each day.

- It is a time-consuming process to change schedules on multiple time clocks.

Energy management systems, or sometimes called electronic time clocks, are designed to overcome these problems plus provide increased control of building operations.

ENERGY MANAGEMENT SYSTEMS

Recent advances in digital technology, dramatic decreases in the cost of this technology and increased energy awareness have resulted in the increased application of computer-based controllers (i.e., energy management systems) in commercial buildings and industrial plants. These devices can control anywhere from one to a virtually unlimited number of items of equipment.

By concentrating the control of many items of equipment at a single point, the EMS allows the building operator to tailor building operation to precisely satisfy occupant needs. This ability to maximize energy conservation, while preserving occupant comfort, is the ultimate goal of an energy engineer.

Microprocessor Based

Energy management systems can be placed in one of two broad, and sometimes overlapping, categories referred to as microprocessor-based and mini-computer based.

Microprocessor-based systems can control from 1 to 40 input/out-points and can be linked together for additional loads. Programming is accomplished by a keyboard or hand-held console, and an LED display is used to monitor/review operation of the unit. A battery maintains the programming in the event of power failure. (See Figures 4-11 and 4-12.)

Capabilities of this type of EMS are generally pre-programmed so that operation is relatively straightforward. Programming simply in-

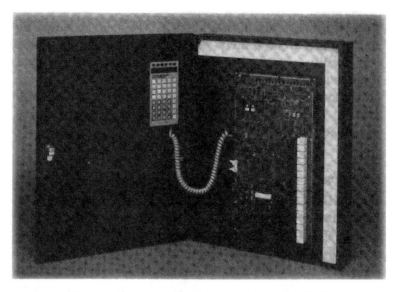

Figure 4-11. Microprocessor-based Programmable EMS (*Photograph Courtesy Control Systems International*)

Figure 4-12. Keyboard Used to Access EMS (*Photo Courtesy Control Systems International*)

volves entering the appropriate parameters (e.g., the point number and the on and off times) for the desired function. Microprocessor-based EMS can have any or all of the following capabilities:

- Scheduling
- Duty Cycling
- Demand Limiting
- Optimal Start
- Monitoring
- Direct Digital Control

Scheduling

Scheduling with an EMS is very much the same as it is with a time clock. Equipment is started and stopped based on the time of day and the day of week. Unlike a time clock, however, multiple start/stops can be accomplished very easily and accurately (e.g., in a classroom, lights can be turned off during morning and afternoon break periods and during lunch). It should be noted that this single function, if accurately programmed and depending on the type of facility served, can account for the largest energy savings attributable to an EMS.

Additionally, holiday dates can be entered into the EMS a year in advance. When the holiday occurs, regular programming is overridden and equipment can be kept off.

Duty Cycling

Most HVAC fan systems are designed for peak load conditions; consequently, these fans are usually moving much more air than is needed. They can sometimes be shut down for short periods each hour, typically 15 minutes, without affecting occupant comfort. Turning equipment off for pre-determined periods of time during occupied hours is referred to as duty cycling, and can be accomplished very easily with an EMS. Duty cycling saves fan and pump energy but does not reduce the energy required for space heating or cooling since the thermal demand must still be met.

The more sophisticated EMS's monitor the temperature of the conditioned area and use this information to automatically modify the duty cycle length when temperatures begin to drift. If, for example, the desired temperature in an area is 70° and at this temperature equipment is cycled 50 minutes on and 10 minutes off, a possible temperature-compensated EMS may respond as shown in Figure 4-13. As the space temperature increases above (or below if so programmed) the setpoint, the equipment off time is reduced until, at 80° in this example, the equipment operates continuously.

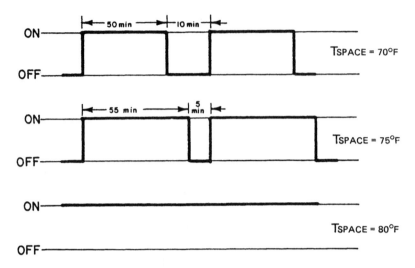

Figure 4-13. Temperature Compensated Duty Cycling

Duty cycling is best applied in large, open-space offices which are served by a number of fans. Each fan could be programmed so that the off times do not coincide, thereby assuring adequate air flow to the offices at all times.

Duty cycling of fans which provide the only air flow to an area should be approached carefully to insure that ventilation requirements are maintained and that varying equipment noise does not annoy the occupants. Additionally, duty cycling of equipment imposes extra stress on motors and associated equipment. Care should be taken, particularly with motors over 20 hp, to prevent starting and stopping of equipment in excess of what is recommended by the manufacturer.

Demand Charges

Electrical utilities charge commercial customers based not only on the amount of energy used (kWh) but also on the peak demand (kW) for each month. Peak demand is very important to the utility so that they may properly size the required electrical service and insure that sufficient peak generating capacity is available to that given facility.

To determine the peak demand during the billing period, the utility establishes short periods of time called the demand interval (typically 15, 30, or 60 minutes). The billing demand is defined as the highest average demand recorded during any one demand interval within the billing period. (See Figure 4-14.) Many utilities now utilize "ratchet" rate charges. A "ratchet" rate means that the billed demand for the month is based on the highest demand in the previous 12 months, or an average of the current month's peak demand and the previous highest demand in the past year.

Depending on the facility, the demand charge can be a significant portion, as much as 20%, of the utility bill. The user will get the most electrical energy per dollar if the load is kept constant, thereby minimizing the demand charge. The objective of demand control is to even out the peaks and valleys of consumption by deferring or rescheduling the use of energy during peak demand periods.

A measure of the electrical efficiency of a facility can be found by calculating the load factor. The load factor is defined as the ratio of energy usage (kWh) per month to the peak demand (kW) × the facility operating hours.

Example Problem 4-4

What is the load factor of a continuously operating facility that consumed 800,000 kWh of energy during a 30-day billing period and established a peak demand of 2000 kW?

Answer

$$\text{Load Factor} = \frac{800{,}000 \text{ kWh}}{2000 \text{ kW} \times 30 \text{ days} \times 24 \text{ hours/day}} = 0.55$$

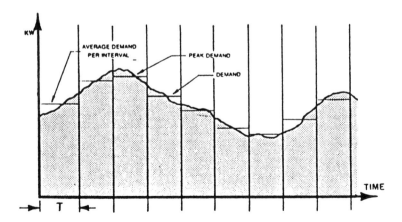

T - DEMAND INTERVAL

Figure 4-14. Peak Demand

The ideal load factor is 1.0, at which demand is constant; therefore, the difference between the calculated load factor and 1.0 gives an indication of the potential for reducing peak demand (and demand charges) at a facility.

Demand Limiting

Energy management systems with demand limiting capabilities utilize either pulses from the utility meter or current transformers to predict the facility demand during any demand interval. If the facility demand is predicted to exceed the user-entered setpoint, equipment is "shed" to control demand. Figure 4-15 illustrates a typical demand chart before and after the actions of a demand limiter.

Electrical load in a facility consists of two major categories: essential loads which include most lighting, elevators, escalators, and most production machinery; and non-essential ("sheddable") loads such as electric heaters, air conditioners, exhaust fans, pumps, snow melters, compressors

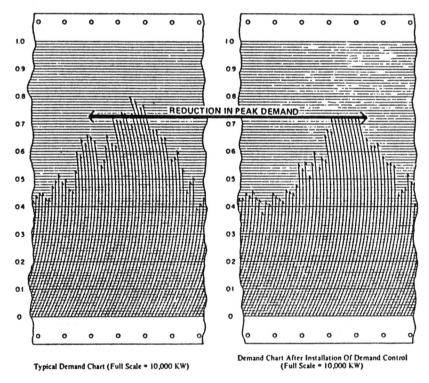

Typical Demand Chart (Full Scale = 10,000 KW)

Demand Chart After Installation Of Demand Control (Full Scale = 10,000 KW)

Figure 4-15. Demand Limiting Comparison

and water heaters. Sheddable loads will not, when turned off for short periods of time to control demand, affect productivity or comfort.

To prevent excessive cycling of equipment, most energy management systems have a deadband that demand must drop below before equipment operation is restored (See Figure 4-16). Additionally, minimum on and maximum off times and shed priorities can be entered for each load to protect equipment and insure that comfort is maintained.

It should be noted that demand shedding of HVAC equipment in commercial office buildings should be applied with caution. Since times of peak demand often occur during times of peak air conditioning loads, excessive demand limiting can result in occupant discomfort.

Time of Day Billing

Many utilities are beginning to charge their larger commercial users based on the time of day that consumption occurs. Energy and demand during peak usage periods (i.e., summer weekday afternoons and winter weekday evenings) are billed at much higher rates than consumption during other times. This is necessary because utilities must augment the power production of their large power plants during periods of peak demand with small generators which are expensive to operate. Some of the more sophisticated energy management systems can now account for

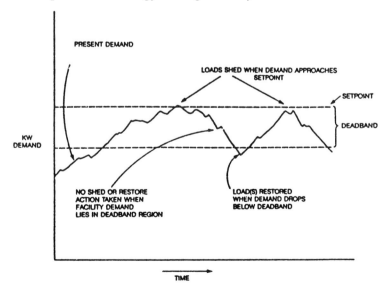

Figure 4-16. Demand Limiting Actions

these peak billing periods with different demand setpoints based on the time of day and day of week.

Optimal Start

External building temperatures have a major influence on the amount of time it takes to bring the building temperature up to occupied levels in the morning. Buildings with mechanical time clocks usually start HVAC equipment operation at an early enough time in the morning (as much as 3 hours before occupancy time) to bring the building up to temperature on the coldest day of the year. During other times of the year when temperatures are not as extreme, building temperatures can be up to occupied levels several hours before it is necessary, and consequently unnecessary energy is used. (See Figure 4-17.)

Energy management systems with optimal start capabilities, however, utilize indoor and outdoor temperature information, along with learned building characteristics, to vary start time of HVAC equipment

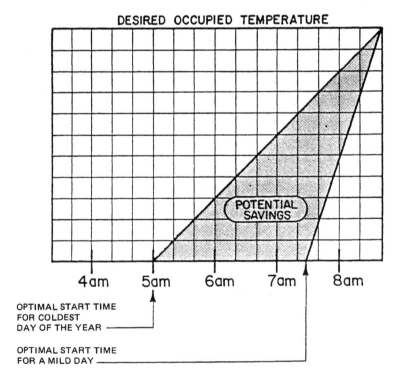

DESIRED OCCUPIED TEMPERATURE

4am   5am   6am   7am   8am

OPTIMAL START TIME
FOR COLDEST
DAY OF THE YEAR

OPTIMAL START TIME
FOR A MILD DAY

POTENTIAL
SAVINGS

Figure 4-17. Typical Variation in Building Warm -up Times

so that building temperatures reach desired values just as occupancy occurs. Consequently, if a building is scheduled to be occupied at 8:00 a.m., on the coldest day of the year, the HVAC equipment may start at 5:00 a.m. On milder days, however, equipment may not be started until 7:00 a.m. or even later, thereby saving significant amounts of energy.

Most energy management systems have a "self-tuning" capability to allow them to learn the building characteristics. If the building is heated too quickly or too slowly on one day, the start time is adjusted the next day to compensate.

Monitoring

Microprocessor-based EMS can usually accomplish a limited amount of monitoring of building conditions including the following:

• Outside air temperature.
• Several indoor temperature sensors.
• Facility electrical energy consumption and demand.
• Several status input points.

The EMS can store this information to provide a history of the facility. Careful study of these trends can reveal information about facility operation that can lead to energy conservation strategies that might not otherwise be apparent.

Direct Digital Control

The most sophisticated of the microprocessor-based EMSs provide a function referred to as direct digital control (DDC). This capability allows the EMS to provide not only sophisticated energy management but also basic temperature control of the building's HVAC systems.

Direct digital control has taken over the majority of all process control application and is now becoming an important part of the HVAC industry. Traditionally, pneumatic controls were used in most commercial facilities for environmental control.

The control function in a traditional facility is performed by a pneumatic controller which receives its input from pneumatic sensors (i.e., temperature, humidity) and sends control signals to pneumatic actuators (valves, dampers, etc.). Pneumatic controllers typically perform a single, fixed function which cannot be altered unless the controller itself is changed or other hardware is added. (See Figure 4-18 for a typical pneumatic control configuration.)

With direct digital control, the microprocessor functions as the primary controller. Electronic sensors are used to measure variables such as temperature, humidity and pressure. This information is used, along with the appropriate application program, by the microprocessor to determine the correct control signal, which is then sent directly to the controlled device (valve or damper actuator). (See Figure 4-18 for a typical DDC configuration.)

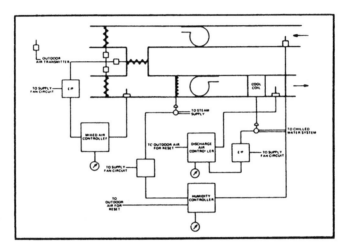

CONVENTIONAL PNEUMATIC CONTROL SYSTEM

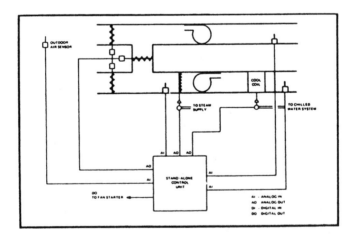

DIRECT DIGITAL CONTROL SYSTEM

Figure 4-18. Comparison of Pneumatic and DDC Controls

Direct digital control (DDC) has the following advantage over pneumatic controls:

- Reduces overshoot and offset errors, thereby saving energy.
- Flexibility to easily and inexpensively accomplish changes of control strategies.
- Calibration is maintained more accurately, thereby saving energy and providing better performance.

To program the DDC functions, a user programming language is utilized. This programming language uses simple commands in English to establish parameters and control strategies.

Mini-computer Based

Mini-computer based EMS can provide all the functions of the microprocessor based EMS, as well as the following:

- Extensive graphics.
- Special reports and studies.
- Fire and security monitoring and detection.
- Custom programs.

These devices can control and monitor from 50 to an unlimited number of points and form the heart of a building's (or complex's) operations.

Figure 4-19 shows a typical configuration for this type of system. The "central processing unit" (CPU) is the heart of the EMS. It is a mini-computer with memory for the operating system and applications software. The CPU performs arithmetic and logical decisions necessary to perform central monitoring and control.

Data and programs are stored or retrieved from the memory or mass storage devices (generally a disk storage system). The CPU has programmed I/O ports for specific equipment, such as printers and cathode ray tube (CRT) consoles. During normal operation, it coordinates operation of all other EMS components.

A cathode ray tube console (CRT), either color and/or black and white, with a keyboard is used for operator interaction with the EMS. It accepts operator commands, displays data and graphically displays systems controlled or monitored by the EMS. A "printer," (or printers) provides a permanent copy of system operations and historical data.

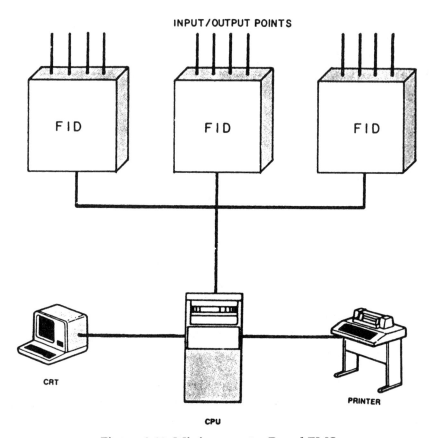

Figure 4-19. Mini-computer Based EMS

A "field interface device" (FID) provides an interface to the points which are monitored and controlled, performs engineering conversions to or from a digital format, performs calculations and logical operations, accepts and processes CPU commands and is capable, in some versions, of stand-alone operations in the event of CPU or communications link failure.

The FID is essentially a microprocessor based EMS as described in the previous section. It mayor may not have a keyboard/display unit on the front panel.

The FIDs are generally located in the vicinity of the points to be monitored and/or controlled and are linked together and to the CPU by a single twisted pair of wires which carries multiplexed data (i.e., data from a number of sources combined on a single channel) from the FID to the CPU and back. In some versions, the FIDs can communicate directly with each other.

Early versions of mini-computer based EMS used the CPU to perform all of the processing with the FID used merely for input and output. A major disadvantage of this type of "centralized" system is that the loss of the CPU disables the entire control system. The development of "intelligent" FIDs in a configuration known as "distributed processing" helped to solve this problem. This system, which is becoming prevalent today, utilizes microprocessor-based FIDs to function as remote CPUs. Each panel has its own battery pack to insure continued operation should the main CPU fail.

Each intelligent panel sends signals back to the main CPU only upon a change of status rather than continuously transmitting the same value as previous "centralized" systems have done. This streamlining of data flow to the main CPU frees it to perform other functions such as trend reporting. The CPU's primary function becomes one of directing communications between various FID panels, generating reports and graphics and providing operator interface for programming and monitoring.

Features

The primary difference in operating functions of the minicomputer based EMS is its increased capability to monitor building operations. For this reason, these systems are sometimes referred to as energy monitoring and control systems (EMCS). Analog inputs such as temperature and humidity can be monitored, as well as digital inputs such as pump or valve status.

The mini-computer based EMS is also designed to make operator interaction very easy. Its operation can be described as "user friendly" in that the operator, working through the keyboard, enters information in English in a question and response format. In addition, custom programming languages are available so that powerful programs can be created specifically for the building through the use of simplified English commands.

The graphics display CRT can be used to create HVAC schematics, building layouts, bar charts, etc. to better understand building systems operation. These graphics can be "dynamic" so that values and statuses are continuously updated.

Many mini-computer based EMS can also easily incorporate fire and security monitoring functions. Such a configuration is sometimes referred to as a Building Automation System (BAS). By combining these functions with energy management, savings in initial equipment costs can be achieved. Reduced operating costs can be achieved as well by having a single operator for these systems.

The color graphics display can be particularly effective in pinpointing alarms as they occur within a building and guiding quick and appropriate response to that location. In addition, management of fan systems to control smoke in a building during a fire is facilitated with a system that combines energy management and fire monitoring functions.

Note, however, that the incorporation of fire, security and energy management functions into a single system increases the complexity of that system. This can result in longer start-up time for the initial installation and more complicated troubleshooting if problems occur. Since the function of fire monitoring is critical to building operation, these disadvantages must be weighed against the previously mentioned advantages to determine if a combined BAS is desired.

DATA TRANSMISSION METHODS

A number of different transmission systems can be used in an EMS for communications between the CPU and FID panels. These transmission systems include telephone lines, coaxial cables, electrical power lines, radio frequency, fiber optics and microwave. Table 4-6 compares the various transmission methods.

Twisted Pair
One of the most common data transmission methods for an EMS is a twisted pair of wires. A twisted pair consists of two insulated conductors twisted together to minimize interference from unwanted signals.

Twisted pairs are permanently hardwired lines between the equipment sending and receiving data that can carry information over a wide range of speeds, depending on line characteristics. To maintain a particular data communication rate, the line bandwidth, time delay or the signal-to-noise ratio may require adjustment by conditioning the line.

Data transmission in twisted pairs, in most cases, is limited to 1200 bps (bits per second) or less. By using signal conditioning, operating speeds up to 9600 bps may be obtained.

Voice Grade Telephone Lines
Voice grade lines used for data transmission are twisted pair circuits

Table 4-6. Transmission Method Comparisons

Method	First Cost	Scan Rates	Reliability	Maint. Effort	Expandibility	Compatibility with Future Requirements
Coaxial	high	fast	excellent	min.	unlimited	unlimited
Twisted pair	low	med.	very good	min.	unlimited	unlimited
RF	med.	fast but limited	low	high	very limited	very limited
Microwave	very high	very fast	excellent	high	unlimited	unlimited
Telephone	very low	slow	low to high	min.	limited	limited
Fiber optics	high	very fast	excellent	min.	unlimited	unlimited
Power Line Carrier	med.	med.	med.	high	limited	limited

defined as type 3002 by the Bell Telephone Company. The local telephone company charges a small connection fee for this service, plus a monthly equipment lease fee. Maintenance is included in the monthly lease fee, with a certain level of service guaranteed.

Two of the major problems involve the quality of telephone pairs provided to the installer and the transmission rate. The minimum quality of each line intended for use in the trunk wiring system must be clearly defined.

The most common voice grade line used for data communication is the unconditioned type 3002 allowing transmission rates up to 1200-bps. The 3002 type line may be used for data transmission up to 9600 bps with the proper line conditioning. Voice grade lines must be used with the same constraints and guidelines for twisted pairs.

Coaxial Cable

Coaxial cable consists of a center conductor surrounded by a shield. The center conductor is separated from the shield by a dielectric. The shield protects against electromagnetic interference. Coaxial cables can operate at data transmission rates in the megabits per second range, but attenuation becomes greater as the data transmission rate increases.

The transmission rate is limited by the data transmission equipment and not by the cable. Regenerative repeaters are required at specific intervals depending on the data rate, nominally every 2000 feet to maintain the signal at usable levels.

Power Line Carrier

Data can be transmitted to remote locations over electric power lines using carrier current transmission that superimposes a low power RF (radio frequency) signal, typically 100 kHz, onto the 60 Hz power distribution system. Since the RF carrier signal cannot operate across transformers, all communicating devices must be connected to the same power circuit (same transformer secondary and phase) unless RF couplers are installed across transformers permitting the transmitters and receivers to be connected over a wider area of the power system. Transmission can be either one-way or two-way.

Note that power line carrier technology is sometimes used in microprocessor based EMS retrofit applications to control single loads in a facility where hard wiring would be difficult and expensive (e.g., wiring between two buildings). Figure 4-20 shows a basic power line carrier system configuration.

Radio Frequency

Modulated RF signals, usually VHF or FM radio, can be used as a data transmission method with the installation of radio receivers and transmitters. RF systems can be effectively used for two-way communication between CPU and FID panels where other data transmission methods are not available or suitable for the application. One-way RF systems can be effectively used to control loads at remote locations such as warehouses and unitary heaters and for family housing projects.

The use of RF at a facility, however, must be considered carefully to avoid conflict with other existing or planned facility RF systems. Additionally, there may be a difficulty in finding a frequency on which to transmit, since there are a limited number available.

The kinds of signals sent over an FM radio system are also limited as are the distances over which the signals can be transmitted. The greater the distance, the greater the likelihood that erroneous signals will be received.

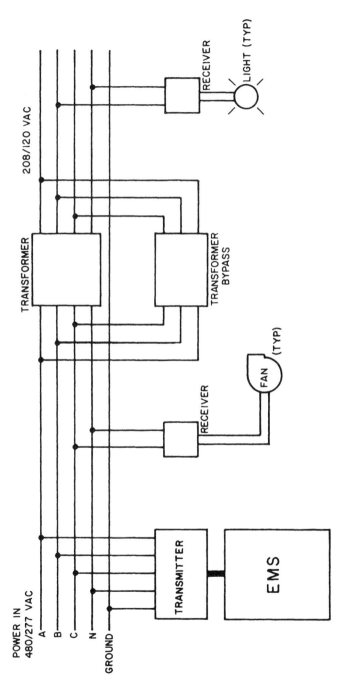

Figure 4-20. Power Line Carrier System Configuration

Fiber Optics

Fiber optics uses the wideband properties of infrared light traveling through transparent fibers. Fiber optics is a reliable communications media which is rapidly becoming cost competitive when compared to other high-speed data transmission methods.

The bandwidth of this media is virtually unlimited, and extremely high transmission rates can be obtained. The signal attenuation of high-quality fiber optic cable is lower than the best coaxial cables. Repeaters required nominally every 2000 feet, for coaxial cable, are 3 to 6 miles apart in fiber optics systems.

Fiber optics terminal equipment selection is limited to date, and there is a lack of skilled installers and maintenance personnel familiar with this media. Fiber optics must be carefully installed and cannot be bent at right angles.

Microwave Transmission

For long distance transmission, a microwave link can be used. The primary drawback of microwave links is first cost. Receivers/ transmitters are needed at each building in a multi-facility arrangement.

Microwave transmission rates are very fast and are compatible with present and future data requirements. Reliability is excellent, too, but knowledgeable maintenance personnel are required. The only limit on expansion is cost.

SUMMARY

The term energy management system denotes equipment whose functions can range from simple time clock control to sophisticated building automation. Two broad and overlapping categories of these systems are microprocessor and mini-computer based.

Capabilities of EMS can include scheduling, duty cycling, demand limiting, optimal start, monitoring, direct digital control, fire detection and security. Direct digital control capability enables the EMS to replace the environmental control system so that it directly manages HVAC operations.

5

Waste Heat Recovery

INTRODUCTION

Waste heat is heat which is generated in a process but then "dumped" to the environment even though it could still be reused for some useful and economic purpose.

The essential quality of heat is not the amount but rather its "value."

The strategy of how to recover this heat depends in part on the temperature of the waste heat gases and the economics involved.

This chapter will present the various methods involved in traditionally recovering waste heat.

Portions of material used in Chapters 5 and 6 are based upon the *Waste Heat Management Guidebooks* published by the U.S. Department of Commerce/National Bureau of Standards. The authors express appreciation to Kenneth G. Kreider; Michael B. McNeil; W.M. Rohrer, Jr.; R. Ruegg; B. Leidy and W. Owens who have contributed extensively to this publication.

SOURCES OF WASTE HEAT

Sources of waste energy can be divided according to temperature into three temperature ranges. The high temperature range refers to temperatures above 1200°F. The medium temperature range is between 450°F and 1200°F, and the low temperature range is below 450°F.

High and medium temperature waste heat can be used to produce process steam. If high temperature waste heat exists, instead of producing steam directly, consider the possibility of using the high temperature energy to do useful work before the waste heat is extracted. Both gas and steam turbines are useful and fully developed heat engines.

In the low temperature range, waste energy which would be otherwise useless can sometimes be made useful by application of mechanical work through a device called the heat pump.

HIGH TEMPERATURE HEAT RECOVERY

The combustion of hydrocarbon fuels produces product gases in the high temperature range. The maximum theoretical temperature possible in atmospheric combustors is somewhat under 3500°F, while measured flame temperatures in practical combustors are just under 3000°F. Secondary air or some other dilutant is often admitted to the combustor to lower the temperature of the products to the required process temperature (for example, to protect equipment) thus lowering the practical waste heat temperature.

Table 5-1 gives temperatures of waste gases from industrial process equipment in the high temperature range. All of these result from direct fuel fired processes.

Table 5-1

Type of Device	*Temperature F*
Nickel refining furnace	2500-3000
Aluminum refining furnace	1200-1400
Zinc refining furnace	1400-2000
Copper refining furnace	1400-1500
Steel heating furnaces	1700-1900
Copper reverberatory furnace	1650-2000
Open hearth furnace	1200-1300
Cement kiln (Dry process)	1150-1350
Glass melting furnace	1800-2800
Hydrogen plants	1200-1800
Solid waste incinerators	1200-1800
Fume incinerators	1200-2600

MEDIUM TEMPERATURE HEAT RECOVERY

Table 5-2 gives the temperatures of waste gases from process equipment in the medium temperature range. Most of the waste heat in this temperature range comes from the exhausts of directly fired process units. Medium temperature waste heat is still hot enough to allow consideration of the extraction of mechanical work from the waste heat, by a steam or gas turbine. Gas turbines can be economically utilized in some cases at inlet pressures in the range of 15 to 30 lb/in² g. Steam can be generated at almost any desired pressure and steam turbines used when economical.

Table 5-2

Type of Device	Temperature F
Steam boiler exhausts	450-900
Gas turbine exhausts	700-1000
Reciprocating engine exhausts	600-1100
Reciprocating engine exhausts (turbocharged)	450-700
Heat treating furnaces	800-1200
Drying and baking ovens	450-1100
Catalytic crackers	800-1200
Annealing furnace cooling systems	800-1200

LOW TEMPERATURE HEAT RECOVERY

Table 5-3 lists some heat sources in the low temperature range. In this range it is usually not practical to extract work from the source, though steam production may not be completely excluded if there is a need for low pressure steam. Low temperature waste heat may be useful in a supplementary way for preheating purposes. Taking a common example, it is possible to use economically the energy from an air-conditioning condenser operating at around 90°F to heat the domestic water supply. Since the hot water must be heated to about 160°F, obviously the air-conditioner waste heat is not hot enough. However, since the cold water enters the domestic water system at about 50°F, energy interchange can take place to raise the water to something less than 90°F. Depending upon the relative air-conditioning load and hot water requirements, any excess condenser heat can be rejected, and the additional energy required by the hot water can be provided by the usual electrical or fired heater.

Table 5-3

Source	Temperature F
Process steam condensate	130-190
Cooling water from:	
Furnace doors	90-130
Bearings	90-190
Welding machines	90-190
Injection molding machines	90-190
Annealing furnaces	150-450
Forming dies	80-190
Air compressors	80-120
Pumps	80-190
Internal combustion engines	150-250
Air conditioning and	
refrigeration condensers	90-110
Liquid still condensers	90-190
Drying, baking and curing ovens	200-450
Hot processed liquids	90-450
Hot processed solids	200-450

WASTE HEAT RECOVERY APPLICATIONS

To use waste heat from sources such as those above, one often wishes to transfer the heat in one fluid stream to another (e.g., from flue gas to feedwater or combustion air). The device which accomplishes the transfer is called a heat exchanger. In the discussion immediately below is a listing of common uses for waste heat energy and, in some cases, the name of the heat exchanger that would normally be applied in each particular case.

The equipment that is used to recover waste heat can range from something as simple as a pipe or duct to something as complex as a waste heat boiler.

Some applications of waste heat are as follows:
- Medium to high temperature exhaust gases can be used to preheat the combustion air for:
  — Boilers using air preheaters.
  — Furnaces using recuperators.

— Ovens using recuperators.
— Gas turbines using regenerators.

• Low to medium temperature exhaust gases can be used to preheat boiler feedwater or boiler makeup water using *economizers,* which are simply gas-to-liquid water heating devices.

• Exhaust gases and cooling water from condensers can be used to preheat liquid and/or solid feedstocks in industrial processes. Finned tubes and tube-in-shell *heat exchangers* are used.

• Exhaust gases can be used to generate steam in *waste heat boilers* to produce electrical power, mechanical power, process steam, and any combination of above.

• Waste heat may be transferred to liquid or gaseous process units directly through pipes and ducts or indirectly through a secondary fluid such as steam or oil.

• Waste heat may be transferred to an intermediate fluid by heat exchangers or waste heat boilers, or it may be used by circulating the hot exit gas through pipes or ducts. Waste heat can be used to operate an absorption cooling unit for air conditioning or refrigeration.

THE WASTE HEAT RECOVERY SURVEY

In order to identify sources of waste heat, a survey is usually made. Figure 5-1 illustrates a survey form which can be used for the waste heat audit. It is important to record flow and temperature of waste gases.

Composition data are required for heat recovery and system design calculations. Be sure to note contaminants since this factor could limit the type of heat recovery equipment to apply. Contaminants can foul or plug heat exchangers.

Operation schedule affects the economics and type of equipment to be specified. For example, an incinerator that is only used one shift per day may require a different method of recovering discharges than if it were used three shifts a day. A heat exchanger used for waste heat recovery in this service would soon deteriorate due to metal fatigue. A different type of heat recovery incinerator utilizing heat storage materials such as rock or ceramic would be more suitable.

SURVEY FORM FOR INDUSTRIAL PROCESS UNITS

NAME OF PROCESS UNIT ____________ INVENTORY NUMBER ____________

LOCATION OF PROCESS UNIT, PLANT NAME ____________ BUILDING ____________

MANUFACTURER ____________ MODEL ________ SERIAL NUMBER ____________

	NAME	FIRING RATE	HHV	TEMPERATURE OF: COMB. AIR	TEMPERATURE OF: FUEL	TEMPERATURE OF: STACK	FLUE GAS COMPOSITION % VOLUME: CO_2	O_2	CO	CH	N_2
PRIMARY FUEL											
FIRST ALTERNAT.											
SECOND ALTERNAT.											

	FLOW PATH 1	FLOW PATH 2	FLOW PATH 3	FLOW PATH 4
FLUID COMPOSITION				
FLOW RATE				
INLET TEMPERATURE				
OUTLET TEMPERATURE				
DESCRIPTION				

ANNUAL HOURS OPERATION ____________ ANNUAL CAPACITY FACTOR, % ____________

ANNUAL FUEL CONSUMPTION: PRIMARY FUEL ______; FIRST ALTERN. ______ SEC. ALTERN. ______

PRESENT FUEL COST: PRIMARY FUEL ______; FIRST ALTERN. ______ SEC. ALTERN. ______

ANNUAL ELECTRICAL ENERGY CONSUMPTION, KWHR. ____________

PRESENT ELECTRICAL ENERGY RATE ____________

CONTAMINANTS ____________

Figure 5-1. Waste Heat Survey

WASTE HEAT RECOVERY CALCULATIONS

From the heat balance (Chapter 6), the heat recovered from the source is determined by Formula 5-1.

$$q = m\, c_p\, \Delta T \qquad\qquad \textit{Formula (5-1)}$$

Where

q = heat recovered, Btu/hr
m = mass flow rate lb/hr
c_p = specific heat of fluid, Btu/lb°F
ΔT = temperature change of gas or liquid during heat recovery °F

If the flow is air, then Formula 5-1 can be expressed as

$$q = 1.08\ \text{cfm}\ \Delta T \qquad\qquad \textit{Formula (5-2)}$$

Where cfm = volume flow rate in standard cubic feet per minute

If the flow is water, then Formula 5-1 can be expressed as

$$q = 500\ \text{gpm}\ \Delta T \qquad\qquad \textit{Formula (5-3)}$$

Where gpm = volume flow rate in gallons per minute

Example Problem 5-1

A waste heat audit survey indicates 10,000 lb/hr of water at 190°F is discharged to the sewer. How much heat can be saved by utilizing this fluid as makeup to the boiler instead of the 70°F feedwater supply? Fuel cost is $6 per million Btu, boiler efficiency .8, and hours of operation 4000.

Analysis

$q = mc_p\, \Delta T = 10{,}000 \times (190{-}70) = 1.2 \times 10^6$ Btu/hr
Savings $= 1.2 \times 10^6 \times 4000 \times \$6/10^6/.8 = \$36{,}000$

Heat Transfer by Convection

Convection is the transfer of heat to or from a fluid, gas, or liquid. Formula 5-4 is indicative of the basic form of convective heat transfer. U_0, in this case, represents the convection film conductance, $Btu/ft^2 \cdot hr \cdot °F$.

Heat transferred for heat exchanger applications is predominantly a combination of conduction and convection expressed as

$$q = U_0 A \, \Delta T_m \qquad \qquad \text{Formula (5-4)}$$

Where
$\quad q \; = \;$ rate of heat flow by convection, Btu/hr
$\quad U_0 \; = \;$ is the overall heat transfer coefficient $Btu/ft^2 \cdot hr \cdot °F$.
$\quad A \; = \;$ is the area of the tubes in square feet
$\quad \Delta T_m \; = \;$ is the logarithmic mean temperature difference and represents the situation where the temperature of two fluids change as they transverse the surface.

$$\Delta T_m = \frac{\Delta T_1 - \Delta T_2}{Log_e \, [\Delta T_1 / \Delta T_2]} \qquad \qquad \text{Formula (5-5)}$$

To understand the different logarithmic mean temperature relationships, Figure 5-2 should be used. Referring to Figure 5-2, the ΔT_m for the counterflow heat exchanger is

$$\Delta T_m = \frac{(t_1 - t'_2) - (t_2 - t'_1)}{Log_e \, [(t_1 - t'_2 / t_2 - t'_1)]} \qquad \qquad \text{Formula (5-6)}$$

The ΔT_m for the parallel flow heat exchanger is

$$\Delta T_m = \frac{(t_1 - t'_1) - (t_2 - t'_2)}{Log_e \, [t_1 - t'_2 / t_2 - t'_2]} \qquad \qquad \text{Formula (5-7)}$$

HEAT TRANSFER BY RADIATION

Radiation is the transfer of heat energy by electromagnetic means between two materials whose surfaces "see" each other. The governing equation is known as the Stefan-Boltzmann equation, and is written

$$q = \sigma F_e \, F_a A \, (T^4_{aba_1} - T^4_{aba_2}) \qquad \qquad \text{Formula (5-8)}$$

A. COUNTERFLOW

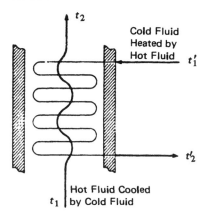

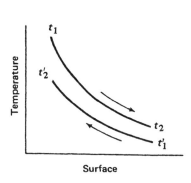

B. PARALLEL FLOW

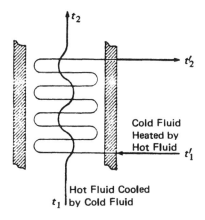

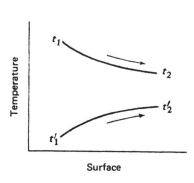

Figure 5-2. Temperature Relationships for Heat Exchangers

Again, q and A are as defined for conduction and convection, and T_{aba_1} and T_{aba_2} are the absolute temperatures of the two surfaces involved. The factor F_e is a function of the condition of the radiation surfaces and in some cases the areas of the surfaces. The factor F_a is the configuration factor and is a function of the areas and their positions. Both F_e and F_a are dimensionless. The Stefan-Boltzmann constant σ is equal to $0.1714\,\text{Btu}/\text{h}\bullet\text{ft}^2\bullet R^4$. Although radiation is usually associated with solid surfaces, certain gases can emit and absorb radiation. These include the so-called nonpolar molecular gases such as H_2O, CO_2, CO, SO_2, NH_3 and the hydrocarbons. Some of these gases are present in every combustion process.

It is convenient in certain calculations to express the heat transferred by radiation in the form of

$$q = b_r A \, (T_1 - T_2) \qquad \qquad Formula \ (5\text{-}9)$$

where the coefficient of radiation, h_r, is defined as

$$h_r = \frac{\sigma F_e \, F_a \, (T^4 \, aba_1 - T^4 \, aba_2)}{(T_1 - T_1)} \qquad \qquad Formula \ (5\text{-}10)$$

Note that b_r is still dependent on the factors F_e and F_a and also on the absolute temperatures to the fourth power. The main advantage is that in Formula 5-10 the rate of heat flow by radiation is a function of the temperature difference and can be combined with the coefficient of convection to determine the total heat flow to or from a surface.

WASTE HEAT RECOVERY EQUIPMENT

Industrial heat exchangers have many pseudonyms. They are sometimes called recuperators, regenerators, waste heat steam generators, condensers, heat wheels, temperature and moisture exchangers, etc. Whatever name they may have, they all perform one basic function: the transfer of heat.

Heat exchangers are characterized as single or multipass gas to gas, liquid to gas, liquid to liquid, evaporator, condenser, parallel flow, counterflow, or crossflow. The terms single or multipass refer to the heating or cooling media passing over the heat transfer surface once or a number of times. Multipass flow involves the use of internal baffles. The next three terms refer to the two fluids between which heat is transferred in the heat exchanger and imply that no phase changes occur in those fluids. Here the term "fluid" is used in the most general sense. Thus, we can say that these terms apply to non-evaporator and noncondensing heat exchangers. The term evaporator applies to a heat exchanger in which heat is transferred to an evaporating (boiling) liquid, while a condenser is a heat exchanger in which heat is removed from a condensing vapor. A parallel flow heat exchanger is one in which both fluids flow in approximately the same direction, whereas in counterflow the two fluids move in opposite directions. When the two fluids move at right angles to each other, the heat exchanger is considered to be of the crossflow type.

The principal methods of reclaiming waste heat in industrial plants make use of heat exchangers. The heat exchanger is a system which separates the stream containing waste heat and the medium which is to absorb it but which allows the flow of heat across the separation boundaries. The reasons for separating the two streams may be any of the following:

(1)   A pressure difference may exist between the two streams of fluid. The rigid boundaries of the heat exchanger can be designed to withstand the pressure difference.

(2)   In many, if not most, cases the one stream would contaminate the other, if they were permitted to mix. The heat exchanger prevents mixing.

(3)   Heat exchangers permit the use of an intermediate fluid better suited than either of the principal exchange media for transporting waste heat through long distances. The secondary fluid is often steam, but another substance may be selected for special properties.

(4)   Certain types of heat exchangers, specifically the heat wheel, are capable of transferring liquids as well as heat. Vapors being cooled in the gases are condensed in the wheel and later re-evaporated into the gas being heated. This can result in improved humidity and/or process control, abatement of atmospheric air pollution and conservation of valuable resources.

The various names or designations applied to heat exchangers are partly an attempt to describe their function and partly the result of tradition within certain industries. For example, a recuperator is a heat exchanger which recovers waste heat from the exhaust gases of a furnace to heat the incoming air for combustion. This is the name used in both the steel and the glass making industries. The heat exchanger performing the same function in the steam generator of an electric power plant is termed an air preheater, and in the case of a gas turbine plant, a regenerator.

However, in the glass and steel industries the word regenerator refers to two chambers of brick checkerwork which alternately absorb heat from the exhaust gases and then give up part of that heat to the incoming air. The flows of flue gas and of air are periodically reversed by valves so that one chamber of the regenerator is being heated by the products of combustion while the other is being cooled by the incoming air. Regenerators are often more expensive to buy and more expensive to maintain than are recuperators, and their application is primarily in glass melt tanks and in open hearth steel furnaces.

It must be pointed out, however, that although their functions are similar, the three heat exchangers mentioned above may be structurally quite different as well as different in their principal modes of heat transfer. A more complete description of the various industrial heat exchangers follows later in this chapter, and details of their differences will be clarified.

The specification of an industrial heat exchanger must include the heat exchange capacity, the temperatures of the fluids, the allowable pressure drop in each fluid path, and the properties and volumetric flow of the fluids entering the exchanger. These specifications will determine construction parameters and thus the cost of the heat exchanger. The final design will be a compromise between pressure drop, heat exchanger effectiveness, and cost. Decisions leading to that final design will balance out the cost of maintenance and operation of the overall system against the fixed costs in such a way as to minimize the total. Advice on selection and design of heat exchangers is available from vendors.

The essential parameters which should be known in order to make an optimum choice of waste heat recovery devices are:

•   Temperature of waste heat fluid.

•   Flow rate of waste heat fluid.

•   Chemical composition of waste heat fluid.

•   Minimum allowable temperature of waste heat fluid.

•   Temperature of heated fluid.

•   Chemical composition of heated fluid.

•   Maximum allowable temperature of heated fluid.

•   Control temperature, if control required.

In the rest of this chapter, some common types of waste heat recovery devices are discussed in some detail.

GAS TO GAS HEAT EXCHANGERS

Recuperators

The simplest configuration for a heat exchanger is the metallic radiation recuperator which consists of two concentric lengths of metal tubing as shown in Figure 5-3.

The inner tube carries the hot exhaust gases while the external annulus carries the combustion air from the atmosphere to the air inlets of the furnace burners. The hot gases are cooled by the incoming combustion air which now carries additional energy into the combustion chamber. This is energy which does not have to be supplied by the fuel; consequently, less fuel is burned for a given furnace loading. The saving in fuel also means a decrease in combustion air and therefore stack losses are decreased not only by lowering the stack gas temperatures but also by discharging smaller quantities of exhaust gas. This particular recuperator gets its name from the fact that a substantial portion of the heat transfer from the hot gases to the surface of the inner tube takes place by radiative heat transfer. The cold air in the annulus, however, is almost transparent

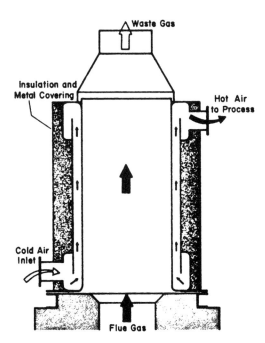

Figure 5-3. Diagram of Metallic Radiation Recuperator

to infrared radiation so that only convection heat transfer takes place to the incoming air. As shown in the diagram, the two gas flows are usually parallel, although the configuration would be simpler and the heat transfer more efficient if the flows were opposed in direction (or counterflow). The reason for the use of parallel flow is that recuperators frequently serve the additional function of cooling the duct carrying away the exhaust gases and consequently extending its service life.

The inner tube is often fabricated from high temperature materials such as stainless steels of high nickel content. The large temperature differential at the inlet causes differential expansion, since the outer shell is usually of a different and less expensive material. The mechanical design must take this effect into account. More elaborate designs of radiation recuperators incorporate two sections: the bottom operating in parallel flow and the upper section using the more efficient counterflow arrangement. Because of the large axial expansions experienced and the stress conditions at the bottom of the recuperator, the unit is often supported at the top by a free-standing support frame with an expansion joint between the furnace and recuperator.

A second common configuration for recuperators is called the tube type or convective recuperator. As seen in the schematic diagram of Figure 5-4, the hot gases are carried through a number of parallel small diameter tubes, while the incoming air to be heated enters a shell surrounding the tubes and passes over the hot tubes one or more times in a direction normal to their axes.

If the tubes are baffled to allow the gas to pass over them twice, the

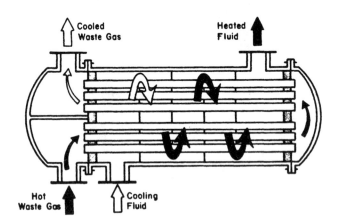

Figure 5-4. Diagram of Convective-type Recuperator

heat exchanger is termed a two-pass recuperator; if two baffles are used, a three-pass recuperator, etc. Although baffling increases both the cost of the exchanger and the pressure drop in the combustion air path, it increases the effectiveness of heat exchange. Shell- and tube-type recuperators are generally more compact and have a higher effectiveness than radiation recuperators, because of the larger heat transfer area made possible through the use of multiple tubes and multiple passes of the gases.

The principal limitation on the heat recovery of metal recuperators is the reduced life of the liner at inlet temperatures exceeding 2000°F. At this temperature, it is necessary to use the less efficient arrangement of parallel flows of exhaust gas and coolant in order to maintain sufficient cooling of the inner shell. In addition, when furnace combustion air flow is dropped back because of reduced load, the heat transfer rate from hot waste gases to preheat combustion air becomes excessive, causing rapid surface deterioration. Then, it is usually necessary to provide an ambient air bypass to cool the exhaust gases.

In order to overcome the temperature limitations of metal recuperators, ceramic tube recuperators have been developed whose materials allow operation on the gas side to 2800°F and on the preheated air side to 2200°F on an experimental basis and to 1500°F on a more or less practical basis. Early ceramic recuperators were built of tile and joined with furnace cement, and thermal cycling caused cracking of joints and rapid deterioration of the tubes. Later developments introduced various kinds of short silicon carbide tubes which can be joined by flexible seals located in the air headers. This kind of patented design illustrated in Figure 5-5 maintains the seals at comparatively low temperatures and has reduced the seal leakage rates to a few percent.

Earlier designs had experienced leakage rates from 8 to 60 percent. The new designs are reported to last two years with air preheat temperatures as high as 1300°F, with much lower leakage rates.

An alternative arrangement for the convective type recuperator, in which the cold combustion air is heated in a bank of parallel vertical tubes that extend into the flue gas stream, is shown schematically in Figure 5-6. The advantage claimed for this arrangement is the ease of replacing individual tubes, which can be done during full capacity furnace operation. This minimizes the cost, the inconvenience and possible furnace damage due to a shutdown forced by recuperator failure.

For maximum effectiveness of heat transfer, combinations of radiation type and convective type recuperators are used, with the convective type

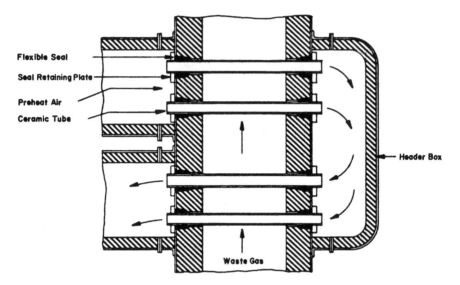

Figure 5-5. Ceramic Recuperator

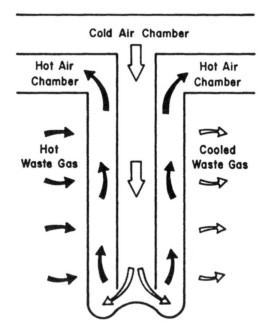

Figure 5-6. Diagram of Vertical Tube-within-tube Recuperator

always following the high temperature radiation recuperator. A schematic diagram of this arrangement is seen in Figure 5-7.

Although the use of recuperators conserves fuel in industrial furnaces and although their original cost is relatively modest, the purchase of the unit is often just the beginning of a somewhat more extensive capital improvement program. The use of a recuperator, which raises the temperature of the incoming combustion air, may require purchase of high temperature burners, larger diameter air lines with flexible fittings to allow for expansion, cold air lines for cooling the burners, modified combustion controls to maintain the required air/fuel ratio despite variable recuperator heating, stack dampers, cold air bleeds, controls to protect the recuperator during blower failure or power failures and larger fans to overcome the additional pressure drop in the recuperator. It is vitally important to protect the recuperator against damage due to excessive temperatures, since the

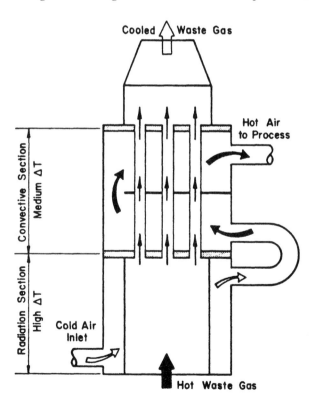

Figure 5-7. Diagram of Combined Radiation and Convective Type Recuperator

cost of rebuilding a damaged recuperator may be as high as 90 percent of the initial cost of manufacture, and the drop in efficiency of a damaged recuperator may easily increase fuel costs by 10 to 15 percent.

Figure 5-8 shows a schematic diagram of one radiant tube burner fitted with a radiation recuperator. With such a short stack, it is necessary to use two annuli for the incoming air to achieve reasonable heat exchange efficiencies.

Recuperators are used for recovering heat from exhaust gases to heat other gases in the medium to high temperature range. Some typical applications are in soaking ovens, annealing ovens, melting furnaces, afterburners and gas incinerators, radiant-tube burners, reheat furnaces, and other gas to gas waste heat recovery applications in the medium to high temperature range.

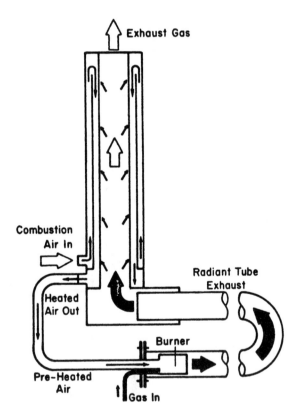

Figure 5-8. Diagram of a Small Radiation-Type Recuperator Fitted to a Radiant Tube Burner

Heat Wheels

A rotary regenerator (also called an air preheater or a heat wheel) is finding increasing applications in low to medium temperature waste heat recovery. Figure 5-9 is a sketch illustrating the application of a heat wheel. It is a sizable porous disk, fabricated from some material having a fairly high heat capacity, which rotates between two side-by-side ducts: one a cold gas duct, the other a hot gas duct. The axis of the disk is located parallel to, and on the partition between, the two ducts. As the disk slowly rotates, sensible heat (and, in some cases, moisture that contains latent heat) is transferred to the disk by the hot air and, as the disk rotates, from the disk to the cold air. The overall efficiency of sensible heat transfer for this kind of regenerator can be as high as 85 percent. Heat wheels have been built as large as 70 feet in diameter with air capacities up to 40,000 ft^3/min. Multiple units can be used in parallel. This may help to prevent a mismatch between capacity requirements and the limited number of sizes available in packaged units. In very large installations such as those required for preheating combustion air in fixed station electrical generating stations, the units are custom designed.

The limitation on temperature range for the heat wheel is primarily due to mechanical difficulties introduced by uneven expansion of the

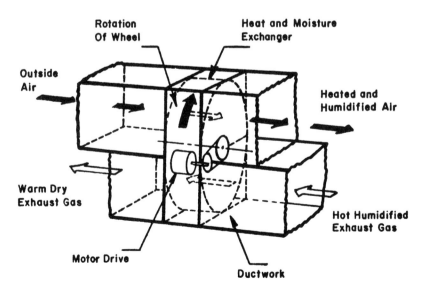

Figure 5-9. Heat and Moisture Recovery Using a Heat Wheel Type Regenerator

rotating wheel when the temperature differences mean large differential expansion, causing excessive deformations of the wheel and thus difficulties in maintaining adequate air seals between duct and wheel.

Heat wheels are available in four types. The first consists of a metal frame packed with a core of knitted mesh stainless steel or aluminum wire, resembling that found in the common metallic kitchen pot scraper; the second, called a laminar wheel, is fabricated from corrugated metal and is composed of many parallel flow passages; the third variety is also a laminar wheel but is constructed from a ceramic matrix of honeycomb configuration. This type is used for higher temperature applications with a present-day limit of about 1600°F. The fourth variety is of laminar construction, but the flow passages are coated with a hygroscopic material so that latent heat may be recovered. The packing material of the hygroscopic wheel may be any of a number of materials. The hygroscopic material is often termed a desiccant.

Most industrial stack gases contain water vapor (since water vapor is a product of the combustion of all hydrocarbon fuels and since water is introduced into many industrial processes) and part of the process water evaporates as it is exposed to the hot gas stream. Each pound of water requires approximately 1000 Btu for its evaporation at atmospheric pressure; thus each pound of water vapor leaving in the exit stream will carry 1000 Btu of energy with it. This latent heat may be a substantial fraction of the sensible energy in the exit gas stream. A hygroscopic material is one such as lithium chloride (LiCl) which readily absorbs water vapor. Lithium chloride is a solid which absorbs water to form a hydrate, $LiCl \cdot H_2O$, in which one molecule of lithium chloride combines with one molecule of water. Thus, the ratio of water to lithium chloride in $LiCl \cdot H_2O$ is 3/7 by weight. In a hygroscopic heat wheel, the hot gas stream gives up part of its water vapor to the coating; the cool gases which enter the wheel to be heated are drier than those in the inlet duct, and part of the absorbed water is given up to the incoming gas stream. The latent heat of the water adds directly to the total quantity of recovered waste heat. The efficiency of recovery of water vapor can be as high as 50 percent.

Since the pores of heat wheels carry a small amount of gas from the exhaust to the intake duct, cross contamination can result. If this contamination is undesirable, the carryover of exhaust gas can be partially eliminated by the addition of a purge section where a small amount of clean air is blown through the wheel and then exhausted to the atmosphere, thereby clearing the passages of exhaust gas. Figure 5-

10 illustrates the features of an installation using a purge section. Note that additional seals are required to separate the purge ducts. Common practice is to use about six air changes of clean air for purging. This limits gas contamination to as little as 0.04 percent and particle contamination to less than 0.2 percent in laminar wheels, and cross contamination to less than 1 percent in packed wheels. If inlet gas temperature is to be held constant, regardless of heating loads and exhaust gas temperatures, then the heat wheel must be driven at variable speed. This requires a variable speed drive and a speed control system using an inlet air temperature sensor as the control element. This feature, however, adds considerably to the cost and complexity of the system. When operating with outside air in periods of high humidity and sub-zero temperatures, heat wheels may require preheat systems to prevent frost formation. When handling gases which contain water-soluble, greasy or adhesive contaminants or large concentrations of process dust, air filters may be required in the exhaust system upstream from the heat wheel.

One application of heat wheels is in space heating situations where unusually large quantities of ventilation air are required for health or safety reasons. As many as 20 or 30 air changes per hour may be required to remove toxic gases or to prevent the accumulation of explosive mixtures. Comfort heating for that quantity of ventilation air is frequently expensive enough to make the use of heat wheels economical. In the summer season the heat wheel can be used to cool the incoming air from the cold exhaust air, reducing the air-conditioning load by as much as 50 percent. It should be pointed out that in many circumstances where large ventilating

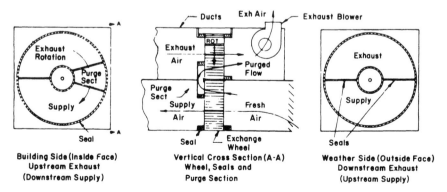

Figure 5-10. Heat Wheel Equipped with Purge Section to Clear Contaminants from the Heat Transfer Surface

requirements are mandatory, a better solution than the installation of heat wheels may be the use of local ventilation systems to reduce the hazards and/or the use of infrared comfort heating at principal work areas.

Heat wheels are finding increasing use for process heat recovery in low and moderate temperature environments. Typical applications would be curing or drying ovens and air preheaters in all sizes for industrial and utility boilers.

Air Preheaters

Passive gas to gas regenerators, sometimes called air preheaters, are available for applications which cannot tolerate any cross contamination. They are constructed of alternate channels (see Figure 5-11) which put the flows of the heating and the heated gases in close contact with each other, separated only by a thin wall of conductive metal. They occupy more volume and are more expensive to construct than are heat wheels, since a much greater heat transfer surface area is required for the same efficiency. An advantage, besides the absence of cross contamination, is the decreased mechanical complexity since no drive mechanism is required. However, it becomes more difficult to achieve temperature control with the passive regeneration, and, if this is a requirement, some of the advantages of its basic simplicity are lost.

Gas to gas regenerators are used for recovering heat from exhaust gases to heat other gases in the low to medium temperature range. A list

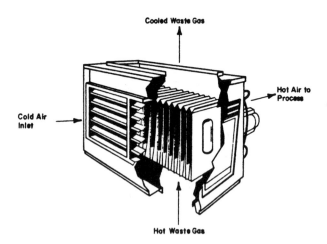

Figure 5-11. A Passive Gas to Gas Regenerator

of typical applications follows:

- Heat and moisture recovery from building heating and ventilation systems.
- Heat and moisture recovery from moist rooms and swimming pools.
- Reduction of building air-conditioner loads.
- Recovery of heat and water from wet industrial processes.
- Heat recovery from steam boiler exhaust gases.
- Heat recovery from gas and vapor incinerators.
- Heat recovery from baking, drying, and curing ovens.
- Heat recovery from gas turbine exhausts.
- Heat recovery from other gas to gas applications in the low through high temperature range.

Heat-Pipe Exchangers

The heat pipe is a heat transfer element that has only recently become commercial, but it shows promise as an industrial waste heat recovery option because of its high efficiency and compact size. In use, it operates as a passive gas to gas finned-tube regenerator. As can be seen in Figure 5-12, the elements form a bundle of heat pipes which extend through the exhaust and inlet ducts in a pattern that resembles the structured finned coil heat exchangers. Each pipe, however, is a separate sealed element consisting of an annular wick on the inside of the full length of the tube, in which an appropriate heat transfer fluid is entrained.

Figure 5-13 shows how the heat absorbed from hot exhaust gases evaporates the entrained fluid, causing the vapor to collect in the center core. The latent heat of vaporization is carried in the vapor to the cold end of the heat pipe located in the cold gas duct. Here the vapor condenses, giving up its latent heat. The condensed liquid is then carried by capillary (and/or gravity) action back to the hot end where it is recycled. The heat pipe is compact and efficient because: (1) the finned-tube bundle is inherently a good configuration for convective heat transfer in both gas ducts, and (2) the evaporative-condensing cycle within the heat tubes is a highly efficient way of transferring the heat internally. It is also free from cross contamination. Possible applications include:

- Drying, curing and baking ovens.
- Waste steam reclamation.
- Air preheaters in steam boilers.
- Air dryers.

- Brick kilns (secondary recovery).
- Reverberatory furnaces (secondary recovery).
- Heating, ventilating and air-conditioning systems.

GAS OR LIQUID TO LIQUID REGENERATORS

Finned-Tube Heat Exchangers

When waste heat in exhaust gases is recovered for heating liquids for purposes such as providing domestic hot water, heating the feedwater for steam boilers, or for hot water space heating, the finned-tube heat

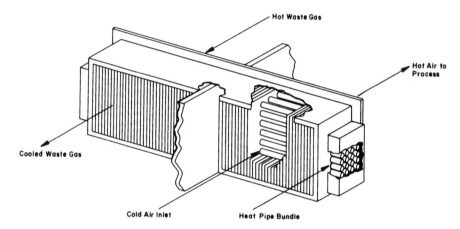

Figure 5-12. Heat Pipe Bundle Incorporated in Gas-to-gas Regenerator

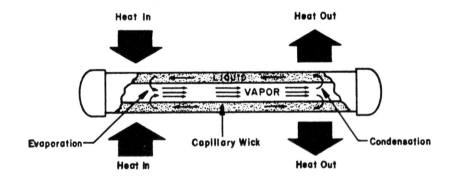

Figure 5-13. Heat Pipe Schematic

exchanger is generally used. Round tubes are connected together in bundles to contain the heated liquid, and fins are welded or otherwise attached to the outside of the tubes to provide additional surface area for removing the waste heat in the gases.

Figure 5-14 shows the usual management for the finned-tube exchanger positioned in a duct and details of a typical finned-tube construction. This particular type of application is more commonly known as an economizer. The tubes are often connected all in series but can also be arranged in series-parallel bundles to control the liquid side pressure drop. The air side pressure drop is controlled by the spacing of the tubes and the number of rows of tubes within the duct.

Finned-tube exchangers are available prepackaged in modular sizes or can be made up to custom specifications very rapidly from standard components. Temperature control of the heated liquid is usually provided by a bypass duct arrangement which varies the flow rate of hot gases over the heat exchanger. Materials for the tubes and the fins can be selected to

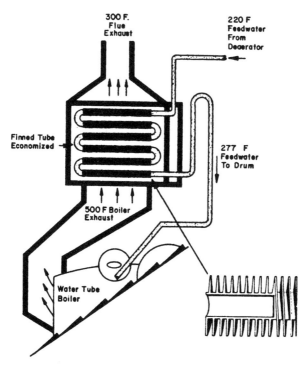

Figure 5-14. Finned-tube Gas to Liquid Regenerator (Economizer)

withstand corrosive liquids and/or corrosive exhaust gases.

Finned-tube heat exchangers are used to recover waste heat in the low to medium temperature range from exhaust gases for heating liquids. Typical applications are domestic hot water heating, heating boiler feedwater, hot water space heating, absorption-type refrigeration or air conditioning and heating process liquids.

Shell and Tube Heat Exchanger

When the medium containing waste heat is a liquid or a vapor which heats another liquid, then the shell and tube heat exchanger must be used since both paths must be sealed to contain the pressures of their respective fluids. The shell contains the tube bundle, and usually internal baffles, to direct the fluid in the shell over the tubes in multiple passes. The shell is inherently weaker than the tubes so that the higher pressure fluid is circulated in the tubes while the lower pressure fluid flows through the shell. When a vapor contains the waste heat, it usually condenses, giving up its latent heat to the liquid being heated. In this application, the vapor is almost invariably contained within the shell. If the reverse is attempted, the condensation of vapors within small diameter parallel tubes causes flow instabilities. Tube and shell heat exchangers are available in a wide range of standard sizes with many combinations of materials for the tubes and shells.

Typical applications of shell and tube heat exchangers include heating liquids with the heat contained by condensates from refrigeration and air-conditioning systems; condensate from process steam; coolants from furnace doors, grates, and pipe supports; coolant from engines, air compressors, bearings, and lubricants; and the condensates from distillation processes.

Waste Heat Boilers

Waste heat boilers are ordinarily water tube boilers in which the hot exhaust gases from gas turbines, incinerators, etc., pass over a number of parallel tubes containing water. The water is vaporized in the tubes and collected in a steam drum from which it is drawn off for use as heating or processing steam.

Figure 5-15 indicates one arrangement that is used, where the exhaust gases pass over the water tubes twice before they are exhausted to the air. Because the exhaust gases are usually in the medium temperature range and in order to conserve space, a more compact boiler can be produced if the water tubes are finned in order to increase the effective heat transfer area on the gas side. The diagram shows a mud drum, a set of tubes over which the hot gases make a double pass, and a steam drum which collects the steam generated above the water surface. The pressure at which the steam is generated and the rate of steam production depend on the temperature of the hot gases entering the boiler, the flow rate of the hot gases, and the efficiency of the boiler. The pressure of a pure vapor in the presence of its liquid is a function of the temperature of the liquid from

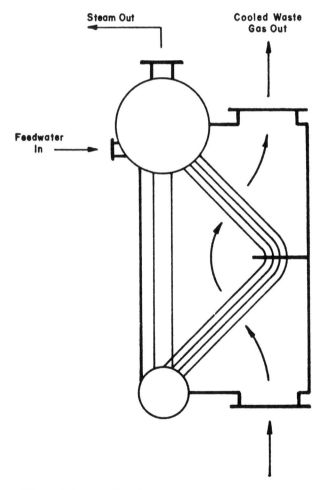

Figure 5-15. Waste Heat Boiler for Heat Recovery from Gas Turbines or Incinerators

which it is evaporated. The steam tables tabulate this relationship between saturation pressure and temperature. Should the waste heat in the exhaust gases be insufficient for generating the required amount of process steam, it is sometimes possible to add auxiliary burners which burn fuel in the waste heat boiler or to add an afterburner to the exhaust gas duct just ahead of the boiler. Waste heat boilers are built in capacities from less than a thousand to almost a million ft^3/min. of exhaust gas.

Typical applications of waste heat boilers are to recover energy from the exhausts of gas turbines, reciprocating engines, incinerators, and furnaces.

Gas and Vapor Expanders

Industrial steam and gas turbines are in an advanced state of development and readily available on a commercial basis. Recently, special gas turbine designs for low pressure waste gases have become available; for example, a turbine is available for operation from the top gases of a blast furnace. In this case, as much as 20 MW of power could be generated, representing a recovery of 20 to 30 percent of the available energy of the furnace exhaust gas stream. Maximum top pressures are of the order of 40 lb/in^2 g.

Perhaps of greater applicability than the last example are steam turbines used for producing mechanical work or for driving electrical generators. After removing the necessary energy for doing work, the steam turbine exhausts partially spent steam at a lower pressure than the inlet pressure. The energy in the turbine exhaust stream can then be used for process heat in the usual ways. Steam turbines are classified as back-pressure turbines, available with allowable exit pressure operation above 400 lb/in^2 g, or condensing turbines which operate below atmospheric exit pressures. The steam used for driving the turbines can be generated in direct fired or waste heat boilers. A list of typical applications for gas and vapor expanders follows:

- Electrical power generation.
- Compressor drives.
- Pump drives.
- Fan drives.

Heat Pumps

In the commercial options previously discussed in this chapter, we find waste heat being transferred from a hot fluid to a fluid at a lower temperature. Heat must flow spontaneously "downhill," that is, from a system at high temperature to one at a lower temperature. This can be expressed scientifically in a number of ways—all the variations of the statement of the second law of thermodynamics. The practical impact of these statements is that energy, as it is transformed again and again and transferred from system to system, becomes less and less available for use. Eventually that energy has such low intensity (resides in a medium at such low temperature) that it is no longer available at all to perform a useful function. It has been taken as a general rule of thumb in industrial operations that fluids with temperatures less than 250°F are of little value for waste heat extraction; flue gases should not be cooled below 250°F (or, better, 300°F to provide a safe margin), because of the risk of condensation of corrosive liquids. However, as fuel costs continue to rise, such waste heat can be used economically for space heating and other low temperature applications. It is possible to reverse the direction of spontaneous energy flow by the use of a thermodynamic system known as a heat pump.

This device consists of two heat exchangers, a compressor and an expansion device. A liquid or a mixture of liquid and vapor of a pure chemical species flows through an evaporator, where it absorbs heat at low temperature and, in doing so, is completely vaporized. The low temperature vapor is compressed by a compressor which requires external work. The work done on the vapor raises its pressure and temperature to a level where its energy becomes available for use. The vapor flows through a condenser where it gives up its energy as it condenses to a liquid. The liquid is then expanded through a device back to the evaporator where the cycle repeats. The heat pump was developed as a space heating system where low temperature energy from the ambient air, water, or earth is raised to heating system temperatures by doing compression work with an electric motor-driven compressor. The performance of the heat pump is ordinarily described in terms of the coefficient of performance or COP, which is defined as

$$COP = \frac{\text{Heat transferred in condenser}}{\text{Compressor work}} \qquad \textit{Formula (5-11)}$$

which in an ideal heat pump is found as

$$COP = \frac{T_H}{T_H - T_L} \qquad \textit{Formula (5-12)}$$

where T_L is the temperature at which waste heat is extracted from the low temperature medium and T_H is the high temperature at which heat is given up by the pump as useful energy. The coefficient of performance expresses the economy of heat transfer.

In the past, the heat pump has not been applied generally to industrial applications. However, several manufacturers are now redeveloping their domestic heat pump systems as well as new equipment for industrial use. The best applications for the device in this new context are not yet clear, but it may well make possible the use of large quantities of low-grade waste heat with relatively small expenditures of work.

SUMMARY

Table 5-4 presents the collation of a number of significant attributes of the most common types of industrial heat exchangers in matrix form. This matrix allows rapid comparisons to be made in selecting competing types of heat exchangers. The characteristics given in the table for each type of heat exchanger are allowable temperature range, ability to transfer moisture, ability to withstand large temperature differentials, availability as packaged units, suitability for retrofitting, and compactness and the allowable combinations of heat transfer fluids.

Table 5-4. Operation and Application Characteristics of Industrial Heat Exchangers

SPECIFICATIONS FOR WASTE RECOVERY UNIT / COMMERCIAL HEAT TRANSFER EQUIPMENT	Low Temperature Sub-Zero – 250°F	Intermediate Temp. 250°F – 1200°F	High Temperature 1200°F – 2000°F	Recovers Moisture	Large Temperature Differentials Permitted	Packaged Units Available	Can Be Retrofit	No Cross-Contamination	Compact Size	Gas-to-Gas Heat Exchange	Gas-to-Liquid Heat Exchanger	Liquid-to-Liquid Heat Exchanger	Corrosive Gases Permitted with Special Construction
Radiation Recuperator			●		●	1	●	●		●			●
Convection Recuperator		●	●		●	●	●	●		●			●
Metallic Heat Wheel	●	●		2		●	●	3	●	●			●
Hygroscopic Heat Wheel	●			●		●	●	3	●	●			
Ceramic Heat Wheel		●	●		●	●	●		●	●			●
Passive Regenerator	●	●			●	●	●	●		●			●
Finned-Tube Heat Exchanger	●	●			●	●	●	●	●		●		4
Tube Shell-and-Tube Exchanger	●	●			●	●	●	●			●	●	
Waste Heat Boilers	●	●				●	●	●			●		4
Heat Pipes	●	●			5	●	●	●	●	●			●

1. Off-the-shelf items available in small capacities only.
2. Controversial subject. Some authorities claim moisture recovery. Do not advise depending on it.
3. With a purge section added, cross-contamination can be limited to less than 1% by mass.
4. Can be constructed of corrosion-resistant materials, but consider possible extensive damage to equipment caused by leaks or tube ruptures.
5. Allowable temperatures and temperature differential limited by the phase equilibrium properties of the internal fluid.

6

Utility System Optimization

BASIS OF THERMODYNAMICS

Thermodynamics deals with the relationships between heat and work. It is based on two basic laws of nature: the first and second laws of thermodynamics. The principles are used in the design of equipment such as steam engines, turbines, pumps, and refrigerators, and in practically every process involving a flow of heat or a chemical equilibrium.

First Law: The first law states that energy can neither be created nor destroyed, thus, it is referred to as the law of conservation of energy. Formula 6-1 expresses the first law for the steady state condition.

$$E_2 - E_1 = Q - W \qquad \text{Formula (6-1)}$$

Where
$E_2 - E_1$ is the change in stored energy at the boundary states 1 and 2 of the system
   Q is the heat added to the system
   W is the work done by the system

Figure 6-1 illustrates a thermodynamic process where mass enters and leaves the system. The potential energy (Z) and the kinetic energy $(V^2/64.2)$ plus the enthalpy represent the stored energy of the mass. Note, Z is the elevation above the reference point in feet, and V is the velocity of the mass in ft/sec. In the case of the steam turbine, the change in Z, V, and Q are small in comparison to the change in enthalpy. Thus, the energy equation reduces to

$$W/778 = h_1 - h_2 \qquad \text{Formula (6-2)}$$

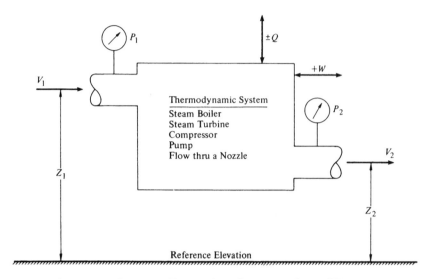

Figure 6-1. System Illustrating Conservation of Energy

Where
  W is the work done in ft • lb/lb
  h_1 is the enthalpy of the entering steam, Btu/lb
  h_2 is the enthalpy of the exhaust steam, Btu/lb
And 1 Btu equals 778 ft • lb

Second Law: The second law qualifies the first law by discussing the conversion between heat and work. All forms of energy, including work, can be converted to heat, but the converse is not generally true. The Kelvin-Planck statement of the second law of thermodynamics says essentially the following: Only a portion of the heat from a heat work cycle, such as a steam power plant, can be converted to work. The remaining heat must be rejected as heat to a sink of lower temperature (to the atmosphere, for instance).

The Clausius statement, which also deals with the second law, states that heat, in the absence of some form of external assistance, can only flow from a hotter to a colder body.

THE CARNOT CYCLE

The Carnot cycle is of interest because it is used as a comparison of the efficiency of equipment performance. The Carnot cycle offers the maximum thermal efficiency attainable between any given temperatures of heat source and sink. A thermodynamic cycle is a series of processes forming a closed curve on any system of thermodynamic coordinates. The Carnot cycle is illustrated on a temperature-entropy diagram, Figure 6-2A, and on the Mollier diagram for superheated steam, Figure 6-2B.

The cycle consists of the following:

1. Heat addition at constant temperature, resulting in expansion work and changes in enthalpy.
2. Adiabatic isentropic expansion (change in entropy is zero) with expansion work and an equivalent decrease in enthalpy.
3. Constant temperature heat rejection to the surroundings, equal to the compression work and any changes in enthalpy.
4. Adiabatic isentropic compression returning to the starting temperature with compression work and an equivalent increase in enthalpy.

The Carnot cycle is an example of a reversible process and has no counterpart in practice. Nevertheless, this cycle illustrates the principles of thermodynamics. The thermal efficiency for the Carnot cycle is illustrated by Formula (6-3).

$$\text{Thermal efficiency} = \frac{T_1 - T_2}{T_1} \qquad \textit{Formula (6-3)}$$

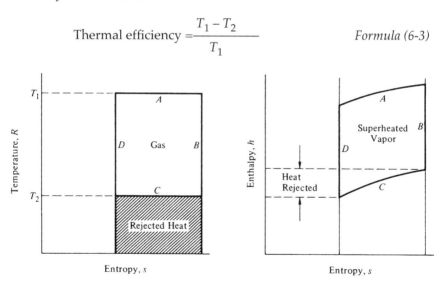

Figure 6-2A. Temperature-entropy diagram; gas.

Figure 6-2B. Mollier diagram; superheated vapor.

Figure 6-2. Carnot Cycles

Where

T_1 = Absolute temperature of heat source, °R (Rankine)
T_2 = Absolute temperature of heat sink, °R

and absolute temperature is given by Formula 6-4.

Absolute temperature = 460 + temperature in Fahrenheit.

Formula (6-4)

T_2 is usually based on atmospheric temperature, which is taken as 500°R.

PROPERTIES OF STEAM PRESSURE AND TEMPERATURE

Water boils at 212°F when it is in an open vessel under atmospheric pressure equal to 14.7 psia (pounds per square inch, absolute). Absolute pressure is the amount of pressure exerted by a system on its boundaries and is used to differentiate it from gage pressure. A pressure gage indicates the difference between the pressure of the system and atmospheric pressure.

psia = psig + atmospheric pressure in psia   *Formula (6-5)*

Changing the pressure of water changes the boiling temperature. Thus, water can be vaporized at 170°F, at 300°F, or any other temperature, as long as the applied pressure corresponds to that boiling point.

SOLID, LIQUID AND VAPOR STATES OF A LIQUID

Water, as well as other liquids, can exist in three states: solid, liquid, and vapor. In order to change the state from ice to water or from water to steam, heat must be added. The heat required to change a solid to liquid is called the *latent heat of fusion*. The heat required to change a liquid to a vapor is called the *latent heat of vaporization*.

In condensing steam, heat must be removed. The quantity is exactly equal to the latent heat that went into the water to change it to steam.

Heat supplied to a fluid, during the change of state to a vapor, will not cause the temperature to rise; thus, it is referred to as the latent heat of vaporization. Heat given off by a substance when it condenses from steam to a liquid is called *sensible heat*. Physical properties of water, such as the latent heat of vaporization, also change with variations in pressure.

USE OF THE STEAM TABLES

Steam properties are illustrated in the Appendix by Tables A-13 and A-14. Table A-13 is referred to as the steam table for saturated steam. Steam properties are shown and correlated to temperature and pressure. (Pressure must be converted to psia.) The properties of superheated steam are indicated by Table A-14.

The term h_{fg} is the latent heat or enthalpy of vaporization. Thus, from Table A-13, at 15.9 psia, the latent heat of vaporization is h_{fg}, 967.8 Btu/lb. At 205.2 psia, the latent heat of vaporization is 840.8 Btu/lb.

The enthalpy h_f represents the amount of heat required to raise one pound of water from 32°F to a liquid state at another temperature. As an example, from Table A-13, to raise water from 32°F to 170°F will require 137.9 Btu/lb.

Example Problem 6-1

How much heat is required to raise 100 pounds of water at 126°F to 170°F?

Answer

From Table A-13:

At 126°F – h_f = 93.9 Btu/lb

At 170°F – h_f = 137.9 Btu/lb

$Q = 100(137.9 - 93.9) = 44 \times 10^2$ Btu.

USE OF THE SPECIFIC HEAT CONCEPT

Another physical property of a material is the *specific heat*. The specific heat is defined as the amount of heat required to raise a unit of mass of a substance one degree. For water, it can be seen from the previous example that one Btu of heat is required to raise one lb water 1°F; thus, the specific heat of water $C_p = 1$. Specific heats for other materials are illustrated in Table 6-1. This leads to two equations.

$$Q = wC_p \Delta T \qquad \qquad \text{Formula (6-6)}$$

Where

Q = quantity of heat, Btu
w = weight of substance, lb
C_p = specific heat of substance, Btu per lb ° F
ΔT = temperature change of substance ° F

$$Q = MC_p \Delta T \qquad \qquad \text{Formula (6-7)}$$

Where

Q = quantity of heat, Btu/hr (Btu/hr is sometimes abbreviated as Btuh)
M = flow rate, lbs/hr
C_p = specific heat, Btu per lb °F

Table 6-1. Specific Heat of Various Substances

SUBSTANCE	Specific Heat Btu/lb°F	SUBSTANCE	Specific Heat Btu/lb°F
SOLIDS		**LIQUIDS**	
ALUMINUM	0.230	ALCOHOL	0.600
ASBESTOS	0.195	AMMONIA	1.100
BRASS	0.086	BRINE. CALCIUM (20% SOLUTION)	0.730
BRICK	0.220	BRINE. SODIUM (20% SOLUTION)	0.810
BRONZE	0.086	CARBON TETRACHLORIDE	0.200
CHALK	0.215	CHLOROFORM	0.230
CONCRETE	0.270	ETHER	0.530
COPPER	0.093	GASOLINE	0.700
CORK	0.485	GLYCERINE	0.576
GLASS. CROWN	0.161	KEROSENE	0.500
GLASS. FLINT	0.117	MACHINE OIL	0.400
GLASS. THERMOMETER	0.199	MERCURY	0.033
GOLD	0.030	PETROLEUM	0.500
GRANITE	0.192	SULPHURIC ACID	0.336
GYPSUM	0.259	TURPENTINE	0.470
ICE	0.480	WATER	1.000
IRON. CAST	0.130	WATER. SEA	0.940
IRON. WROUGHT	0.114		
LEAD	0.031	**GASES**	
LEATHER	0.360	AIR	0.240
LIMESTONE	0.216	AMMONIA	0.520
MARBLE	0.210	BROMINE	0.056
MONEL METAL	0.128	CARBON DIOXIDE	0.200
PORCELAIN	0.255	CARBON MONOXIDE	0.243
RUBBER	0.481	CHLOROFORM	0.144
SILVER	0.055	ETHER	0.428
STEEL	0.118	HYDROGEN	3.410
TIN	0.045	METHANE	0.593
WOOD	0.330	NITROGEN	0.240
ZINC	0.092	OXYGEN	0.220
		SULPHUR DIOXIDE	0.154
		STEAM (SUPERHEATED. 1 PSI)	0.450

ΔT = temperature change of substance °F

Example Problem 6-2
Check answer to Example Problem 6-1 using Formula 6-6.

Answer

$Q = WC_p \Delta T$
$= 100 \times 1 \times (170 - 126) = 44 \times 10^2$ Btu.

The saturated vapor enthalpy h_g represents the amount of heat necessary to change water at 32°F to steam at a specified temperature and pressure.

Example Problem 6-3
How much heat is required to raise 100 pounds of water at 126°F to 15.8 psig steam?

Answer

At 126°F – h_f = 93.9 Btu/lb
15 psig corresponds to 30 psia at 250°F
At 30 psia, h_g = 1164.1 Btu/lb
$Q = 100(1164.1 - 93.9) = 1070.2 \times 10^2$ Btu.

USE OF THE LATENT HEAT CONCEPT

When water is raised from 32°F to 212°F steam, the total heat required is composed of two components:
(a) The heat required to raise the temperature of water from 32°F to 212°F; h_f = 180.17 Btu/lb from Table A-13.
(b) The heat required to evaporate the water at 212°F; h_{fg} = 970.3 Btu/lb from Table A-13.

Thus, $h_g = h_f + h_{fg} = 180.17 + 970.3 = 1150.5$ Btu/lb, which agrees with the value of h_g found in Steam Table A-13.

Example Problem 6-4
30 psig steam is used for a heat exchanger and returns to the system as 30 psig condensate. What amount of heat is given off to the process fluid?

Answer

30 psig = 45 psia. This corresponds to an h_{fg} of approximately 928 Btu/lb.

THE USE OF THE SPECIFIC VOLUME CONCEPT

Another property of water is the specific volume v_f of water and the specific volume v_g of steam. *Specific volume* is defined as the space occupied by one pound of a material. For example, at 50 psia, water occupies 0.01727 ft^3 per lb and steam occupies 8.515 ft^3 per lb.

The *specific weight* is simply the weight of one cubic foot of a material and is the reciprocal of the specific volume.

THE MOLLIER DIAGRAM

A visual tool for understanding and using the properties of steam is illustrated by the Mollier diagram, Figure 6-3. The Mollier diagram enables one to find the relationship between temperature, pressure, enthalpy, and entropy for steam. Constant temperature and pressure curves illustrate the effect of various processes on steam.

For a constant temperature process (isothermal), the change in entropy is equal to the heat added (or subtracted) divided by the temperature at which the process is carried out. This is a simple way of explaining the physical meaning of entropy. The *change* in entropy is of interest. Referring to Table A-13, the value of entropy at 32°F is zero. Increases in entropy are a measure of the portion of heat in a process which is unavailable for conversion to work. Entropy has a close relationship to the second law of thermodynamics, discussed earlier in this chapter.

Another item of interest from the Mollier diagram is the saturation line. The saturation line indicates temperature and pressure relationships corresponding to saturated steam. Below this curve, steam contains a % moisture, as indicated by the % moisture curves. Steam at temperatures above the saturation curve is referred to as superheated. As an example, this chart indicates that at 212°F and 14.696 psia, the enthalpy h_g is 1150 Btu/lb, which agrees with steam table A-14 in the Appendix.

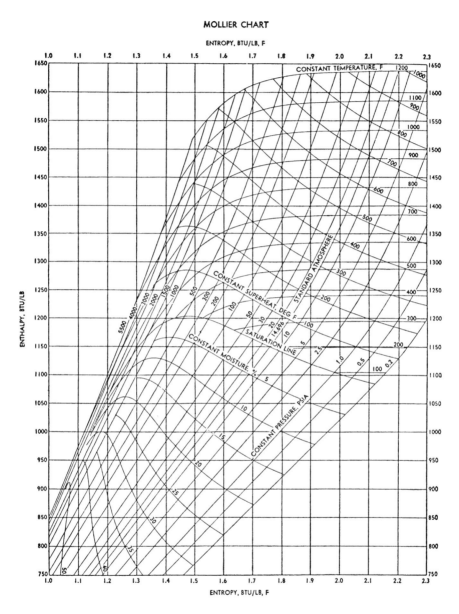

Figure 6-3. Mollier Diagram (*Courtesy Babcock & Wilcox Company and ASME***)**

SUPERHEATED STEAM

If additional heat is added to raise the temperature of steam over the point at which it was evaporated, the steam is termed superheated. Thus, steam at the same temperature as boiling water is saturated steam. Steam at temperatures higher than boiling water, at the same pressure, is superheated steam.

Example Problem 6-5
Using the Mollier diagram, find the enthalpy of steam at 14.696 psia and 300°F.

Answer
Follow the constant pressure line until it intersects the 300°F curve. The answer to the right is 1192 Btu/lb.

In addition to the Mollier diagram, the superheated steam table A-14 is helpful.

Steam cannot be superheated in the presence of water because the heat supplied will only evaporate the water. Thus, the water will be evaporated prior to becoming superheated. Superheated steam is condensed by first cooling it down to the boiling point corresponding to its pressure. When the steam has been de-superheated, further removal of heat will result in its condensation.

In the generation of power, superheated steam has many uses.

HEAT BALANCE

A heat balance is an analysis of a process which shows where all the heat comes from and where it goes. This is a vital tool in assessing the profit implications of heat losses and proposed waste heat utilization projects. The heat balance for a steam boiler, process furnace, air conditioner, etc. must be derived from measurements made during actual operating periods. Chapter 3 provides information on the instrumentation available to make these measurements. The measurements that are needed to get a complete heat balance involve energy inputs, energy losses to the environment and energy discharges.

Energy Input
Energy enters most process equipment either as chemical energy in

the form of fossil fuels, or sensible enthalpy of fluid streams, or latent heat in vapor streams, or as electrical energy.

For each input it is necessary to meter the quantity of fluid flowing or the electrical current. This means that if accurate results are to be obtained, submetering for each flow is required (unless all other equipment served by a main meter can be shut down so that the main meter can be used to measure the inlet flow to the unit). It is not necessary to continuously submeter every flow since temporary installations can provide sufficient information. In the case of furnaces and boilers that use pressure ratio combustion controls, the control flow meters can be utilized to yield the correct information. It should also be pointed out that for furnaces and boilers only the fuel should be metered. Tests of the exhaust products provide sufficient information to derive the oxidant (usually air) flow if accurate fuel flow data are available. For electrical energy inflows, the current is measured with an ammeter, or a kilowatt hour meter may be installed as a submeter. Ammeters using split core transformers are available for measuring alternating current flow without opening the line. These are particularly convenient for temporary installations.

In addition to measuring the flow for each inlet stream, it is necessary to know the chemical composition of the stream. For air, water, and other pure substances no tests for composition are required, but for fossil fuels the composition must be determined by chemical analysis or secured from the fuel supplier. For vapors one should know the quality—this is the mass fraction of vapor present in the mixture of vapor and droplets. Measurement of quality is made with a vapor calorimeter which requires only a small sample of the vapor stream.

Other measurements that are required are the entering temperatures of the inlet stream of fluid and the voltages of the electrical energy entering (unless kilowatt-hour meters are used).

The testing routines discussed above involve a good deal of time, trouble and expense. However, they are necessary for accurate analyses and may constitute the critical element in the engineering and economic analyses required to support decisions to expend capital on waste heat recovery equipment.

Energy Losses

Energy loss from process equipment to the ambient environment is usually by radiative and convective heat transfer. Radiant heat transfer, that is, heat transfer by light or other electromagnetic radiation, is discussed in the section of Chapter 3 dealing with infrared thermography. Convective heat transfer, which takes place by hot gas at the surface of the hot material being displaced by cooler gas, may be analyzed using Newton's law of cooling.

$$Q = UA\,(T_s - T_0) \qquad\qquad Formula\ (6\text{-}8)$$

Where
  Q = rate of heat loss in energy units Btu/h
  U = heat transfer coefficient in Btu/h•ft^2•F
  A = area of surface losing heat in ft^2
  T_s = surface temperature
  T_0 = ambient temperature

Although heat flux meters are available, it is usually easier to measure the quantities above and derive the heat loss from the equation. The problems encountered in using the equation involve the measurement of surface temperatures and the finding of accurate values for the heat transfer coefficient.

Unfortunately, the temperature distribution over the surface of a process unit can be very nonuniform so that an estimate of the overall average is quite difficult. New infrared measurement techniques, which are discussed in Chapter 3, make the determination somewhat more accurate. The heat transfer coefficient is not only a strong function of surface and ambient temperatures but also depends on geometric considerations and surface conditions. Thus for given surface and ambient temperatures a flat, vertical plate will have a different h_{cr} value than will a horizontal or inclined plate.

Energy Discharges
    The composition, discharge rate and temperature of each outflow from the process unit are required in order to complete the heat balance. For a fuel-fired unit, only the composition of the exhaust products, the flue gas temperature and the fuel input rate to the unit are required to derive:
(1) air input rate
(2) exhaust gas flow rate
(3) energy discharge rate from exhaust stack

    The composition of the exhaust products can be determined from an Orsat analysis, a chromatographic test or, less accurately, from a determination of the volumetric fraction of oxygen or CO_2, Figure 6-4 can be used for determining the quantity of excess or deficiency of air in the combustible mixture. It is based on the fact that chemical reactions occur with fixed ratios of reactants to form given products.

For example, natural gas with the following composition:

$$
\begin{array}{lcl}
CO_2 & - & 0.7\% \text{ volume} \\
O_2 & - & 0.0 \\
CH_4 & - & 92.0 \\
C_2H_6 & - & 6.8 \\
N_2 & - & \underline{0.5} \\
& & 100.00
\end{array}
$$

is burned to completion with the theoretical amount of air indicated in the volume equation below:

FUEL: $0.92\ CH_4 + 0.068\ C_2\ H_6$
(1 ft^3) $+\ 0.007\ CO_2 + 0.005\ N_2$

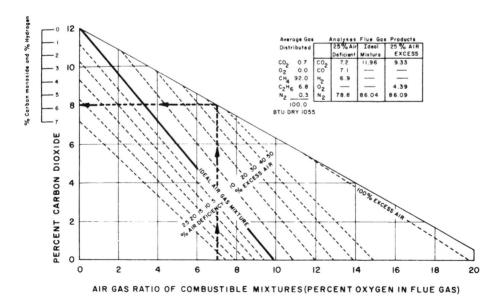

Figure 6-4. Natural Gas Combustion Chart

plus
AIR: $2.078 O_2 + 7.813 N_2 \rightarrow$
(9.891 ft^3)
yields
DRY PRODUCTS: $1.063 CO_2 + 7.818 N_2$
(10.925 ft^3) $+ 2.044 H_2O$ *Formula (6-9)*

The equation is based upon the laws of conservation of mass and elemental chemical species. The ratio of N_2 and O_2 in the combustible mixture results from the approximate volumetric ratio of N_2 to O_2 in air, i.e.,

$$20.9\% \ O_2, 79.1\% \ N_2 \text{ or} \quad \frac{79.1}{20.9} = 3.76 = \frac{\text{Volume } N_2}{\text{Volume } O_2}$$

For gases the coefficients of the chemical equation represent relative volumes of each species reacting. Ordinarily, excess air is provided to the fuel so that every fuel molecule will react with the necessary number of oxygen molecules even though the physical mixing process is imperfect. If 10 percent excess air were supplied, this mixture of reactants and products would give a chemical equation appropriately modified as given below:

FUEL: $0.92 CH_4 + 0.068 C_2 H_6$
(1 ft^3) $+ 0.007 CO_2 + 0.005 N_2$
plus
AIR: $2.286 O_2 + 8.594 N_2 \rightarrow$
(10.880 ft^3)
yields
DRY PRODUCTS: $1.063 CO_2 + 0.208 O_2 + 8.599 N_2$
(11.914 ft^3) $+ 2.044 H_2O$ *Formula (6-10)*

where an additional term representing the excess oxygen appears in the products, along with a corresponding increase in the nitrogen.

As an example let us assume that an oxygen meter has indicated a reading of 7% for the products of combustion from the natural gas whose composition was given previously and that the exhaust gas temperature was measured as 700°F. Figure 6-4 is used as indicated to determine that 45% excess air is mixed with fuel. The combustion equation then becomes:

FUEL: $0.92\ CH_4 + 0.068\ C_2\ H_6$
(1 ft³) $+ 0.007\ CO_2 + 0.005\ N_2$
plus
AIR: $3.013\ O_2 + 11.329\ N_2 \rightarrow$
(14.342 ft³)
yields
DRY PRODUCTS: $1.063\ CO_2 + 0.935\ O_2 + 11.334\ N_2$
(15.376 ft³) $+ 2.044\ H_2O$

For each 1 ft³ of fuel, 14.342 ft³ of air is supplied and 15.376 ft³ of exhaust products (at mixture temperature) is formed. Each cubic foot of fuel contains 1055 Btu of energy, so the fuel energy input is 250,000ft³/h × 1055 = 263,750,000 Btu/h. From Figure 6-5 we compute the exhaust gas losses at 700°F as

$$
\begin{array}{llll}
CO_2: & 250,000 \times & 1.063 \times 17.5 = & 4,651,000\ \text{Btu/h} \\
H_2O: & 250,000 \times & 2.044 \times 14.0 = & 7,154,000 \\
O_2: & 250,000 \times & 0.935 \times 12.3 = & 2,875,000 \\
N_2: & 250,000 \times & 11.334 \times 11.5 = & 32,585,000 \\
\end{array}
$$

$$47,265,000\ \text{Btu/h}$$

or 18% of the fuel energy supplied. Some of this could be recovered by suitable waste heat equipment.

Heat Balance on a Boiler
    Let us consider a further example, a heat balance on a boiler. A process steam boiler has the following specifications:

- Natural gas fuel with HHV = 1001.2 Btu/ft³
- Gas firing rate = 2126.5 ft³/min.
- Steam discharge at 150 lb/in² g saturated
- Steam capacity of 100,000 lb/h
- Condensate returned at 180°F

    The heat balance on the burner is derived from measurements made *after* the burner controls had been adjusted for an optimum air/fuel ratio corresponding to 10% excess air. All values of the heat content of the fluid streams are referred to a base temperature of 60°F. Consequently the com-

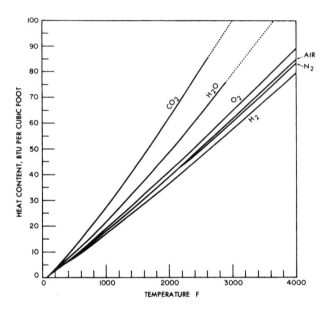

Figure 6-5. Heat Content of Gases Found in Flue Products, Based on Gas Volumes at 60°F (dashed line indicates dissociation)

putations for each fluid stream entering or leaving the boiler are made by use of the equation below.

$$\dot{H} = \dot{m}\left(b - b_o\right)$$ *Formula (6-11)*

Where

$\dot{H}$ = the enthalpy rate for entrance or exit fluids

$\dot{m}$ = mass flow rates for entrance or exit fluids

h = specific enthalpy at the fluid temperature of the fluid entering or leaving the entrance or exit

h_o = specific enthalpy of that fluid at the reference temperature T_o = 60°F

The first law of thermodynamics for the boiler is expressed as: Sum of all the enthalpy rates of substances entering = Sum of all enthalpy rates of substances leaving + q where q is the rate of heat loss to the surroundings. This can be expressed as

$$\Sigma_{in} H_i = \Sigma_{out} H_i + q$$ *Formula (6-12)*

where Σ is the summation sign, and H_i is the enthalpy rate of substance i.

For gaseous fuels the computation for the heat content of the gases is more conveniently expressed in the form

$$H = J_o \, C_{pm} \, (T - T_o) \qquad\qquad \textit{Formula (6-13)}$$

Where

J_o = the volume rate of the gas stream corrected back to 1.0 atmosphere and 60°F ($T_o = 60°F$)

C_{pm} = the specific heat given on the basis of a standard volume of gas averaged over the temperature range $(T - T_o)$ and the gas mixture components.

$$C_{pm} = \Sigma_i X_i C_{pi} \qquad\qquad \textit{Formula (6-14)}$$

Where

X_i = percent by volume of a component in one of the flow paths

C_{pi} = average specific heat over temperature range for each component

From Formula 6-10 we derive the volume fractions of each gas component as follows:

Component	X
CO_2	8.9
H_2O	17.2
N_2	72.2
O_2	1.7
	100.0

The average specific heat is found (using Figure 6-6).

$$
\begin{aligned}
C_{pm} \;=\; & .089 \times 0.0275 = 0.00245 \\
& 0.172 \times 0.0220 = 0.00378 \\
& 0.722 \times 0.0186 = 0.01344 \\
& 0.017 \times 0.0195 = \underline{0.00033} \\
& \qquad\qquad\quad 0.02 \text{ Btu/Scf} \cdot \text{F}
\end{aligned}
$$

The combustion equation also tells that when the products are at standard conditions 11.88 ft³ fuel and air generate 11.915 ft³ of products and that for every cubic foot of gas burned, 10.880 ft³ of air is introduced. Thus for a firing rate of 2126.5 ft³/min. the air required is 2126.5 × 10.88 = 23,136 ft³/min. or 23,136 × 60 = 1,388,179 ft³/h. The total fuel and air flow rate is then almost exactly equal to 1,388,179 + 2126.5 × 60 = 1,515,769 ft³/h. This corresponds to a flue gas discharge rate of 1,519,128 ft³/h.

For a flue gas discharge rate of 1,519,128 ft³/h and a temperature of 702°F, the total exhaust heat rate is found as

$$\dot{H}_{EX\,GAS} = 1,519,128 \, \frac{ft^3}{h} \cdot 0.02 \, \frac{Btu}{ft^3 \cdot F}$$

$$\times \, (702–60)F = 19,505,604 \, Btu/h$$

For the steam leaving the boiler (100,000 lb/h at 150 lb/in² g saturated) the energy flow rate is found using the steam tables in Chapter 15 and the following equation:

$$\dot{H}_{Steam} = \dot{m}b – b_o$$
$$= 100,000 \, lb/h \, (1195.6–28.08) \, Btu/lb$$
$$= 116,752,000 \, Btu/h$$

where 1195.6 is the specific enthalpy of saturated steam at 150 lb/in² g and 28.08 is the specific enthalpy of saturated liquid water at 60°F, since we are using that temperature as our standard reference temperature for the heat balance. We have used 60°F as a reference temperature; this is not universal practice and in the boiler industry 70°F is more common, whereas in other areas 25°C is normal.

CHEMICAL ENERGY IN FUEL

To determine the heat content of the chemical energy in fuel, find the higher heating value (HHV) for the fuel and multiply it by the volumetric flow rate for a gaseous fuel or the mass flow rate for a liquid or solid fuel. The assumed higher heating value for the natural gas used in the boiler of our example is 1001.1 Btu/ft³ and the heat content rate is then

$$\dot{H} = 2126.5 \, \frac{ft^3}{min.} \times 60 \, \frac{min.}{h} \times 1001.1 \, \frac{Btu}{ft^3}$$

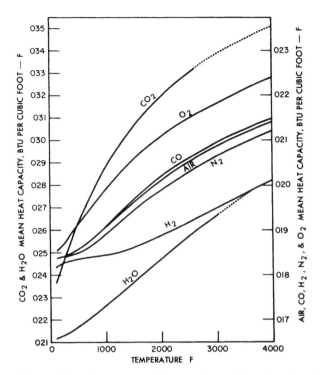

Figure 6-6. Mean Heat Capacity of Gases Found in Flue Products from 60 to $T°$F (dashed line indicates dissociation)

$$= 127{,}743{,}000 \quad \frac{\text{Btu}}{\text{h}}$$

The enthalpy rates for the condensate return and make-up water are derived from data in the steam tables where the specific enthalpy of the compressed liquids is taken to be almost exactly equal to the specific enthalpy of the *saturated* liquid found at the same *temperature.*

The complete heat balance derived in the manner detailed above is presented in Figure 6-7.

Waste heat is available from the combustion products leaving the stack. It amounts to 19,505,604 Btu/h at a temperature of 702°F. Some energy is also available from the condensers.

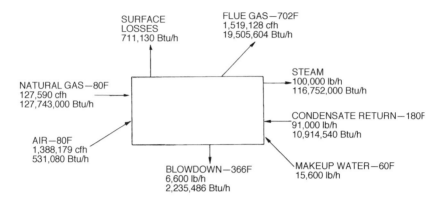

Figure 6-7. Heat Balance for a Simple Steam Generator Burning Natural Gas with 10% Excess Air

WASTE HEAT RECOVERY

The energy exhausted to the atmosphere should not be discarded. A portion of it can be recovered by using a heat exchanger. Any requirement for energy at a temperature in excess of 200°F can be satisfied. It is necessary to identify the prospective uses for the waste energy, make an economic analysis of the costs and savings involved in each of the options, and decide among those options on the basis of the economics of each. An important option in every case is that of rejecting all options if none proves economic. For the illustrative example, let us assume that the following uses for the waste heat have been identified:

- Preheating the combustion air.
- Preheating the boiler feedwater.
- Heating the domestic hot water supply (370 gal/hr from 50° to 170°F).
- A combination of the preceding.

The first two are practices which are standard in energy-intensive high technology industries (e.g., electric companies).

Two rapid calculations give the waste heat available in the exhaust gases between 702°F and 220°F and the heat requirements for the domestic hot water supply to be 14,063,000 Btu/h and 369,900 Btu/h respectively. Since the latter constitutes only 2.6% of the available waste energy, we should reject that option. The remaining options are to preheat the combustion air and the feedwater or to divide the waste energy between those 2 options. For a small boiler we probably would not find the purchase of two separate heat exchangers an economic option, so that we shall limit ourselves to one of the other of the first two options. Without going into a detailed analysis, we may note that preheating the combustion air or the feedwater provides a double benefit. Since the preheated air or water requires less fuel to produce the same steam capacity, a direct fuel saving results. But the smaller quantity of fuel means a smaller air requirement, and this in turn means a smaller quantity of exhaust products and thus smaller stack loss at a given stack temperature.

The economic benefits can be estimated as follows: The air preheater is estimated to save 6% of the fuel. With an average boiler loading of 60% for the 8760 h in a year, this amounts to an annual fuel saving of

$$.6 \times 0.06 \times 8760 \times 127{,}590 = 40{,}237 \ \frac{ft^3}{yr}$$

which is worth, at an average rate of $5.50/1000 ft^3, an annual dollar saving of

$$\frac{40{,}237}{1000} \times 5.50 = \$221{,}000/yr$$

If the costs for installing the air preheater are assumed to be

Cost of preheater	$ 52,000
Cost of installation	57,200
New burners, air piping, controls and fan	56,600
	$165,800

For the feedwater heater (or economizer) the fuel savings is 9.2% or a total annual fuel saving of

$$0.092 \times 127{,}600 \times 8760 \times 0.60 = 61{,}696{,}000 \ \frac{ft^3}{yr}$$

and the economic benefit is

$$\frac{61,696,000 \frac{ft^3}{yr}}{1000} \cdot \$5.50 = \$340,000/yr$$

The cost of the economizer installed is estimated to be $134,000. This is clearly the best option for this particular boiler, especially as there is no need in this case for modifications to the boiler and accessories beyond the heat exchanger retrofit. There are several reasons for its superiority. The first is that in the case of air preheating, we are exchanging the waste heat in the gases to the incoming air which has almost the same mass flow rate and almost the same specific heat. Thus we can expect that the final temperature of the preheater air will be almost the arithmetic mean of the ambient air temperature and the exhaust gas temperature entering the economizer. In the case of the feedwater, the mass flow rate times specific heat is over four and one-quarter times that of the combustion air. Thus we can expect to transfer more energy to the water which results in a lower flue gas temperature leaving the stack.

The second reason is that preheating the air quite often (as in this case) affects the boiler accessories which require additional modifications and thus related capital expenditures.

In the preceding example of the process steam boiler, we are analyzing an efficient process unit. The heat available in the product (process steam) constituted a large percentage of the energy introduced in the fuel. The efficiency in percentage terms is computed as

$$\eta = \frac{\text{Useful output}}{\text{Energy input}} = \frac{Q_{steam}}{Q_{fuel}} \times 100$$

$$\eta = \frac{100,000 \text{ lb/h } (1195.6-148) \times 100}{127,590 \text{ ft}^3/\text{h} \times 1001.2 \text{ Btu/ft}^3}$$

$$= \frac{1.0476 \times 10^{10}}{1.2775 \times 10^8} = 0.82, \text{ or } 82\%$$

HEAT RECOVERY IN STEEL TUBE FURNACE

As a second and very different example, note the steel tube furnace illustrated below in Figure 6-8.

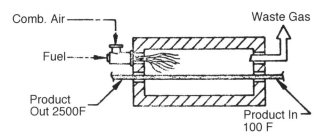

Figure 6-8. Continuous Steel Tube Furnace

The tubing enters the furnace from the right at a temperature of 100°F.

The specifications for the steel tube heating furnace are

Product capacity – 50 ton/h

Product specifications – 0.23% carbon steel

Final product temperature – 2000°F

Air/Fuel inlet temperature – 100°F

Air/Fuel mixture –10% excess air

Fuel – No. 5 fuel oil gravity API° 16

Fuel firing rate 48.71 gpm at 240°F (factory usage)

Utilization factor – 0.62

The useful heat leaves the furnace in the steel at 2000°F. This is called the useful furnace output and equals

$$Q_{prod} = \dot{m}_{prod} C_p (T_{out} - T_{in})$$

$$= 50 \ \frac{ton}{h} \ \times 2200 \ lb/ton$$

$$\times .179 \ Btu/lb \times (2000{-}100) \ F$$

$$= 0.3741 \times 10^8 \ \frac{Btu}{h}$$

0.179 is used as the average specific heat of steel over the 100°F-2000°F range. The heat input to the furnace is the chemical energy in the fuel oil. The heating value for No. 5 fuel oil is found from Table 6-2.

The heat input is determined by

$$Q_{fuel} = 48.71 \text{ gpm} \times 142,300 \ \frac{\text{Btu}}{\text{gal}}$$

$$= 4.159 \times 10^8 \text{ Btu/h}$$

The percent efficiency of the furnace is:

$$\eta = \frac{\text{output} \times 100}{\text{input}}$$

$$= \frac{\text{Enthalpy added to steel} \times 100}{\text{Enthalpy entering with fuel}}$$

$$= \frac{0.3741 \times 10^8}{4.159 \times 10^8} \times 100 = 9\%$$

This means that of the 415.9 MBtu introduced per hour to the furnace, that 378.4 million are released to the atmosphere. At the present average cost of $2/gal. for No. 5 fuel oil, the heat wasted is

$$Q_{waste} = 48.71 \text{ gpm} \times (1-0.09) \ \frac{\text{gal wasted}}{\text{gal use}} \times .62$$

$$\times 60 \text{ min./h} \times \$2/\text{gal} = \$3297/\text{h}$$
$$= \$3297/\text{h} \times 8760\text{h/yr} = \$28,885,000/\text{yr}$$

In order to construct the combustion equation we must note from Table 6-3 that the carbon-hydrogen ratio is about 7.3. Thus

$$C_{6.08}H_{10} + 9.44 \ O_2 + 35.49 \ N_2 \rightarrow 6.08 \ CO_2$$
$$+ 5 \ H_2O + 0.86 \ O_2 + 35.49 \ N_2$$

Again referring to Table 6-2 the density of the liquid fuel is 7.99 lb/gal. Therefore, the above equation represents the reaction for 100 lb fuel/7.00 lb/gal = 12.51 gal and for 35.49 + 9.44 or 44.93 lb mol of air. 44.93 lb mol of air weighs 1301lb (using the molecular weight for air found in

Table 6-2. Physical Properties of Fuel Oil at 60°F

Fuel oil (CS-12-48) Grade No.	Grav-ity, API	Sp gr	Lb per gal	Btu per lb	Net Btu per gal
6	3	1.0520	8.76	18,190	152,100
6	4	1.0443	8.69	18,240	151,300
6	5	1.0366	8.63	18,290	149,400
6	6	1.0291	8.57	18,340	148,800
6	7	1.0217	8.50	18,390	148,100
6	8	1.0143	8.44	18,440	147,500
6	9	1.0071	8.39	18,490	146,900
6	10	1.0000	8.33	18,540	146,200
6	11	0.9930	8.27	18,590	145,600
6	12	.9861	8.22	18,640	144,900
6, 5	14	.9725	8.10	18,740	143,600
6, 5	16	.9593	7.99	18.840	142,300
5	18	.9465	7.89	18,930	140,900
4, 5	20	.9340	7.78	19,020	139,600
4, 5	22	.9218	7.68	19,110	138,300
4, 5	24	.9100	7.58	19,190	137,100
4, 2	26	.8984	7.49	19,270	135,800
4, 2	28	.8871	7.39	19,350	134,600
2	30	.8762	7.30	19,420	133,300
2	32	.8654	7.21	19,490	132,100
2	34	.8550	7.12	19,560	130,900
1, 2	36	.8448	7.04	19,620	129,700
1, 2	38	.8348	6.96	19,680	128,500
1	40	.8251	6.87	19,750	127,300
1	42	0.8156	6.79	19,810	126,200

The relation between specific gravity and degrees API is expressed by the formula:

$$\frac{141.5}{131.5 + API} = \text{sp gr at 60 F.}$$

For each 10 F above 60 F add 0.7 API.

For each 10 F below 60 F subtract 9.7 AP.

Table 6-3. Typical Properties of Commercial Petroleum Products Sold in the East and Midwest

| | Premium Fuels | | | | | Fuel Oils | | | | |
| | Gasolines | | No. 1 fuel, light diesel, or stove oil | Diesel or "gas house" gas oil | Reduced crude, heavy gas oil or premium residuum* | No. 2 | No. 3 | Cold No. 5 | No. 5 | Bunker C or No. 6 |
	12 psia Reid vapor pressure natural gasoline	Straight run gasoline								
Gravity, °API	79	63	38–42	35–38	19–22	32–35	28–32	20–25	16–24	6–14
Viscosity	...	...	34–36†	34–36†	15–50‡	34–36†	35–45†	<20‡	20–40‡	50–300‡
Conradson carbon, wt %	None	None	Trace	Trace	3–7	Trace	<0.15	1–2	2–4	6–15
Pour point, °F	<0	<0	<0	<10	...	<10	<20	<15	15–60	>50
Sulfur, wt %	<0.1	<0.2	<0.1	0.1–0.6	<2.0	0.1–0.6	0.2–1.0	0.5–2.0	0.5–2.0	1–4
Water and sed., wt %	None	None	Trace	Trace	<1.0	<0.05	<0.10	0.1–0.5	0.5–1.0	0.5–2.0
Distillation, °F										
10%	110	150	390–410	400–440	550–650	420–440	450–500	<600	<600	600–700
50%	150	230	440–460	490–510	800–900	490–530	540–560	...	...	850–950
90%	235	335	490–520	580–600	...	580–620	600–670	<700	>700	...
End point	315	370	510–560	630–660	...	630–660	650–700	...	...	...
Flash point, °F	<100	<100	100–140	110–170	>150	130–170	>130	>130	>130	>150
Aniline point, °F	145	125	140–160	150–170	...	120–140	120–140	...	...	...
C to H ratio	5.2	5.6	6.1–6.4	6.2–6.5	6.9–7.3	6.6–7.1	7.0–7.3	7.3–7.7	7.3–7.7	7.7–9.0
Average MBtu recovered/gal of oil:										
Carbureted water gas	101	101	102	101	93	91	86	82	83	74
High-Btu oil gas	90	89	90	90	82	80	76	72	73	64
Tar + carbon, wt % of oil										
Carbureted water gas	...	...	20	22	35	31	36	42	42	52
High-Btu oil gas	...	...	32	34	45	42	46	52	52	60

(Data on gasolines and data below C/H ratio not given in reports.)
* The 6 wt % Conradson carbon oil used in the Hall High Btu Oil Gas Tests (A.G.A. Gas Production Research Committee. Hall High Btu Oil Gas Process, New York, 1949.) and "New England Gas Enriching Oil" are typical examples of this group.
† Saybolt Universal seconds at 100 F.
‡ Saybolt Furol seconds at 122 F.
Remarks:
No. 1 Fuel Oil—A distillate oil intended for vaporizing pot-type burners. A volatile fuel.
No. 2 Fuel Oil—For general purpose domestic heating; for use in burners not requiring No. 1 oil. Moderately volatile.
No. 3 Fuel Oil—Formerly a distillate oil for use in burners requiring low viscosity oil. Now incorporated as a part of No. 2.
No. 4 Fuel Oil—For burner installations not equipped with preheating facilities.
No. 5 Fuel Oil—A residual type oil. Requires preheating to 170–220 F.
No. 6 Fuel Oil—Preheating to 220–260 F suggested. A high viscosity oil.

Table 6-4). The specific volume of air at standard conditions is found from the same table as 378.5 ft³/lb mol. Therefore, the volume of air required to burn 12.51 gal of fuel is 44.93 × 378.5 = 17,000 ft³ at standard temperature and pressure. The air input is then

$$m = \frac{17{,}000 \text{ ft}^3}{12.51 \text{ gal}} \cdot 48.71 \text{ gpm} = 66{,}200 \text{ ft}^3/\text{m}$$

or

$$m_{air} = 66{,}200 \times 60 = 3{,}972{,}000 \text{ ft}^3/\text{h}$$
$$Q_{air} = 1.2 \text{ Btu}/\text{ft}^3 \times 3{,}972{,}000$$
$$= 4{,}766{,}000 \text{ Btu}/\text{h}$$

The volume ratio of stack gases at standard conditions to air inlet

Table 6-4. Gas Constants and Volume of the Pound-Mol for Certain Gases

	$R' = R/M$ specific gas constant, ft per F	M, mol wt. lb per lb-mol	R, universal gas constant, ft·lb per (lb-mol)(F)	Mc, cu ft per lb·mol*
Hydrogen	767.04	2.016	1546	378.9
Oxygen	48.24	32.000	1544	378.2
Nitrogen	55.13	28.016	1545	378.3
Nitrogen, "atmospheric"†	54.85	28.161	1545	378.6
Air	53.33	28.966	1545	378.5
Water vapor	85.72	18.016	1544	378.6
Carbon dioxide	34.87	44.010	1535	376.2
Carbon monoxide	55.14	28.010	1544	378.3
Hydrogen sulfide	44.79	34.076	1526	374.1
Sulfur dioxide	23.56	64.060	1509	369.6
Ammonia	89.42	17.032	1523	373.5
Methane	96.18	16.042	1543	378.2
Ethane	50.82	30.068	1528	374.5
Propane	34.13	44.094	1505	368.7
n·Butane	25.57	58.120	1486	364.3
iso·Butane	25.79	58.120	1499	367.4
Ethylene	54.70	28.052	1534	376.2
Propylene	36.01	42.078	1515	379.1

*At 60 F, 30 in. Hg, dry.
†Includes other inert gases in trace amounts.

rate is found from the combustion equation as

$$\frac{6.08 + 5 + 0.86 + 35.49}{9.44 + 35.49} = \frac{47.43}{44.93} = 1.056$$

$$\begin{aligned} Q_{ex} &= 1.056 \times 3{,}972{,}000 \\ &= 4{,}194{,}000 \text{ ft}^3/\text{h} \end{aligned}$$

The volume fractions of the stack gas components are found from the combustion equation as

	X
CO_2	0.128
H_2O	0.106
O_2	0.018
N_2	0.748
	1.000

The specific enthalpy of the stack gas components may be calculated by use of Figure 6-5 for the exhaust temperature of 2200°F leading to the average specific enthalpy as indicated below.

					h
CO_2	0.128	×	69.3	=	8.870
H_2O	0.106	×	54.0	=	57.24
O_2	0.018	×	45.4	=	0.817
N_2	0.748	×	43.4	=	32.463
					47.874

Therefore

$$Q_{ex} = 4,194,000 \text{ ft}^3/\text{h} \times 47.874 \ \frac{\text{Btu}}{\text{ft}^3}$$

$$= 00,784,000 \text{ h} \ \frac{\text{Btu}}{\text{h}}$$

The heat losses from the surface equal the fuel enthalpy input less the exhaust gas enthalpy and the produce enthalpy.

$$Q = 415,880,000 + 4,766,000 - 200,784,000$$

$$- 37,410,000 = 182,451,270 \ \frac{\text{Btu}}{\text{h}}$$

The complete heat balance diagram is shown in Figure 6-9. Now suppose that we install a metallic recuperator and preheat the combustion air to 750°F. The new heat balance is shown in Figure 6-10.

The economic analysis of this situation is as follows:

The waste heat saved is equal to the fuel savings times the heating value of the fuel.

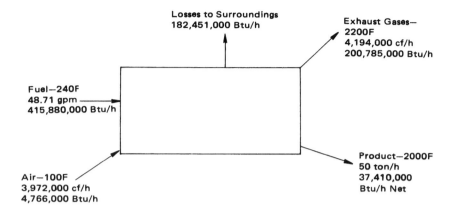

Figure 6-9. Heat Balance for a Simple Continuous Steel Tube Furnace

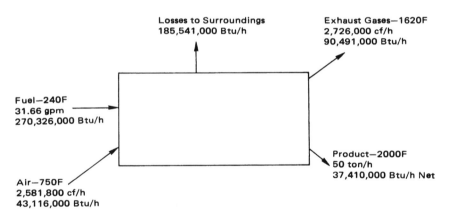

Figure 6-10. Heat Balance for a Continuous Steel Tube Furnace Equipped with a Recuperator Preheating Combustion Air to 750°F

$$Q_{saved} = (48.71 - 31.66) \text{ gpm} \times 142,300 \ \frac{Btu}{gal}$$

$$= 145,554,000 \text{ or } 35\% \text{ saving}$$

which equals

$$0.35 \times 48.71 \text{ gpm} \times 60 \times \$2/\text{gal} = \$2,045.82/\text{h}$$

In using Figures 6-11 and 6-12 remember that the combustion air volume (in SCF) is reduced directly proportional to the reduction in firing rate. If we maintain the same air/fuel ratio for the furnace (10% excess air), the air rate is reduced from the initial value of 3,972,000 ft³/h at a firing rate of 48.71 gpm to a rate of 31.66 gpm/48.71 gpm × 3,972,000 ft³/h = 2,581,800 ft³/h with the lower firing rate of 31.66 gpm when the recuperator is used.

If, as in our example, the furnace is used 0.62 × 8760 or 6531 h/yr, then the annual saving is 5431 × \$2,045.80 = \$11.10 × 10⁶/yr. Since a metallic recuperator can be purchased for a fraction of the savings, the savings will allow a payoff in less than a year.

BOILERS

Boiler Configurations and Components

Industrial boiler designs are influenced by fuel characteristics and firing method, steam demand, steam pressures, firing characteristics and the individual manufacturers. Industrial boilers can be classified as either firetube or watertube, indicating the relative position of the hot combustion gases with respect to the fluid being heated.

Firetube Boilers

Firetube units pass the hot products of combustion through tubes submerged in the boiler water. Conventional units generally employ from 2 to 4 passes to increase the surface area exposed to the hot gases and thereby increase efficiency. Multiple passes, however, require greater fan power, increased boiler complexity and larger shell dimensions. (Refer to Figure 6-13.) Maximum capacity of firetube units is currently limited to 25,000 lbs of steam per hour (750 boiler hp) with an operating pressure of 250 psi due to economic factors related to material strength and thickness.

Advantages of firetube units include: (1) ability to meet wide and sudden load fluctuations with only slight pressure changes, (2) low initial costs and maintenance and (3) simple foundation and installation procedures.

Watertube Boilers

Watertube units circulate the boiler water inside the tubes and the flue gases outside. Water circulation is generally provided by the density

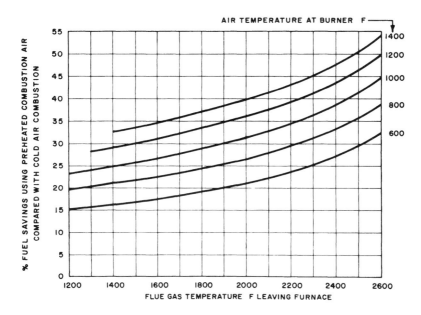

Figure 6-11. Fuel Savings as a Function of Flue Gas Temperature °F Leaving Furnace

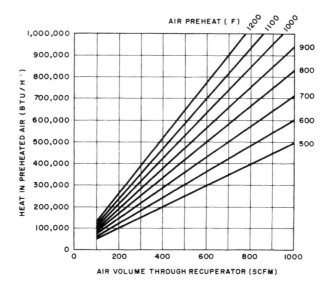

Figure 6-12. Heat Recovered as a Function of Air Volume through Recuperator (SCFM)

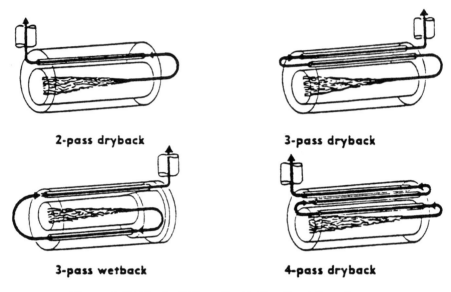

2-pass dryback   3-pass dryback

3-pass wetback   4-pass dryback

Figure 6-13. Typical Firetube Boiler Gas Flow Patterns

variation between cold feed water and the hot water/steam mixture in the riser as illustrated in Figure 6-14. Watertube boilers may be subclassified into different groups by tube shape, by drum number and location and by capacity. Refer to Figure 6-15.

Another important determination is "field" versus "shop" erected units. Many engineers feel that shop assembled boilers can meet closer tolerance than field assembled units and therefore may be more efficient; however, this has not been fully substantiated. Watertube units range in size from as small as 1000 lbs of steam per hour to the giant utility boilers in the 1000 MW class. The largest industrial boilers are generally taken to be about 500,000 lbs of steam per hour. Important elements of a steam generator include the firing mechanism, the furnace water walls, the superheaters, convective regions, the economizer and air preheater and the associated ash and dust collectors.

FUEL HANDLING AND FIRING SYSTEMS

Gas Fired

Natural gas fuel is the simplest fuel to burn in that it requires little preparation and mixes readily with the combustion air supply. Industrial

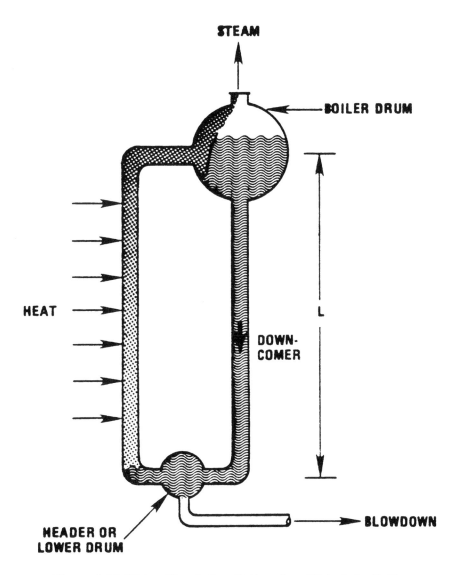

Figure 6-14. Water Circulation Pattern in a Watertube Boiler

boilers generally use low-pressure burners operating at a pressure of 1/8 to 4 psi. Gas is generally introduced at the burner through several orifices that generate gaseous jets that mix rapidly with the incoming combustion air supply. There are many designs in use that differ primarily in the orientation of the burner orifices and their locations in the burner housing.

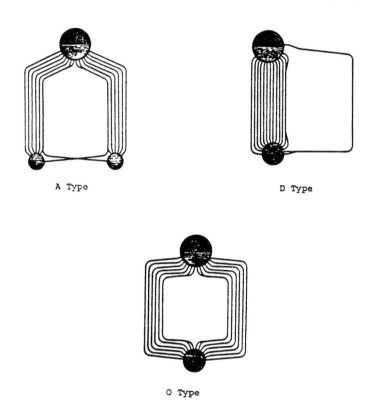

Figure 6-15. Classification of Watertube Boilers by Basic Tube Arrangement

Oil Fired

Oil fuels generally require some type of pretreatment prior to delivery to the burner. This may include the use of strainers to remove solid foreign material and tank and flow line preheaters to assure the proper viscosity. Oil must be atomized prior to vaporization and mixing with the combustion air supply. This generally requires the use of either air, steam or mechanical atomizers. The oil is introduced into the furnace through a gun fitted with a tip that distributes the oil into a fine spray that allows mixing between the oil droplets and the combustion air supply. Oil cups that spin the oil into a fine mist are also employed on small units. An oil burner may be equipped with diffusers that act as flame holders by inducing strong recirculation patterns near the burner. In some burners, primary air nozzles are employed.

Pulverized Coal Fired

The pulverizer system provides four functions: pulverizing, drying, classifying to the required fineness and transporting the coal to the burner's main air stream. The furnace may be designed for dry ash removal in the hopper bottom or for molten ash removal as in a slag tap furnace. The furnace is dependent on the burning and ash characteristics of the coal as well as the firing system and type of furnace bottom. The primary objectives are to control furnace ash deposits and provide sufficient cooling of the gases leaving the furnace to reduce the buildup of slag in the convective regions. Pulverized coal fired systems are generally considered to be economical for units with capacities in excess of 200,000 lbs of steam per hour.

Stoker Fired

Coal stoker units are characterized by bed combustion on the boiler grate with the bulk of the combustion air supplied through the grate. Several stoker firing methods currently in use on industrial sized boilers include underfed, overfed and spreader. In underfed and overfed stokers, the coal is transferred directly on to the burning bed. In a spreader stoker the coal is hurled into the furnace when it is partially burned in suspension before lighting on the grate. Several grate configurations can be used with overfed and spreader stokers including stationary, chain, traveling, dumping and vibrating grates. Each grate configuration has its own requirements as to coal fineness and ash characteristics for optimum operation. Spreader stoker units have the advantage that they can burn a wide variety of fuels including waste products. Underfed and overfed units have the disadvantage that they are relatively slow to respond to load variations. Stoker units can be designed for a wide range of capacities from 2,000 to 350,000 lbs of steam per hour. Spreader stoker units are generally equipped with overfire air jets to induce turbulence for improved mixing and combustible burnout. Stoker units are also equipped with ash reinjection systems that allow collected ash that contains a significant portion of unburned carbon to be reintroduced into the furnace for burning.

COMBUSTION CONTROL SYSTEMS

Combustion controls have two purposes: (1) maintain constant steam conditions under varying loads by adjusting fuel flow, and (2) maintain an appropriate combustion air-to-fuel flow.

Classification

Combustion control systems can be classified as series, parallel and series/parallel.

In series control, either the fuel or air is monitored and the other is adjusted accordingly. For parallel control systems, changes in steam conditions result in a change in both air and fuel flow. In series/parallel systems, variations in steam pressure affect the rate of fuel input, and simultaneously the combustion air flow is controlled by the steam flow.

Combustion controls can be also classified as positioning and metering controls.

Positioning controls respond to system demands by moving to a present position. In metering systems, the response is controlled by actual measurements of the fuel air flows.

Application

The application and degree of combustion controls vary with the boiler size and is dictated by system costs. The parallel positioning jackshaft system has been extensively applied to industrial boilers based on minimum system costs. The combustion control responds to changes in steam pressure and can be controlled by a manual override. The control linkage and cam positions for the fuel and air flow are generally calibrated on start-up.

Improved control of excess air can be obtained by substituting electric or pneumatic systems for the mechanical linkages. In addition, relative position of fuel control and combustion air dampers can be modified. More advanced systems are pressure ratio control of the fuel and air pressure, direct air and fuel metering and excess air correction systems using flue gas O_2 monitoring. Factors that have limited the application of the most sophisticated control systems to industrial boilers include cost, reliability and maintenance.

SAFETY

It is essential that energy engineers conducting boiler evaluation tests and tune-ups understand and be aware of boiler safety devices. Occasionally operators who don't understand safety have been known to bypass safety features in order to keep a unit operating.

Summary of requirements found in the NATIONAL BOILER SAFETY

CODES:

* PREPURGE—4 to 8 air changes to insure no fuel vapors or gases remain in the boiler which could ignite or explode when the pilot light is off.

* PILOT PROVING—10 to 15 second proving period pilot must ignite and be proven before the main fuel valves open.

* MAIN FLAME TRIAL—usually 10 to 15 second trial period for natural gas and oil after main fuel valves open.

* FLAME FAILURE RESPONSE TIME—usually 3 to 4 seconds after the main flame goes out the fuel shut-off valves should automatically close to prevent a build up of explosive vapor concentrations within the boiler.

* INTERLOCKS—automatically shut the boiler down if certain safety features are not in the safe condition.

Loss of atomizing means (steam/air)

High/Low gas pressure

Low fire for light off

Low fuel oil temperature

Low fuel oil pressure

Main combustion air

Low water level

* FUEL VALVES—Safety Shut Off Valves (SSOV). Usually two valves; closing time 1 to 5 seconds depending on fuel and size of boiler.

* FLAME DETECTORS—should be positioned and capable of detecting flame only and not sparks from the spark ignitor or hot refractory so as not to give a false flame signal.

OPERATING GUIDES

Consult the plant engineer and the boiler manufacturers technical manuals for complete boiler operating procedures. Typical items to check before conducting efficiency checks are as follows:

Oil Burners

Make sure the atomizer is of the proper design and size and the burner is centered with dimensions according to manufacturer's drawings.

Inspect oil-tip passages and orifices for wear (use proper size drill as a feeler gage) and remove any coke or gum deposits to assure the proper oil-spray pattern.

Verify proper oil pressure and temperature at the burner.

Verify proper atomizing-steam pressure.

Make sure that the burner diffuser (impeller) is not damaged and is properly located to the oil-gun tip.

Check to see that the oil gun is positioned properly within the burner throat and that the throat refractory is in good condition.

Gas Burners

Inspect gas-ingestion orifices and verify that all passages are unobstructed. Also, be sure filters and moisture traps are in place, clean and operating properly, to prevent plugging of gas orifices.

Confirm proper location and orientation of diffusers, spuds, etc. Look for any burned off or missing burner parts.

Combustion Controls

Inspect all fuel valves to verify proper movement; clean valve internal surfaces if necessary.

Eliminate "play" in control linkages on dampers. Any play, no matter how slight, will cause a loss of efficiency since a precise tune-up will be impossible.

Make sure fuel-supply inlet pressure to pressure regulators are high enough to assure constant regulator-outlet pressures for all firing rates.

Correct any control elements that fail to respond smoothly to varying steam demand. Unnecessary hunting caused by improperly adjusted regulators or automatic master controllers can waste fuel.

Check that all gages are functioning and are calibrated.

Furnace

Inspect boiler furnace and gas side surfaces for excessive deposits and fouling. These lead to higher stack temperatures and lower boiler efficiencies.

Inspect furnace refractory and insulation for cracks that may cause leaks and missing refractory.

Clean furnace inspection parts and make sure that burner throat, furnace walls, and leading connection passes are visible through them since flame observation is an essential part of efficient boiler operation and testing.

BOILER EFFICIENCY IMPROVEMENT

The boiler plant should be designed and operated to produce the maximum amount of usable heat from a given amount of fuel.

Combustion is a chemical reaction of fuel and oxygen which produces heat. Oxygen is obtained from the input air which also contains nitrogen. Nitrogen is useless to the combustion process. The carbon in the fuel can combine with air to form either CO or CO_2, Incomplete combustion can be recognized by a low CO_2 and high CO content in the stack. Excess air causes more fuel to be burned than required. Stack losses are increased and more fuel is needed to raise ambient air to stack temperatures. On the other hand, if insufficient air is supplied, incomplete combustion occurs and the flame temperature is lowered.

Boiler Efficiency

Boiler efficiency *(E)* is defined as

$$\%E = \frac{\text{Heat Out of Boiler}}{\text{Heat Supplied to Boiler}} \times 100 \qquad \textit{Formula (6-15)}$$

For steam-generating boilers

$$\%E = \frac{\text{Evaporation Ratio} \times \text{Heat Content of Steam}}{\text{Calorific Value of Fuel}} \times 100 \qquad \textit{Formula (6-16)}$$

For hot water boilers

$$\%E = \frac{\text{Rate of Flow from Boiler} \times \text{Heat Output of Water}}{\text{Caloric Value of Fuel} \times \text{Fuel Rate}} \times 100 \qquad \textit{Formula (6-17)}$$

The relationship between steam produced and fuel used is called the evaporation ratio.

Boilers are usually designed to operate at the maximum efficiency when running at rated output. Figure 6-16 illustrates boiler efficiency as a function of time on line.

Full boiler capacity for heating occurs only a small amount of the time. On the other hand, part loading of 60% or less occurs approximately 90% of the time.

Where the present boiler plant has deteriorated, consideration

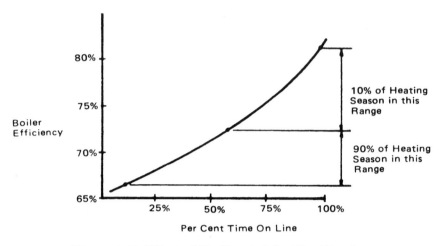

Figure 6-16. Effect of Cycling to Meet Part Loads

should be given to replacement with modular boilers sized to meet the heating load.

The overall thermal efficiency of the boiler and the various losses of efficiency of the system are summarized in Figure 6-17.

To calculate dry flue gas loss, Formula 6-18 is used.

---

1. Overall thermal efficiency . . . . ——

2. Losses due to flue gases

    (a) Dry Flue Gas . . . . ——

The loss due to heat carried up the stack in dry flue gases can be determined, if the carbon dioxide (CO_2) content of the flue gases and the temperatures of the flue gas and air to the furnace are known.

    (b) Moisture % Hydrogen . . . . ——

    (c) Incomplete combustion . . . .

                                                                          ——

3. Balance of account, including radiation and other unmeasured

    losses . . . . ——

            TOTAL 100%

Figure 6-17. Thermal Efficiency of Boiler

$$\text{Flue gas loss} = \frac{K(T - t)}{CO_2} \qquad\qquad \textit{Formula (6-18)}$$

Where

K = constant for type of fuel   = 0.39 Coke
       = 0.37 Anthracite
       = 0.34 Bituminous Coal
       = 0.33 Coal Tar Fuel
       = 0.31 Fuel Oil

T =   temperature of flue gases in °F
t =   temperature of air supply to furnace in °F
CO_2 =   percentage CO_2 content of flue gas measured volumetrically.

It should be noted that this formula does not apply to the combustion of any gaseous fuels such as natural gas, propane, butane, etc. Basic combustion formulas or nomograms should be used in the gaseous fuel case.

The savings in fuel as related to the change in efficiency is given by Formula 6-19.

$$\text{Savings in Fuel} = \frac{\text{New Efficiency-Old Efficiency}}{\text{New Efficiency}}$$

$$\times \text{Fuel Consumption} \qquad\qquad \textit{Formula (6-19)}$$

Figure 6-19 can be used to estimate the effect of flue gas composition, excess air, and stack temperature on boiler efficiency.

To estimate losses due to moisture, Figure 6-18 is used.

BOILER TUNE-UP TEST PROCEDURES

As illustrated by Figures 6-20 and 6-21, either % CO_2 or % O_2 can be used to determine excess air as long as the boiler is not operating on the fuel rise side of the curve.

The detailed procedures which follow illustrate how to tune-up a boiler to get the best air-to-fuel ratio. Note that for natural gas, % CO_2

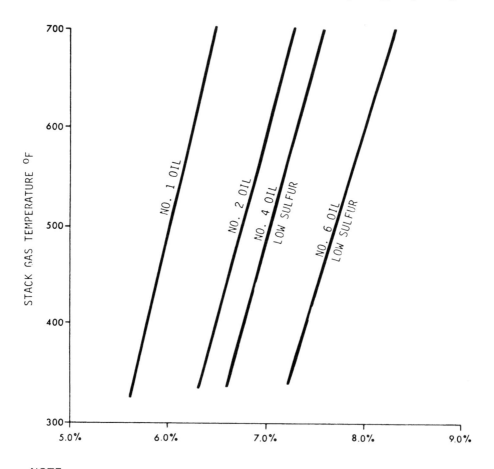

NOTE:
1. The figure gives a simple reference to heat loss in stack gases due to the formation of water in burning the hydrogen in various fuel oils.
2. The graph assumes a boiler room temperature of 80°F.

Figure 6-18. Heat Loss Due to Burning Hydrogen in Fuel (Source: Instructions For Energy Auditors, Volume 1)

must also be measured, while for fuel oil, smoke spot numbers or visual smoke measurements are used.

The principal method used for improving boiler efficiency involves operating the boiler at the lowest practical excess O_2 level with an adequate margin for variations in fuel properties, ambient conditions and the repeatability and response characteristics of the combustion control system.

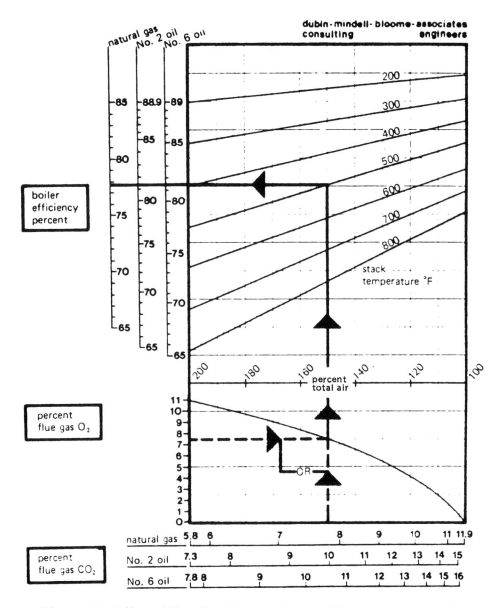

Figure 6-19. Effect of Flue Gas Composition and Temperature on Boiler Efficiency (*Source: Guidelines for Saving Energy in Existing Buildings-Engineers, Architects and Operators Manual, ECM-2*)

THEORETICAL AIR CURVE
(FUEL OIL)

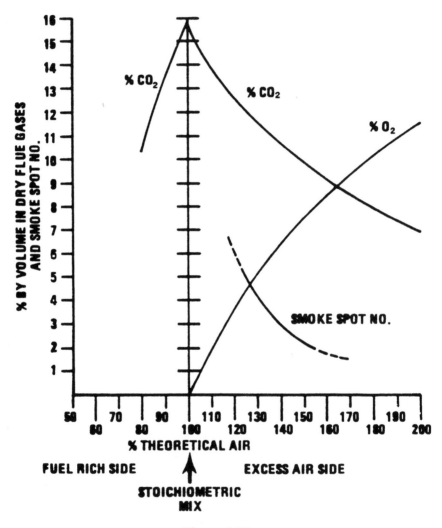

Figure 6-20.

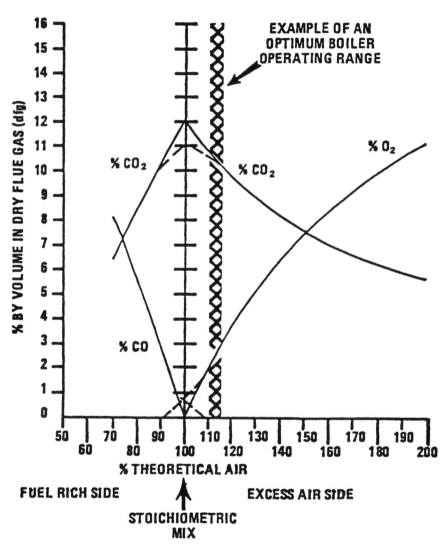

Figure 6-21.

These tests should only be conducted with a through understanding of the test objectives and by following a systematic, organized series of tests.

Cautions

Extremely low excess O_2 operation can result in catastrophic results.

Know at all times the impact of the modification on fuel flow, air flow and the control system.

Observe boiler instrumentation, stack and flame conditions while making any changes.

When in doubt, consult the plant engineering personnel or the boiler manufacturer.

Consult the boiler operation and maintenance manual supplied with the unit for details on the combustion control system or for methods of varying burner excess air.

Test Description

The test series begins with baseline tests that document existing "as-found conditions" for several firing rates over the boiler's normal operating range. At each of these firing rates, variations in excess O_2 level from 1 to 2% above the normal operating point to the "minimum O_2" level are noted. Curves of combustibles as a function of excess O_2 level will be constructed similar to those given in Figures 6-22 and 6-23. As illustrated in these figures, high levels of smoke or CO indicating potentially unstable operation can occur with small changes in excess O_2; thus small changes in excess O_2 should be made for conditions near the smoke or CO limit. It is important to note that the boiler may exhibit a gradual smoke or CO behavior at one firing rate and a steep behavior at another. Minimum excess O_2 will be that at which the boiler just starts to smoke, or the CO emissions rise above 400 ppm or the smoke spot number equals the maximum value as given in Table 6-5. Once minimum excess O_2 levels are established, an appropriate O_2 margin or operating cushion ranging from 0.5 to 2.0% O_2 above the minimum point depending on the particular boiler control system and fuels, should be maintained.

Repeated tests at the same firing condition, approached from both the "high side" and "low side" (i.e., from higher and lower firing rates), can determine whether there is excessive play in the boiler controls.

Record all pertinent data for future comparisons. Readings should

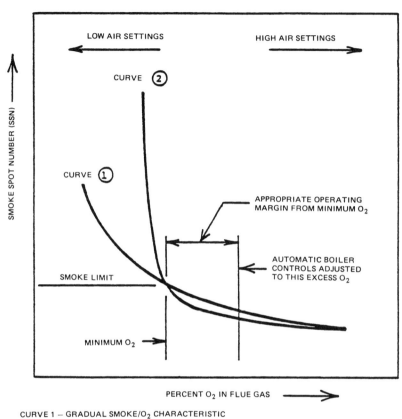

CURVE 1 — GRADUAL SMOKE/O_2 CHARACTERISTIC
CURVE 2 — STEEP SMOKE/O_2 CHARACTERISTIC

Figure 6-22. Typical Smoke-O_2 Characteristic Curves for Coal- or Oil-fired Industrial Boilers

be made only after steady boiler conditions are reached and at normal steam operating conditions.

Step-By-Step Boiler Adjustment Procedure
for Low Excess O_2 Operation
1. Bring the boiler to the test firing rate and put the combustion controls on manual.
2. After stabilizing, observe flame conditions and take a complete set of readings.
3. Raise excess O_2 1 to 2%, allowing time to stabilize and take readings.

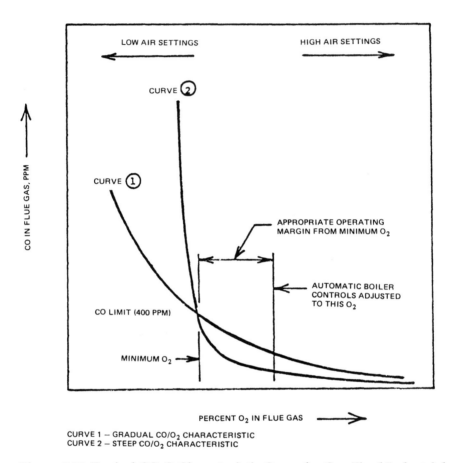

CURVE 1 — GRADUAL CO/O$_2$ CHARACTERISTIC
CURVE 2 — STEEP CO/O$_2$ CHARACTERISTIC

Figure 6-23. Typical CO-O Characteristic Curve for Gas-Fired Industrial Boilers

Table 6-5. Maximum Desirable Smoke Spot Number

Fuel Grade	Maximum Desirable SSN
No. 2	less than 1
No. 4	2
No. 5 (light and heavy), and low-sulfur resid.	3
No. 6	4

4. Reduce excess O_2 in small steps while observing stack and flame conditions. Allow the unit to stabilize following each change and record data.

5. Continue to reduce excess air until a minimum excess O_2 condition is reached.

6. Plot data similar to Figures 6-22 and 6-23.

7. Compare the minimum excess O_2 value to the value provided by the boiler manufacturers. High excess O_2 levels should be investigated.

8. Establish the margin in excess O_2 above the minimum and reset the burner controls to maintain this level.

9. Repeat Steps 1 thru 8 for each firing rate to be considered. Some compromise in optimum O_2 settings may be necessary since control adjustments at one firing rate may affect conditions at other firing rates.

10. After these adjustments have been completed, verify the operation of these settings by making rapid load pick-ups and drops. If undesirable conditions are encountered, reset controls.

11. Low fire conditions may require special consideration.

12. Perform tests on any alternate fuel used. Again, some discretion will be required to resolve differences in control settings for optimum conditions on the two fuels.

Evaluation of the New Low O_2 Settings

Extra attention should be given to furnace and flame patterns for the first month or two following implementation of the new operating modes. Thoroughly inspect the boiler during the next shutdown. To assure high boiler efficiency, periodically make performance evaluations and compare with the results obtained during the test program.

Burner Adjustments

Adjustments to burner and fuel systems can also be made in addition to the low excess O_2 test program previously described. The approach in testing these adjustments is a "trial-and-error" procedure with sufficient organization to allow meaningful comparisons with established data.

Items that may result in lower minimum excess O_2 levels include changes in burner register settings, oil gun tip position, oil gun diffuser position, coal spreader position, fuel oil temperature, fuel and atomizing pressure, and coal particle size. Evaluation of each of these items involves the same general procedures, precautions and data evaluation as outlined previously. The effect of these adjustments on minimum O_2 is variable from boiler to boiler and difficult to predict.

CONCLUSIONS

1. Combustion modifications are potentially catastrophic unless caution is continuously observed.
2. Test programs must be well thought out and planned in advance and anticipated results formulated and checked with obtained data.
3. Any modifications at low excess O_2 levels must be made slowly in small steps and with continuous evaluation of the flame, furnace and stack gas conditions.
4. All data must be recorded to allow future comparisons to be made.
5. Practical operating excess O_2 levels must be given an adequate excess O_2 margin to allow for load variation and fuel property or ambient air condition changes.
6. New recommended operating conditions should be monitored for a sufficient length of time to gain confidence in their long-term use.
7. Periodic efficiency checks can indicate deviations from optimum performance conditions.

AUXILIARY EQUIPMENT FOR INCREASING EFFICIENCY

The efficiency improvement potential of auxiliary equipment modifications is dependent on the existing boiler conditions. Stack gas heat recovery equipment (air preheaters and economizers) are generally the most cost-effective auxiliary equipment additions. Addition of turbulators to firetube boilers can improve operating efficiencies and promote balanced gas flows between tube banks. Advanced combustion control systems and burners generally are less beneficial than stack gas heat recovery equipment on industrial sized units. Insulation and sootblowers, judiciously applied, can have beneficial effects on boiler efficiency. Significant energy savings potentials exist in wastewater heat recovery from blowdown water and returned condensate.

Table 6-6 summarizes the options available to improve boiler efficiency.

Table 6-6. Boiler Efficiency Improvement Equipment

DEVICE	PRINCIPLE OF OPERATION	EFFICIENCY IMPROVEMENT POTENTIAL	SPECIAL CONSIDERATIONS
Air Preheaters	Transfer energy from stack gases to incoming combustion air.	2.5 % for each 100 degree F decrease in stack gas temperature.	* Results in improved combustion condition. * Minimum flue gas temperatures limited by corrosion characteristics of the flue gas. * Application limited by space, duct orientation and maximum combustion air temperatures
Economizers	Transfer energy from stack gases to incoming feedwater.	2.5 % increase for each 100 degree F decrease in stack gas temperature (1 % increase for each 10 degree F increase in feedwater temperature.	* Minimum flue gas temperature limited by corrosion characteristics of the flue gas. * Application limited on low pressure boilers. * Generally preferred over air preheaters for small (50,000 lbs./hr.) units.
Firetube Turbulators	Increases turbulence in the secondary passes of fi-etube units thereby increasing efficiency.	2.5 % increase for each 100 degree F decrease in stack gas temperature.	* Limited to gas and oil fired units. * Properly deployed, they can balance gas flows through the tubes. * Increases pressure drop in the system.
Combustion Control Systems	Regulate the quantity of fuel and air flow.	0.25 % increase for each 1 % decrease in excess O_2 depending on the stack gas temperature.	* Vary in complexity from the simplest jackshaft system to cross limited oxygen correction system. * Can operate either pneumatically or electrically. * Retrofit applications must be compatible with existing burner hardware.
Instrumentation	Provide operational data.		* Provide records so that efficiency comparisions can be made.
Oil and Gas Burners	Promote flame conditions that result in complete combustion at lower excess air levels.	0.25 % increase for each 1 % decrease in excess O_2 depending on the stack gas temperature.	* Operation with the most elaborate low excess air burners require the use of advanced combustion control systems. Flame shape and heat release rate must be compatible with furnace characteristics.

(Continued)

Table 6-6. Boiler Efficiency Improvement Equipment (*Continued*)

Equipment	Function	Value	Comments
			• Flame scanners increase reliability. • Advanced atomizing systems are available. • Close control of oil viscosity improves atomization.
Insulation	Reduce external heat transfer.	Dependent on surface temperature	• Mass type insulation has low thermal conductivity and release heat loss by conduction. • Reflective insulation has smooth, metallic surfaces that reduce heat loss by radiation. • Insulation provides several other advantages including structural strength, reduced noise and fire protection.
Sootblowers	Remove boiler tube deposits that retard heat transfer.	Dependent on the gas temperature	• Can use steam or air as the blowing media. • Fixed position systems are used in low temperature regions whereas retractable "losses" are employed in high temperature areas. • The choice of the cleaning media will depend on the characteristics of the deposits.
Blowdown Systems	Transfer energy from expelled blowdown liquids to incoming feedwater.	1 - 3 % dependent on blowdown quantities and operating pressures	• Quantity of expelled blowdown water is dependent on the boiler and makeup water quality. • Continuous blowdown operation not only decreases expelled liquids but also allows the incorporation of heat recovery equipment.
Condensate Return Systems	Reduce hot water requirements by recovering condensate.	12 - 15 %	• Quantity of condensate returned dependent on process and contamination. • Several systems available range from atmospheric (open) to fully pressurized (closed) systems.

Air preheaters and economizers are common equipment used. A caution should be noted in that these and other options will reduce stack temperature. In order to avoid corrosion problems, exit temperatures should be as illustrated in Table 6-7. Exit gas temperature is determined by the extent of boiler convective surface or the presence of stack gas heat recovery equipment.

Table 6-7. Minimum Exit Gas Temperatures

Oil Fuel (>2.5% S)	390°F
Oil Fuel (<1.0% S)	330°F
Bituminous Coal (>3.5% S)	290°F
Bituminous Coal (<1.5% S)	230°F
Pulverized Anthracite	220°F
Natural Gas (Sulfur-free)	220°F

For boilers without heat recovery equipment, the minimum exit gas temperature is fixed by the boiler operating pressure since this determines the steam temperature. Usual design practices result in an outlet gas temperature 150°F above the saturated steam temperature.

It becomes increasingly expensive to approach boiler saturation temperatures by simply adding convective surface area. As operating pressures increase, the stack gas temperature increases to make heat recovery equipment more desirable. Economizers will permit a reduction in exit gas temperatures since the feedwater is at a lower temperature (220°F) than the steam saturation temperature. Stack gas temperatures of 300°F can be achieved with stack gas heat recovery equipment. Further reductions are achieved using air preheaters. Present design criteria limit the degree of cooling using stack gas heat recovery equipment to a level which will minimize condensation on heat transfer surfaces. The sulfur content of the fuel has a direct bearing on the minimum stack gas temperature since SO_3 combines with condensed water to form sulfuric acid and since the SO_3 concentration in the flue gas also determines the condensation temperature.

7

Heating, Ventilation, Air Conditioning, and Building System Optimization

This chapter will review the basics of Heating, Ventilation and Air Conditioning (HVAC) and building as related to energy engineering.

DEGREE DAYS

Degree days are the summation of the product of the difference in temperature (ΔT) between the *average outdoor* and hypothetical *average indoor* temperatures (65°F) and the number of days (t) the outdoor temperature is below 65°F. Therefore:

$$DD = \Delta T \times t, \text{ therefore } \Delta T = DD/t \qquad \textit{Formula (7-1)}$$

Degree days divided by the total number of days on which degree days were accumulated will yield an average ΔT for the season, based on an assumed indoor temperature of 65°F. To find the average outdoor temperature of the season, this figure must be subtracted from 65°F.

Example Problem 7-1

If there are 6750 degree days recorded over a heating season of 270 days, what is the mean outdoor temperature for that season?

Answer

$$\Delta T = DD/t \qquad \Delta T = \frac{6750DD}{270 \text{ days}} \qquad \Delta T = 25°P$$

The average outdoor temperature can now be found, since
$\Delta T = T$ (avg. indoor) $- T$ (avg. outdoor)
T (avg. outdoor) $= T$ (avg. indoor) $- \Delta T$
T (avg. outdoor) $= 65°F - 25°F$
T (avg. outdoor) $= 40°F$

RESISTANCE (R) TO HEAT FLOW AND CONDUCTANCE (U) AND CONDUCTIVITY (K)

The rate at which heat flows through a material depends on its characteristics. Some materials transmit heat more readily than others. This characteristic of materials which affects the flow of heat through them can be viewed either as their *resistance* to the flow of heat or as their *conductance* allowing the flow of heat.

For a section of a building, such as a wall, the conductance is expressed as the U-value for that wall, that is, the number of Btus that will pass through a one-square-foot section of a building in one hour with a one-degree temperature difference between the two surfaces.

U = Btus per square foot per hour per degree Fahrenheit.

or

$$U = Btu / ft^2 \, h°F$$

R – Value = Thermal Resistance = The unit time for a unit area of a particular body or assembly having defined surfaces with a unit average temperature difference established between the two surfaces per unit of thermal transmission.

$$\frac{hr \cdot ft^2 \cdot °F}{Btu}$$

$$R = 1/U$$

Where

$$U = \frac{1}{1/hi + x_1/K_1 + \dfrac{x_2}{K_2} + \ldots + 1/ho}$$

Formula (7-2)

x = Respective thickness of different materials
K = Respective conductivities of layers
h = Convective heat transfer coefficients

The conductivity of a material as related to conductance and resistance is illustrated by Formula 7-3.

$$U \; = \; \frac{K}{d} \; = \; \frac{1}{R} \qquad\qquad Formula\ (7\text{-}3)$$

where *d* is the thickness of the material.

In buildings, Table 7-1 may be used to find acceptable conductance values for walls.

Table 7-1. Minimum Conductance Values for Walls

Heating Degree Days	Min Acceptable *U*-Value
1000	0.40
2000	0.30
3000	0.30
4000	0.20
5000	0.20
6000	0.15
7000	0.15
8000	0.10
9000	0.10

VOLUME (V) OF AIR

The volume of air within a structure is constant even though the air itself changes—new air enters; old air leaves. Total volume is equal to the volume of space in the conditioned portion of the facility. (Only the volume of conditioned space is considered since air entering and leaving the unconditioned part of the facility does not demand energy to condition it.)

To determine the volume (*V*) of air, multiply the height (*H*) of the space times the width (*W*) of the space times the length (*L*) of the space.*

---

*This is only appropriate for structures with flat ceilings.

While this can be done for the facility as a whole, it is more accurate to calculate it for each room and then add these volumes.

AIR CHANGES PER HOUR (AC)

The rate at which the volume of air in a structure changes per hour differs greatly from building to building. The number of air changes per hour (AC/h) has wide variation due to a number of factors such as:

- *The number and size of openings* in the envelope—around doors and windows and in the siding itself.
- *The average speed* of the wind blowing against the structure and the protection the structure has from this wind.
- *The number and size* of chimneys, vents, and exhaust fans and the frequency of their use.
- *The number of times* that doors and windows are opened.
- *How the structure is used.*

Currently, the general standard, or accepted value, for air changes within an average home is .35 air changes (of entire house) per hour, unless for some reason a room contains toxic fumes, etc.

HEATING CAPACITY OF AIR (HC)

Air can be heated and cooled. A certain amount of heat is necessary to change the temperature of each cubic foot (ft^3) of air one degree Fahrenheit (F). This amount of heat depends on the density of air which varies with temperature and pressure. This figure will generally be within the range of 0.018-0.022 Btus/ft^3 °F.

BUILDING DYNAMICS

The building experiences heat gains and heat losses depending on whether the cooling or heating system is present, as illustrated in Figures 7-1 and 7-2. Only when the total season is considered in conjunction with lighting and heating, ventilation and air conditioning (HVAC) can the en-

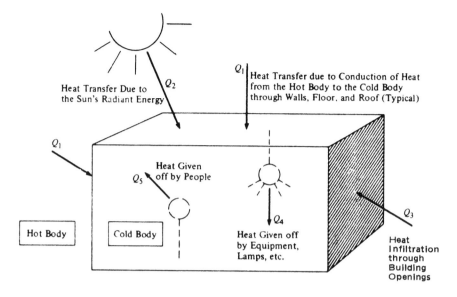

Heat Gain = $Q_1 + Q_2 + Q_3 + Q_4 + Q_5$

Figure 7-1. Heat Gain of a Building

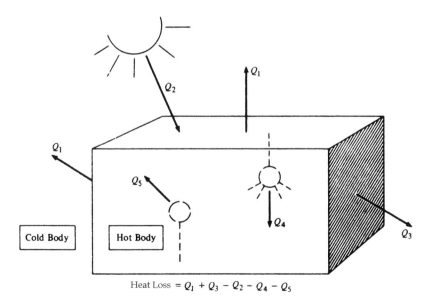

Heat Loss = $Q_1 + Q_3 - Q_2 - Q_4 - Q_5$

Figure 7-2. Heat Loss of a Building

ergy utilization choice be decided. One way of reducing energy consumption of HVAC equipment is to reduce the overall heat gain or heat loss of a building.

CONDUCTION HEAT LOSS

The formula used to determine the amount of heat conducted through the envelope is as follows: Degree days (*DD*) is the product of the difference in temperature (ΔT) and the time (*t*) in days, providing that the days are converted to hours. This is accomplished by multiplying *DD* times 24 hours a day. This will yield the quantity of heat (*Q*) conducted through a particular section of the envelope for the entire heating season.

The formula can be written:

$$Q_{(heating\ season)} = U \times DD \times 24\ \text{hrs/day} \qquad \textit{Formula (7-4)}$$

or

$$Q_{(heating\ season)} = \frac{A \times DD \times 24\ \text{hrs/day}}{R} \qquad \textit{Formula (7-5)}$$

In general, heat flow through a flat surface is defined as

$$Q = U\ A\Delta T \qquad\qquad\qquad \textit{Formula (7-6)}$$

where ΔT is the temperature difference causing the heat flow.

For a composite wall, the heat flow is represented by Figure 7-3. To calculate the overall *U* of conductance value, the resistance of each material is added in series. This is analogous to an electrical circuit.

$$R = R_1 + R_2 + R_3 + R_4 + R_5 \qquad\qquad \textit{Formula (7-7)}$$

FACTORS IN CONDUCTION

There are four factors which affect the conduction of heat from one area to another. They are

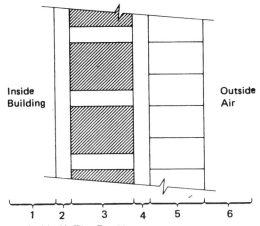

1   Inside Air Film $R = .68$
2   1/2" Plaster Board Interior Finish $R = .44$
3   8" Concrete Block, Sand & Gravel Aggregate $R = 1.1$
4   1/2" Cement Mortar $R = .1$
5   4" Brick Exterior $R = .44$
6   Outside Air Film @ 15 mph Wind $R = .17$

Figure 7-3. Typical Wall Construction

- *The difference in temperature* (ΔT) between the warmer area and the colder area.
- *The length of time* (t) over which the transfer occurs.
- *The area* (A) *in common* between the warmer and the colder area.
- *The resistance* (R) *to heat flow and conduction* (U) between the warmer and the colder area.

In a typical house, the major losses due to conduction occur through the walls, the roof, and the floor.

The Wall
- Major heat loss due to conduction, other losses are minor.
- Major concern is about the U-values of the walls.
- Minimum acceptable R-value for walls should be above 10.
- For winter the most important wall to insulate is north.
- For summer the most important wall to insulate is south.

The Roof
- Major heat loss due to conduction, other losses are minor.

- If the funds for insulation are limited, insulate the roof first, then the north wall, then the south wall.
- Use Table 7-2 as a guide for insulating the roof.

Table 7-2. Minimum Conductance Values for Roof

Heating Degree Days	Min. Acceptable U-Value
1000	0.30
2000	0.20
3000	0.20
4000	0.15
5000	0.15
6000	0.10
7000	0.10
8000	0.06

The Floor
- Major heat loss through conduction along the perimeter of the floor.
- Insulating floor should be the last priority compared to the walls and roof.
- Use Table 7-3 as a guide for insulating floors.

Table 7-3. Minimum Conductance Values for Floors

Heating Degree Days	Min. Acceptable U-Value
1000	0.40
2000	0.35
3000	0.30
4000	0.22
5000	0.22
6000	0.18
7000	0.18
8000	0.12

Difference in Temperature

Heat flows (much as water moves downhill) from warm areas to cold ones. The steeper the gradient between its origin and its destination, the faster it will flow. In fact, the rate at which heat is conducted is directly proportionate to the difference in temperature (ΔT) between the warm area and the colder one.

Length of Time

The longer the heat is allowed to flow across the gradient, the more heat will be conducted. The amount of heat (Btus) is directly proportionate to the time span (*t*) of the transfer.

Btu/h is the amount of heat transferred in one hour.

The Area (*A*) in Common

The larger the area common to the warmer and colder surfaces through which the heat flows, the greater is the rate of conducted heat. For the same material, for the same length of time, at the same ΔT, the amount of heat (Btus) transferred is directly proportionate to the area (A) in common.

Example Problem 7-2

Calculate the heat loss through 20,000 ft² of building wall, as indicated by Figure 7-3. Assume a temperature differential of 17°F.

Answer

Description	Resistance
Outside air film at 15 mph	0.17
4" brick	0.44
Mortar	0.10
Block	1.11
Plaster Board	0.44
Inside film	0.68
Total resistance	2.94

$$U = 1/R = 0.34$$
$$Q = U\,A\,\Delta T$$
$$= 0.34 \times 20{,}000 \times 17 = 115{,}600 \text{ Btu/h}$$

In Figure 7-4 a surface film conductance is introduced.

The surface or film conductance is the amount of heat transferred in Btu per hour from a surface to air or from air to a surface per square foot for one degree difference in temperature. The flow of heat for the composite material can also be specified in terms of the conductivity of the material and the conductance of the air film.

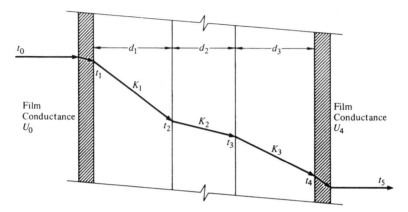

Figure 7-4. Temperature Distribution of the Composite Wall

$$Q = \frac{A\,(t_0 - t_5)}{1/U_o + d_1/K_1 + d_2/K_2 + d_3/K_3 + 1/U_4} \qquad Formula\ (7\text{-}8)$$

LATENT HEAT AND SENSIBLE HEAT

The latent heat gain of a space means that moisture is being added to the air in the space. Moisture in the air is really in the form of super-heated steam. Removing sensible heat from a space through air-conditioning equipment lowers the dry-bulb temperature of the air. On the other hand, removing latent heat from a space changes the substance state from a vapor to a liquid. The latent heat gain of a space is expressed in terms of moisture, heat units (Btu) or grains of moisture per hour (7,000 grains equals one pound), The average value for the latent heat of vaporization for superheated steam in air is 1050.

Example Problem 7-3
2000 grains of moisture are released in a conditioned room each hour. Calculate the heat that must be removed in order to condense this moisture at the cooling coils.

Answer
$$\frac{2000}{7000} \times 1050 = 299.9\ \text{Btuh}$$

Example Problem 7-4

Calculate the quantity of heat (Q) required by infiltration in an 8,000-cubic foot (ft^3) home that has 1.7 air changes per hour (AC/h) when the outside temperature is 48°F and the inside temperature is 68°F ($\Delta T = 20°F$) for one day (24 hours).

Answer

$$Q = V \times AC/h \times 0.020^* \text{ Btu}/\text{ft}^3 \text{ °F} \times \Delta T \times t$$
$$Q = 8,000 \text{ ft}^3/AC \times 1.7 \text{ AC/h} \times 0.020^* \text{ Btu}/\text{ft}^3 \text{ °F} \times 20°F \times 24 \text{ hrs}$$
$$Q = 8,000 \times 1.7 \times 0.020° \times 20 \times 24$$
$$Q = 130,560 \text{ Btus}^{**}$$

INFILTRATION

Leakage or infiltration of air into a building is similar to the effect of additional ventilation. Unlike ventilation, it cannot be controlled or turned off at night. It is the result of cracks, openings around windows and doors, and access openings. Infiltration is also induced into the building to replace exhaust air unless the HVAC balances the exhaust. Wind velocity increases infiltration, and stack effects are potential problems. Air that is pushed out the window and door cracks is referred to as exfiltration.

To estimate infiltration, the air change method or crack method is used.

Air Change Method

The five factors which determine the amount of energy lost through infiltration can be assembled in a formula that states:

The quantity of heat (Q) equals the volume of air (V) times the number of air changes per hour (AC/h) times the amount of heat required to raise the temperature of air one degree Fahrenheit (0.018-0.022 Btus/ft^3°F) times the temperature difference (ΔT) times the length of time (t). This is expressed as follows:

$$Q = V \times AC/h \times 0.020 \text{ Btus} \bullet /\text{ft}^3 \text{ °F} \times \Delta T \times t \qquad \textit{Formula (7-9)}$$

---

*This is a regional variable.

**Once again, all of the units in the formula cancel except Btus, leaving the units for Q as Btus.

The air change method is considered to be a quick estimation method and is not usually accurate enough for air-conditioning design. A second method used to determine infiltration is the crack method.

Crack Method

When infiltration enters a space, it adds sensible and latent loads to the room. To calculate this gain, the following equations (7-10 and 7-11) are used.

Sensible Heat Gain $Q_S = 1.08 \text{ CFM } \Delta T$ *Formula (7-10)*

Latent Heat Gain $Q_L = .7 \text{ CFM } (HR_o - HR_i)$ *Formula (7-11)*

Where
  Q_S = Sensible heat gain Btuh
  Q_L = Latent heat gain Btuh
  CFM   = Air Flow Rate
  ΔT = Temperature differential between outside and inside air, F
  HR_o = Humidity ratio of outside air, grains per lb
  HR_i = Humidity ratio of room air, grains per lb

BODY HEAT

The human body releases sensible and latent heat depending on the degree of activity. Heat gains for typical applications are summarized in Table 7-4.

Table 7-4. Heat Gain from Occupants

Activity	Sensible Heat Btuh	Latent Heat Btuh	Total Heat Gain Btuh
Very Light Work — Seated			
(Offices, Hotels, Apartments)	215	185	400
Moderately Active Work			
(Offices, Hotels, Apartments)	220	230	450
Moderately Heavy Work			
(Manufacturing)	330	670	1000
Heavy Work (Manufacturing)	510	940	1450

Source: ASHRAE—Guide & Data Book

EQUIPMENT, LIGHTING AND MOTOR HEAT GAINS

It is important to include heat gains from equipment, lighting systems and motor heat gain in the overall calculations.

For a manufacturing facility, the major source of heat gains will be from the process equipment. Consideration must be given to all equipment including motors driving supply and exhaust fans.

To convert motor horsepower to heat gain in Btuh, Formula 7-12 is used.

$$Q = \frac{hp \times .746}{\eta} \times 3412 \qquad \text{Formula (7-12)}$$

Where
  hp is the running motor horsepower
  η is the efficiency of the motor
  Q is the heat gain from the motor Btuh

Similarly, the kilowatts of the lighting system can be converted to heat gain.

$$Q = (kW_F + kW_B) \times 3412 \qquad \text{Formula (7-13)}$$

Where
  kW_F is the kilowatts of the lighting fixtures
  kW_B is the kilowatts of the ballast
   Q is the heat gain from the lighting system Btuh

RADIANT HEAT GAIN

Heat from the sun's rays greatly increases heat gain of a building. If the building energy requirements were mainly due to cooling, then this gain should be minimized. Solar energy affects a building in the following ways:

1. *Raises the surface temperature;* thus a greater temperature differential will exist at roofs than at walls.
2. A large percentage of direct solar radiation and diffuse sky radiation *passes through* transparent materials, such as glass.
3. Envelopes can be designed to have individual control loops to stop or enhance the solar radiation into the building.

SURFACE TEMPERATURES

The temperature of a wall of roof depends upon

(a)   the angle of the sun's rays

(b)   the color and roughness of the surface

(c)   the reflectivity of the surface

(d)   the type of construction

When an engineer is specifying building materials, he should consider the above factors, A simple example is color. The darker the surface, the more solar radiation will be absorbed. Obviously, white surfaces have a lower temperature than black surfaces after the same period of solar heating. Another factor is that smooth surfaces reflect more radiant heat than do rough ones.

In order to properly take solar energy into account, the angle of the sun's rays must be known. If the latitude of the facility is known, the angle can be determined.

SUNLIGHT AND GLASS CONSIDERATIONS

A danger in the energy conservation movement is to take steps backward. A simple example would be to exclude glass from building designs because of the poor conductance and solar heat gain factors of clear glass. The engineer needs to evaluate various alternate glass constructions and coating in order to maintain and improve the aesthetic qualities of good design while minimizing energy inefficiencies. It should be noted that the method to reduce heat gain of glass due to conductance is to provide an insulating air space.

To reduce the solar radiation that passes through glass, several techniques are available. Heat absorbing glass (tinted glass) is very popular. Reflective glass is gaining popularity, as it greatly reduces solar heat gains.

To calculate the relative heat gain through glass, a simple method is illustrated below.

$$Q = \pm U A (t_0 - t_1) + A \times S_1 \times S_2 \qquad\qquad \textit{Formula (7-14)}$$

Where

Q is the total heat gain for each glass orientation (Btuh).

U is the conductance of the glass (Btu/h•ft²•°F).

A is the area of glass; the area used should include framing, since it will generally have a poor conductance compared with the surrounding material. (ft²)

$t_0 - t_1$ is the temperature difference between the inside temperature and outside ambient. (°F)

S_1 is the shading coefficient; S_1 takes into account external shades, such as Venetian blinds and draperies, and the qualities of the glass, such as tinting and reflective coatings.

S_2 is the solar heat gain factor. This factor takes into account direct and diffused radiation from the sun. Diffused radiation is basically caused by reflections from dust particles and moisture in the air.

Optimizing Windows
- Use storm windows in winter.
- Use a clear plastic cover to cover windows in the winter so as to allow radiation but stop infiltration.
- Use double or triple glass windows.
- Promote thermal barriers.
- Use solar control devices.

THE PSYCHROMETRIC CHART

Just as the steam table and the Mollier diagram are used to relate the properties of steam, the psychrometric chart is used to illustrate the properties of moist air. The psychrometric chart is a very important tool in the design of air-conditioning systems.

PROPERTIES OF MOIST AIR

Air expands and contracts with temperature. If pressure is held constant, then air expands or contracts at a specified rate with change in temperature as defined by Formula 7-15.

$$V_2 = V_1 \frac{T_2}{T_1}$$ *Formula (7-15)*

Where
V_1 = initial volume of air
V_2 = final volume of air
T_1 = absolute temperature
T_2 = final absolute temperature

At constant temperature,

$$V_2 = V_1 \frac{P_2}{P_1}$$ *Formula (7-15a)*

Where
P_2 = initial pressure, psia
P_1 = final pressure, psia

The change in the volume occupied by air at any temperature can be found by first using Formula 7-15a to calculate the change in volume with pressure and then using Formula 7-15 to calculate the change in volume with temperature.

The temperature at which the water vapor in the atmosphere begins to condense is the *dew point temperature*. It should be noted that the mass of moisture per pound of dry air in a mixture of air and water vapor depends on the dew point temperature alone. If there is no condensation of moisture, the dew point temperature remains constant.

Humidity ratio is defined as the mass of water vapor mixed with one pound of dry air.

Degree of saturation is defined as the actual humidity ratio divided by the humidity ratio at saturation.

Relative humidity is defined as the vapor pressure of air divided by the saturation pressure of pure water at the same temperature.

Sensible heat of an air-vapor mixture is defined as the heat which affects the dry-bulb temperature of the mixture only.

Wet-bulb Temperature can be determined by covering the bulb of a thermometer with a wet wick and holding it in a stream of swiftly moving air. At first the temperature will drop quickly and then reach a stationary point referred to as the wet-bulb temperature. The wet-bulb temperature is lower than the dry-bulb temperature. The amount of water which evaporates from the wet wick into the air depends on the amount of water vapor initially in the air flowing past the wet bulb. A sling psychrometer is a convenient instrument used to measure wet-bulb temperatures.

Total heat is defined as the sum of the sensible and latent heat. Sensible heat depends only upon the dry-bulb temperature, while the latent heat content depends only upon its dew point.

The *enthalpy of an air-water vapor mixture* can be calculated by Formula 7-16.

$$h_{(mix)} = h_{(dry\ air)} + h_{(water\ vapor)}$$

or

$$h_{(mix)} = Cp \times T_{DB} + HR\ hg \qquad \text{Formula (7-16)}$$

Where
$h_{(mix)}$ = enthalpy of the mixture of dry air and water vapor, Btu per lb
Cp = specified heat
T_{DB} = dry bulb temperature
HR = humidity ratio of the mixture
hg = enthalpy of saturated vapor (steam) at the dew point temperature

To determine the properties of air such as the humidity ratio, relative humidity and enthalpy, the psychrometric chart is frequently used. Figure 7-5 illustrates the psychrometric chart.

Example Problem 7-5
Given air at 70°F DB and 50% relative humidity, for the air vapor mixture find

- Wet-bulb Temperature
- Enthalpy
- Humidity Ratio
- Dew Point Temperature

- Specific Volume
- Vapor Pressure
- Percentage Humidity

Answer
From Figure 7-5 find the intersection of 70°F DB and 50% *HR*, point "A."

The WB temperature is found as 58.6 WB, point "B."

The enthalpy is found as 25.5 Btu/lb, point "C."

The humidity ratio is found to be 56 grains of moisture per pound of dry air, point "D," and the dew point temperature is 53°F, point "E."

The specific volume is found to be 13.5 cubic feet per pound of air, point "F," with a vapor pressure of .38 inches of mercury, point "G."

The percentage humidity equals the actual humidity 56, point "D," divided by the humidity ratio at saturation (100% *RH*) which is found to be 110, point "H." Thus % humidity = 56/110 = .50.

Example Problem 7-6

Given 8000 CFM of chilled air at 55°F DB and 50°F WB mixed with 3000 CFM of outside air at 90°F DB and 80°F WB, compute the properties of the mixture.

Answer

From Figure 7-6, the intersection of 55°F DB and 50°F WB is point "A." The specific volume is then 13.1 cubic feet/lb, point "B."

Similarly, for the outside air, the specific volume is 14.3 cubic feet/lb, point "D."

The total weight and dry-bulb temperature of the mixture can be found by the following ratios:

$$\frac{8{,}000}{13.1} = 610.6 \text{ lb/min.}$$

$$\frac{3{,}000}{14.3} = \underline{209.7 \text{ lb/min.}}$$

$$\text{Total} \qquad 820.3 \text{ lb/min. for total weight}$$

The dry-bulb temperature is

$$\frac{610.6}{820.3} \times 55 = 40.93°\text{P}$$

$$\frac{209.7}{820.3} \times 90 = \underline{23.0°\text{F}}$$

$$63.9°\text{F DB}$$

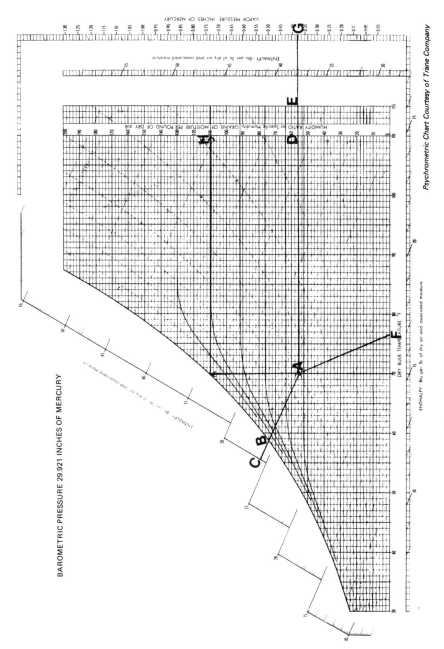

Figure 7-5. Psychrometric Chart

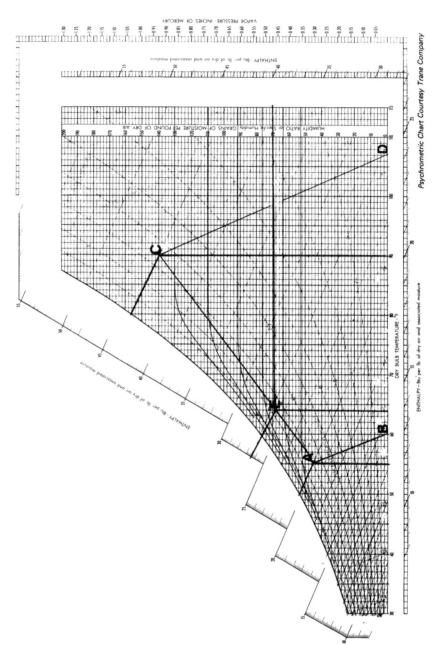

Figure 7-6. Psychrometric Chart for Mixture of Air

The properties of the mixture, point "E," can now be determined from the chart.

$$WB = 59.8°F$$
$$h = 26.6 \text{ Btu}/\text{lb}$$
$$\text{Humidity Ratio} = 70 \text{ gr of moisture}/\text{lb of dry air.}$$

BASICS OF FAN DISTRIBUTION SYSTEMS

In order to distribute conditioned or ventilated air, fans are the chief vehicle used. Several basic types of fans commonly used in industry are illustrated in Figure 7-7 and are listed below.

Centrifugal, airfoil blade—used on large heating, ventilating, and air-conditioning systems. Airfoil fans are used where clean air is handled.

Centrifugal, backward curved blade—used for general heating, ventilating, and air-conditioning systems. Air handling need not be as clean as above.

Centrifugal, radial blade—a rugged, heavy duty fan for high pressure applications. It is designed to handle sand, wood chips, etc.

Centrifugal, forward curved blade—ideal for low pressure applications, such as domestic furnaces or room and packaged air conditioners.

The brake horsepower of a fan is illustrated by Formula 7-17.

$$\text{Brake Horsepower} = \frac{\text{CFM} \times \text{Fan PS}}{6356 \times \eta_F} \qquad \text{Formula (7-17)}$$

Where
CFM is the quantity of air in CFM
η_F is the fan static efficiency
Fan PS is the fan static pressure in inches

To compute the fan static pressure

$$\text{Fan PS} = P_T(O) - P_T(i) - P_v(O) \qquad \text{Formula (7-18)}$$

Where
$P_T(O)$ is the total pressure at fan outlet
$P_T(i)$ is the total pressure at fan inlet

Figure 7-7. Fan Types: (A) Vaneaxial; (B) Backward Curved Blade; (C) Tubeaxial; (D) Radial; (E) Radial Tip Blade; (F) Airfoil Blade. (*Courtesy of Buffalo Forge Company.*)

$P_v(O)$ is the velocity pressures at fan outlet

The excess pressure above the static pressure is known as the velocity pressure and is computed by Formula 7-19 for standard air having value of 13.33 cu ft per lb as

$$P_v = \left(\frac{V}{4005}\right)^2$$

Formula (7-19)

where *V* is the velocity of air in FPM.

Note that the pressure of air in sheet metal ducts is so low that ordinary pressure gage (Bourdon type) cannot be used; thus a V-tube or manometer is used, which measures pressure in inches of water. A pressure of 1 psi will support a column of water 2.31 feet high or 27.7 inches.

Fan Laws

The performance of a fan at varying speeds and air densities may be predicted by certain basic fan laws as illustrated in Table 7-5.

Table 7-5. Fan Laws

1. Fan Law for variation in fan speed at constant air density with a constant system
1.1 Air volume, CFM varies as fan speed
1.2 Static velocity or total pressure varies as the square of fan speed
1.3 Power varies as cube of fan speed
2. Fan Law variation in air density at constant fan speed with a constant system
2.1 Air volume is constant
2.2 Static velocity gr pressure varies as density
2.3 Power varies as density

Example Problem 7-7

An energy audit indicates that the ventilation requirements of a space can be reduced from 15,000 CFM to 12,000 CFM. Comment on the savings in brake horsepower if the fan pulley is changed to reduce the fan speed accordingly.

Answer

From the Fan Laws

$$hp_1 = hp_2 \cdot \left(\frac{CFM\ new}{CFM\ old}\right)^3$$

$$= hp_2 \cdot \left(\frac{12,000^3}{15,000}\right) = hp_2(.8)^3 = .512\ hp_2$$

or a 48.8% savings.

The fan performance is affected by the density of the air that the fan is handling, All fans are rated at standard air with a density of .075 lb per cu ft and a specific volume of 13.33 cu ft per lb. When a fan is tested in a laboratory at different than standard air, the brake horsepower is corrected by using the fan laws.

Fan Performance Curves

Fan performance curves are used to determine the relationship between the quantity of air that a fan will deliver and the pressure it can discharge at various air quantities. For each fan type, the manufacturer can supply fan performance curves which can be used in design and as a tool of determining the fan efficiency.

As illustrated by Ex. Prob. 7-6, one energy engineering technique to reduce fan horsepower is to reduce fan speed. An alternate way is to throttle the air flow by a damper. The fan performance curves can be used to illustrate the best choice of these options. The system characteristics can be plotted on the fan curves to show the static pressure required to overcome the friction loss in the duct system. From the fan laws the system friction loss varies with the square of fan speed; thus, as the air quantity increases, the friction loss will vary as illustrated by Figure 7-8.

For a detailed analysis, the fan performance curve should be used to predict how a specific fan will perform in a desired application.

Example Problem 7-8

Given Fan Performance Curve Figure 7-9. The fan delivers 21,500 CFM at 600 rpm at a brake horsepower of 12.3. Comment on the savings in brake horsepower by reducing air flow to 14,400 CFM by each of the following methods: (a) reducing fan speed to 400 rpm, (b) throttling the air flow by a damper.

(a) From the fan laws the brake horsepower is reduced as follows:

$$hp = (12.3)\left(\frac{400}{600}\right)^3 = 3.64$$

Using the fan performance curve, Figure 7-10, the system characteristic curve "A" is plotted.

By reducing the rpm from 600 to 400, the system operates at point 1 and then moves to point 2. The brake horsepower is found from Figure 7-9 to be 3.7.

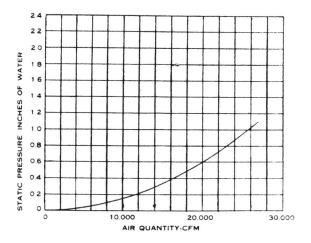

Figure 7-8. System Characteristic Curve

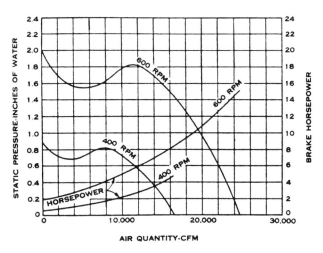

Figure 7-9. Fan Performance Curves

(b) By closing the air damper, the air flow is reduced to 14,400 CFM while still running the fan at 600 rpm. Using the fan performance curve, Figure 7-10, the system operates at point 1 and then moves to point 3. The power to operate the fan at point 3 is 7.2 hp from Figure 7-9. Thus, if the fan speed can be reduced, it is more efficient than throttling the air flow damper.

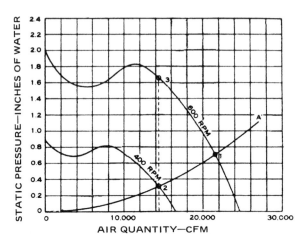

Figure 7-10. Fan Performance and System Characteristic Curves

FLUID FLOW

Pump and piping considerations are extremely important due to the fact that energy transport losses are a part of any distribution system. Losses occur due to friction, and that lost energy must be supplied by pump horsepower.

Centrifugal pumps are commonly used in heating, ventilating and air-conditioning applications as well as utility systems. The output torque for the pump is supplied by a driver such as a motor. Liquid enters the eye of the impeller which rotates. Pressure energy builds up by the action of centrifugal force, which is a function of the impeller vane peripheral velocity.

As with fan systems, pump laws and curves can be used to predict system responsiveness. The affinity laws of a pump are illustrated in Table 7-6.

The horsepower required to operate a pump is illustrated by Formula 7-20.

$$\text{hp} = \frac{\Delta P \text{ GPM}}{1715 \, \eta} \qquad \qquad Formula\ (7\text{-}20)$$

Where
  ΔP is the differential pressure across a pump in psi
  GPM is the required flow rate in gallons per minute
    η is the pump efficiency

Table 7-6. Affinity Laws for Pumps

Impeller Diameter	Speed	Specific Gravity (SG	To Correct for	Multiply by
			Flow	$\left(\dfrac{\text{New Speed}}{\text{Old Speed}}\right)$
Constant	Variable	Constant	Head	$\left(\dfrac{\text{New Speed}}{\text{Old Speed}}\right)^2$
			BHP (or kW)	$\left(\dfrac{\text{New Speed}}{\text{Old Speed}}\right)^3$
			Flow	$\left(\dfrac{\text{New Diameter}}{\text{Old Diameter}}\right)$
Variable		Constant	Head	$\left(\dfrac{\text{New Diameter}}{\text{Old Diameter}}\right)^2$
			BHP (or kW)	$\left(\dfrac{\text{New Diameter}}{\text{Old Diameter}}\right)^3$
Constant		Variable	BHP (or kW)	$\left(\dfrac{\text{New SG}}{\text{Old SG}}\right)$

To convert psi to read in feet use Formula 7-21.

$$\text{Head in Feet} = \frac{\text{psi} \times 2.31}{\text{Specific Gravity of Fluid}} \qquad \textit{Formula (7-21)}$$

Basically, the size of discharge line piping from the pump determines the friction loss through the pipe that the pump must overcome. The greater the line loss, the more pump horsepower required. If the line is short or has a small flow, this loss may not be significant in terms of the total system head requirements. On the other hand, if the line is long and has a large flow rate, the line loss will be significant.

To calculate the pressure loss for water system piping and the corresponding velocity, Formulas 7-22 and 7-23 are used.

$$\Delta P = \frac{= 0.55\, CF^{1.85}}{d^{4.87}} \qquad \textit{Formula (7-22)}$$

$$V = \frac{.41F}{d^2} \qquad\qquad \text{Formula (7-23)}$$

Where

ΔP = pressure loss per 100 feet of pipe, psi
  V = velocity of fluid, ft/sec.
  C = roughness factor
      1 for copper tubing
      1.62 for steel pipe
      .77 for plastic pipe
  F = flow rate in gallons per minute
  d = inside diameter of pipe, inches

The pressure loss due to fittings is determined by Formula 7-24.

$$\Delta P = .0067 \, KV^2 \qquad\qquad \text{Formula (7-24)}$$

where *K is* the loss coefficient.

Options to Reduce Pump Horsepower

There are several alternatives that will significantly reduce pump horsepower. Summarized below are some of the options available.

1. Many pumps are oversized to very conservative design practices. If the pump is oversized, install a smaller impeller to match the load.
2. In some instances heating or cooling supply flow rates can be reduced. To save on pump horsepower, either reduce motor speed or change the size of the motor sheave.
3. Check economics of replacing corroded pipe with a large pipe diameter to reduce friction losses.
4. Consider using variable speed pumps to better match load conditions. Motor drive speed can be varied to match pump flow rate or head requirements.
5. Consider adding a smaller auxiliary pump. During part load situations a larger pump can be shut down and a smaller auxiliary pump used.

The distribution and overall load that is placed on HVAC transportation devices can additionally be optimized by using the following guidelines.

1. Reduce the resistance to flow in ducts and piping.
2. Eliminate leaks in ducts or piping.
3. Improve the efficiency of terminal devices. (Heat exchangers, diffusers, etc.)
4. Reduce or eliminate the opportunity for heat transfer during distribution.
5. Reduce the transport rate. (Flow velocity)

Remember that improvements in the envelope efficiency will automatically reduce the distribution load. This effect should always be considered and optimized by sizing distribution components accordingly.

Common Sense Methods

Do not underestimate the effects of energy saving techniques which require only a change of habit or life style.

Some of the following guidelines can lead to significant savings with little or no initial cost.

1. Set back the thermostat. Some suggested guidelines are 68°F during the heating season and 78°F during the cooling season. When possible, reduce thermostat in unoccupied areas. This is easily done with baseboard electric heat.
2. Use or adjust thermostats to allow for a temperature deadband or range in which no action is taken by HVAC systems.
3. Regulate thermostat settings. (Hide or lock where necessary.)
4. Turn off lights in unoccupied areas and reduce overall lighting levels by illuminating the required area (task area). A good example of the potential for energy savings is illustrated on pages 302-304 of this text.
5. Reduce levels of relative humidity during the heating season. The energy required to vaporize water for humidification purposes can be significant.
6. Shut down ventilation systems during unoccupied hours and reduce levels to acceptable limit during occupied hours.
7. Check and ensure that ventilation dampers open and close as desired by HVAC system.
8. Filter indoor air to reduce outdoor air ventilation requirements.

Filtering systems are available to remove cigarette smoke, cooking smoke and odors, and other common air contaminants.
9. Use storm windows or plastic to reduce infiltration as well as conduction and convection heating/cooling losses of windows.
10. Correct duct inadequacies which cause leakage, obstruction, or allow unnecessary heat transfer.

HVAC SYSTEMS

The summary below illustrates the types of systems frequently encountered in heating arid air-conditioning systems.

Reducing Energy Consumption in HVAC Systems

Variable Air Volume (VAV) System—A variable air volume system provides heated or cooled air at a constant temperature to all zones served. VAV boxes located in each zone or in each space adjust the quantity of air reaching each zone or space depending on its load requirements. Methods for conserving energy consumed by this system include:
1. Reduce the volume of air handled by the system to that point which is minimally satisfactory.
2. Lower hot water temperature and raise chilled water temperature in accordance with space requirements.
3. Lower air supply temperature to that point which will result in the VAV box serving the space with the most extreme load being fully open.
4. Consider installing static pressure controls for more effective regulation of pressure bypass (inlet) dampers.
5. Consider installing fan inlet damper control systems if none now exists.

Constant Volume System—Most constant volume systems either are part of another system—typically dual duct systems—or serve to provide precise air supply at a constant volume. Opportunities for conserving energy consumed by such systems include:
1. Determine the minimum amount of airflow which is satisfactory and reset the constant volume device accordingly.
2. Investigate the possibility of converting the system to variable (step controlled) constant volume operation by adding the necessary controls.

Single Zone System

Single zone systems consist of a mixing, conditioning and fan section. The conditioning section may have heating, cooling, humidifying or a combination of capabilities. Single zone systems can be factory assembled roof-top units or built up from individual components and may or may not have distributing duct work.

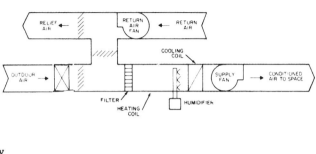

Terminal Reheat System

Reheat systems are modifications of single zone systems. Fixed cold temperature air is supplied by the central conditioning system and reheated in the terminal units when the space cooling load is less than maximum. The reheat is controlled by thermostats located in each conditioned space.

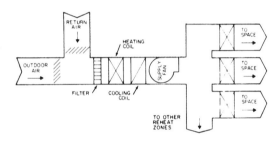

Multizone Systems

Multizone systems condition all air at the central system and mix heated and cooled air at the unit to satisfy various zone loads as sensed by zone thermostats. These systems may be packaged roof-top units or field-fabricated systems.

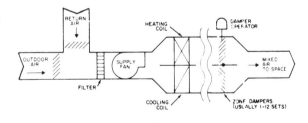

Dual Duct Systems

Dual duct systems are similar
to multizone systems
except heated and cooled
air is ducted to the conditioned spaces and mixed as
required in terminal
mixing boxes.

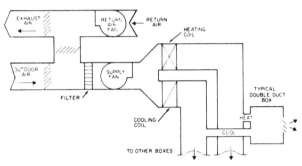

Variable Air Volume Systems

A variable air volume system
delivers a varying amount of air as
required by the conditioned spaces.
The volume control may be by
fan inlet (vortex) damper,
discharge damper or fan speed
control. Terminal sections may be
single duct variable volume units
with or without reheat, controlled
by space thermostats.

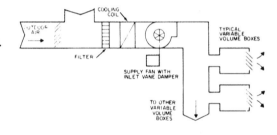

Induction Systems

Induction systems generally have units at the outside
perimeter of conditioned spaces. Conditioned primary
air is supplied to the units where it passes through
nozzles or jets and by induction draws room air
through the induction unit coil. Room temperature
control is accomplished by modulating water flow
through the unit coil.

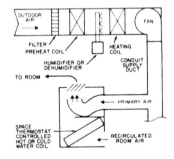

Fan Coil Units

A fan coil unit consists of a cabinet with heating and/or cooling coil,
motor and fan and a filter. The unit may be floor or ceiling mounted
and uses 100% return air to condition a space.

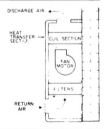

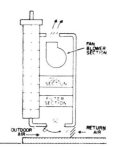

Unit Ventilator

A unit ventilator consists of a cabinet with heating and/or cooling coil, motor and fan, a filter and a return air—outside air mixing section. The unit may be floor or ceiling mounted and uses return and outside air as required by the space.

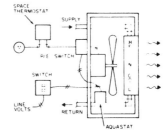

Unit Heater

Unit heaters have a fan and heating coil which may be electric, hot water or steam. They do not have distribution duct work but generally use adjustable air distribution vanes. Unit heaters may be mounted overhead for heating open areas or enclosed in cabinets for heating corridors and vestibules.

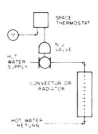

Perimeter Radiation

Perimeter radiation consists of electric resistance heaters or hot water radiators usually within an enclosure but without a fan. They are generally used around the conditioned perimeter of a building in conjunction with other interior systems to overcome heat losses through walls and windows.

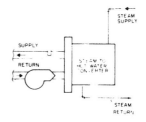

Hot Water Converters

A hot water converter is a heat exchanger that uses steam or hot water to raise the temperature of heating system water. Converters consist of a shell and tubes with the water to be heated circulated through the tubes and the heating steam or hot water circulated in the shell around the tubes.

Source: "Energy Conservation with Comfort," Honeywell

Induction System—Induction systems comprise an air handling unit which supplies heated or cooled primary air at high pressure to induction units located on the outside walls of each space served. The high pressure primary air is discharged within the unit through nozzles inducing room air through a cooling or heating coil in the unit. The resultant mixture of primary air and induced air is discharged to the room at a temperature dependent upon the cooling and heating load of the space. Methods for conserving energy consumed by this system include:

1. Set primary air volume to original design values when adjusting and balancing work is performed.
2. Inspect nozzles. If metal nozzles, common on most older models, are installed, determine if the orifices have become enlarged from years of cleaning. If so, chances are that the volume/pressure relationship of the system has been altered. As a result, the present volume of primary air and the appropriate nozzle pressure required must be determined. Once done, rebalance the primary air system to the new nozzle pressures and adjust individual induction units to maintain airflow temperature. Also, inspect nozzles for cleanliness. Clogged nozzles provide higher resistance to air flow, thus wasting energy.
3. Set induction heating and cooling schedules to minimally acceptable levels.
4. Reduce secondary water temperatures during the heating season.
5. Reduce secondary water flow during maximum heating and cooling periods by pump throttling or, for dual-pump systems, by operating one pump only.
6. Consider manual setting of primary air temperature for heating, instead of automatic reset by outdoor or solar controllers.

Dual-duct System—The central unit of a dual-duct system provides both heated and cooled air, each at a constant temperature. Each space is served by two ducts, one carrying hot air, the other carrying cold air. The ducts feed into a mixing box in each space which, by means of dampers, mixes the hot and cold air to achieve that air temperature required to meet load conditions in the space or zone involved. Methods for improving the energy consumption characteristics of this system include:

1. Lower hot deck temperature and raise cold deck temperature.
2. Reduce air flow to all boxes to minimally acceptable level.
3. When no cooling loads are present, close off cold ducts and shut down the cooling system. Reset hot deck according to heating loads

and operate as a single duct system. When no heating loads are present, follow the same procedure for heating ducts and hot deck. It should be noted that operating a dual-duct system as a single-duct system reduces air flow, resulting in increased energy savings through lowered fan speed requirements.

Single Zone System—A zone is an area or group of areas in a building which experiences similar amounts of heat gain and heat loss. A single zone system is one which provides heating and cooling to one zone controlled by the zone thermostat. The unit may be installed within or remote from the space it serves, either with or without air distribution ductwork.

1. In some systems air volume may be reduced to the minimum required, therefore reducing fan power input requirements. Fan brake horsepower varies directly with the cube of air volume. Thus, for example, a 10% reduction in air volume will permit a reduction in fan power input by about 27% of original. This modification will limit the degree to which the zone serviced can be heated or cooled as compared to current capabilities.

2. Raise supply air temperatures during the cooling season and reduce them during the heating season. This procedure reduces the amount of heating and cooling which a system must provide, but, as with air volume reduction, limits heating and cooling capabilities.

3. Use the cooling coil for both heating and cooling by modifying the piping. This will enable removal of the heating coil, which provides energy savings in two ways. First, air flow resistance of the entire system is reduced so that air volume requirements can be met by lowered fan speeds. Second, system heat losses are reduced because the surface area of cooling coils is much larger than that of heating coils, thus enabling lower water temperature requirements. Heating coil removal is not recommended if humidity control is critical in the zone serviced and alternative humidity control measures will not suffice.

Multizone System—A multizone system heats and cools several zones—each with different load requirements—from a single, central unit. A thermostat in each zone controls dampers at the unit which mix the hot and cold air to meet the varying load requirements of the zone involved. Steps which can be taken to improve energy efficiency of multizone systems include:

1.   Reduce hot deck temperatures and increase cold deck temperatures. While this will lower energy consumption, it also will reduce the system's heating and cooling capabilities as compared to current capabilities.

2.   Consider installing demand reset controls which will regulate hot and cold deck temperatures according to demand. When properly installed, and with all hot deck or cold deck dampers partially closed, the control will reduce hot and raise cold deck temperature progressively until one or more zone dampers is fully open.

3.   Consider converting systems serving interior zones to variable volume. Conversion is performed by blocking off the hot deck, removing or disconnecting mixing dampers, and adding low pressure variable volume terminals and pressure bypass.

Terminal Reheat System—The terminal reheat system essentially is a modification of a single zone system which provides a high degree of temperature and humidity control. The central heating/cooling unit provides air at a given temperature to all zones served by the system. Secondary terminal heaters then reheat air to a temperature compatible with the load requirements of the specific space involved. Obviously, the high degree of control provided by this system requires an excessive amount of energy. Several methods for making the system more efficient include:

1.   Reduce air volume of single zone units.

2.   If close temperature and humidity control must be maintained for equipment purposes, lower water temperature and reduce flow to reheat coils. This still will permit control but will limit the system's heating capabilities somewhat.

3.   If close temperature and humidity control is not required, convert the system to variable volume by adding variable volume valves and eliminating terminal heaters.

THE ECONOMIZER CYCLE

The basic concept of the economizer cycle is to use outside air as the cooling source when it is cold enough. There are several parameters which should be evaluated in order to determine if an economizer cycle is justified. These include:

- Weather
- Building occupancy
- The zoning of the building
- The compatibility of the economizer with other systems
- The cost of the economizer

What Are the Costs of Using The Economizer Cycle

Outside air cooling is accomplished usually at the expense of an additional return air fan, economizer control equipment, and an additional burden on the humidification equipment. Therefore, economizer cycles must be carefully evaluated based on the specific details of the application.

Using outside air to cool a building can result in lower mechanical refrigeration cost whenever outdoor air has a lower total heat content (enthalpy) than the return air. This can be accomplished by an "integrated economizer" or enthalpy control. See Figure 7-11 for a comparison of controls.

Dry-bulb Economizer

Operation of the "integrated economizer" can be made automatic by providing (1) dampers capable of providing 100% outdoor air, and (2) local controls that sequence the chilled water or DX (direct expansion) coil and dampers so that during economizer operation, on a rise in discharge (or space) temperature, the outdoor damper opens first; then on a further rise, the cooling coil is turned "on."

Economizer operation is activated by outside air temperature, for example, 72°F DB*. If outside air is below 72°F, the above described economizer sequence occurs. Above 72°F, outside air cooling is not economical, and the outdoor air damper closes to its minimum position to satisfy ventilation requirements only.

Enthalpy Control

If an economizer system is equipped with enthalpy control, savings will accrue due to a more accurate changeover point. The load on a cooling coil for an air handling system is a function of the total *heat* of air entering the coil. Total heat is a function of *two* measurements, dry-bulb (DB) and relative humidity (RD) or dew point (DP). The enthalpy control measures both conditions (DB and RH) in the return air duct and outdoors.

---

*This varies according to location.

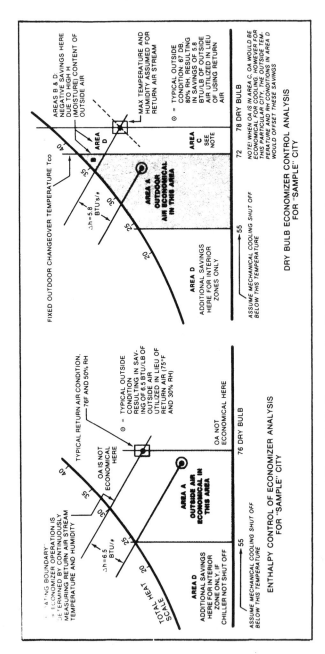

Figure 7-11. Comparative Economizer Controls

It then computes which air source would impose the lowest load on the cooling system. If outside air is the smallest load, the controller enables the economizer cycle. (Dry-bulb economizer control savings will be less than those shown for enthalpy control, except in dry climates.)

Enthalpy Control Savings Calculations

Savings are based on the assumption that the system previously had either (a) no 100% O.A. damper or (b) a fixed minimum outdoor air setting (whenever the fan operated) sufficient for ventilation purposes; it is also assumed that (c) minimum outdoor air has already been reduced, and the minimum damper opening will be at the *new value*.

Step 1. Determine minimum CFM of outdoor air to be used during occupied hours.

Step 2. Calculate annual savings.

$$A\frac{ft^3}{min.}\left(1-\frac{B\%}{100}\right)\cdot K\frac{10^6 Btu}{yr\ 1000\ CFM}\cdot \frac{operating\ hrs/wk}{50}\cdot J\frac{\$}{10^6 Btu}$$

$$= \$ \text{ SAVED PER YEAR} \hspace{4cm} Formula\ (7\text{-}25)$$

Where

A = air handling capacity $\left(\frac{ft^3}{min.}\right)$

B = present ventilation air (%)

J = cost of cooling $\left(\frac{\$}{10^6 Btu}\right)$

K = seasonal cooling savings $\left(\frac{10^6 Btu}{yr\ 1000\ CFM}\right)$

Formula 7-25 is used to calculate the savings resulting from enthalpy control of outdoor air. The calculated savings generally will be greater than the savings resulting from a dry-bulb economizer. To estimate dry-bulb economizer savings, multiply the enthalpy savings by .93.

8

HVAC Equipment

PERFORMANCE RATIOS

Measuring Efficiency by Using the Coefficient of Performance

The coefficient of performance (COP) is the basic parameter used to compare the performance of refrigeration and heating systems. COP for cooling and heating applications is defined as follows:

$$\text{COP (Cooling)} = \frac{\text{Rate of Net Heat Removal}}{\text{Total Energy Input}} \qquad \textit{Formula (8-1)}$$

$$\textit{COP}\text{ (Heating, Heat Pump*)} = \frac{\text{Rate of Useful Heat Delivered*}}{\text{Total Energy Input}}$$
$$\textit{Formula (8-2)}$$

Measuring System Efficiency Using the Energy Efficiency Ratio

The energy efficiency ratio (EER) is used primarily for air-conditioning systems and is defined by Formula 8-3.

$$EER = \frac{COP\,(3412)}{1000}\ \text{Btu}/\text{watt-hr} \qquad \textit{Formula (8-3)}$$

THE HEAT PUMP

The heat pump has gained wide attention due to its high potential *COP*. The heat pump in its simplest form can be thought of as a window air

---

*For heat pump applications, exclude supplemental heating.

conditioner. During the summer, the air on the room side is cooled while air is heated on the outside-air side. If the window air conditioner is turned around in the winter, some heat will be pumped into the room. Instead of switching the air conditioner around, a cycle reversing valve is used to switch functions. This valve switches the function of the evaporator and condenser, and refrigeration flow is reversed through the device. Thus, *the heat pump is heat recovery through a refrigeration cycle.* Heat is removed from one space and placed in another. In Chapter 7, it was seen that the direction of heat flow is from hot to cold. Basically, energy or pumping power is needed to make heat flow "up hill." The mechanical refrigeration compressor "pumps" absorbed heat to a higher level for heat rejection. The refrigerant gas is compressed to a higher temperature level so that the heat absorbed by it, during the evaporation or cooling process, is rejected in the condensing or heating process. Thus, the heat pump provides cooling in the summer and heating in the winter. The source of heat for the heat pump can be from one of three elements: air, water or the ground.

Air to Air Heat Pumps

Heat exists in air down to 460°F below zero. Using outside air as a heat source has its limitations, since the efficiency of a heat pump drops off as the outside air level drops below 55°F. This is because the heat is more dispersed at lower temperatures, or more difficult to capture. Thus, heat pumps are generally sized on cooling load capacities. Supplemental heat is added to compensate for declining capacity of the heat pump. This approach allows for a realistic first cost and an economical operating cost.

An average of two to three times as much heat can be moved for each kW input compared to that produced by use of straight resistance heating. Heat pumps can have a COP of much more than 2 or 3 in industrial processes, depending on the temperature involved. Commercially available heat pumps range in size from 2 to 3 tons for residences up to 40 tons for commercial and industrial users. Figure 8-1 illustrates a simple scheme for determining the supplemental heat required when using an air-to-air heat pump.

Hydronic Heat Pump

The hydronic heat pump is similar to the air to air unit, except the heat exchange is between water and refrigerant instead of air to refrigerant, as illustrated in Figure 8-2. Depending on the position of the reversing valve, the air heat exchanger either cools or heats room air. In

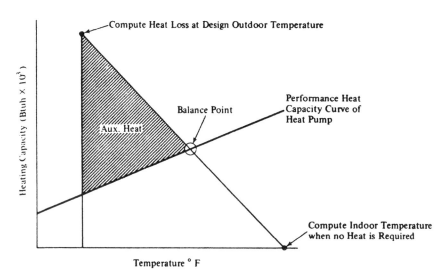

Figure 8-1. Determining Balance Point of Air to Air Heat Pump

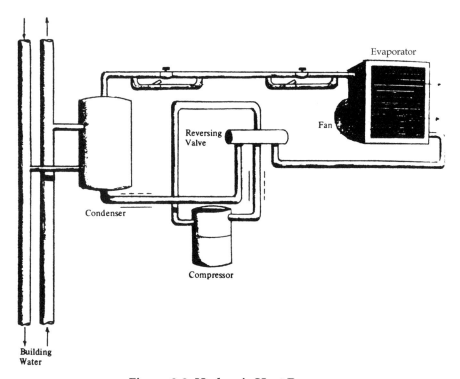

Figure 8-2. Hydronic Heat Pump

the case of cooling, heat is rejected through the water-cooled condenser to the building water. In the case of heating, the reversing valve causes the water to refrigerant heat exchanger to become an evaporator. Heat is then absorbed from the water and discharged to the room air.

Imagine several hydronic heat pumps connected to the same building water supply. In this arrangement, it is conceivable that while one unit is providing cool air to one zone, another is providing hot air to another zone; the first heat pump is providing the heat source for the second unit, which is heating the room. This illustrates the principle of energy conservation. In practice, the heat rejected by the cooling units does not equal the heat absorbed. An additional evaporative cooler is added to the system to help balance the loads. A better heat source would be the water from wells, lakes or rivers which is thought of as a constant heat source. Care should be taken to insure that a heat pump connected to such a heat source does not violate ecological interests.

Liquid Chiller

A liquid chilling unit (mechanical refrigeration compressor) cools water, brine, or any other refrigeration liquid, for air-conditioning or refrigeration purposes. The basic components include a compressor, liquid cooler, condenser, the compressor drive, and auxiliary components. A simple liquid chiller is illustrated in Figure 8-3. The type of chiller usually depends on the capacity required. For example, small units below 80 tons are usually reciprocating, while units above 350 tons are usually centrifugal.

Factors which affect the power usage of liquid chillers are the percent load and the temperature of the condensing water. A *reduced condenser water temperature saves energy*. In Figure 8-4, it can be seen that by reducing the original condenser water temperature by 10 degrees, the power consumption of the chiller is reduced. Likewise, a chiller operating under part load consumes less power. The "ideal" coefficient of performance (COP) is used to relate the measure of cooling effectiveness. Approximately 0.8 kW is required per ton of refrigeration (0.8 kW is power consumption at full load, based on typical manufacturer's data).

$$COP = \frac{1 \text{ ton} \times 12{,}000 \text{ Btu/ton}}{0.8 \text{ kW} \times 3412 \text{ Btu/kW}} = 4.4$$

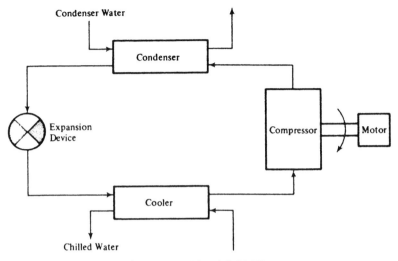

Figure 8-3. Liquid Chiller

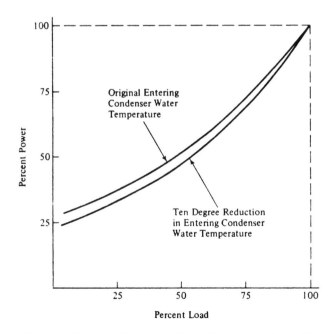

Figure 8-4. Typical Power Consumption Curve for Centrifugal Liquid Chiller

Chillers in Series and in Parallel

Multiple chillers are used to improve reliability, offer standby capacity, reduce inrush currents and decrease power costs at partial loads. Figure 8-5 shows two common arrangements for chiller staging namely, chillers in parallel and chillers in series.

In the parallel chiller arrangement, liquid to be chilled is divided among the liquid chillers and the streams are combined after chilling. Under part load conditions, one unit must provide colder than designed chilled liquid so that when the streams combine, including the one from the off chiller, the supply temperature is provided. The parallel chillers have a lower first cost than the series chillers counterparts but usually consume more power.

In the series arrangement, a constant volume of flow of chilled water passes through the machines, producing better temperature control and better efficiency under part load operation; thus, the upstream chiller requires less kW input per ton output. The waste of energy during the mixing aspect of the parallel chiller operation is avoided. The series chillers, in general, require higher pumping costs. The energy conservation engineer should evaluate the best arrangement, based on load required and the partial loading conditions.

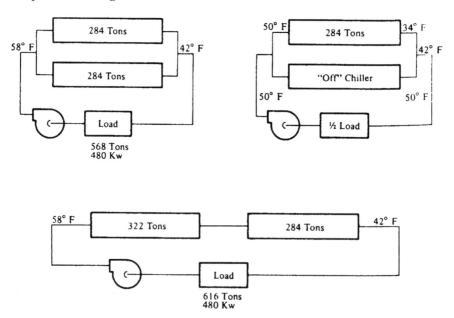

Figure 8-5. Multiple Chiller Arrangements

The Absorption Refrigeration Unit

Any refrigeration system uses external energy to "pump" heat from a low temperature level to a high temperature. Mechanical refrigeration compressors pump absorbed heat to a higher temperature level for heat rejection. Similarly, absorption refrigeration changes the energy level of the refrigerant (water) by using lithium bromide to alternately absorb it at a low temperature level and reject it at a high level by means of a concentration-dilution cycle.

The single-stage absorption refrigeration unit uses 10 to 12 psig steam as the driving force. Whenever users can be found for low pressure steam, energy savings will be realized. A second aspect for using absorption chillers is that they are compatible for use with solar collector systems. Several manufacturers offer absorption refrigeration equipment which uses high temperature water (160°-200°F) as the driving force.

A typical schematic for a single-stage absorption unit is illustrated in Figure 8-6. The basic components of the system are the evaporator, absorber, concentrator and condenser. These components can be grouped in a single or double shell. Figure 8-6 represents a single-stage arrangement.

Evaporator—Refrigerant is sprayed over the top of the tube bundle to provide for a high rate of transfer between water in the tubes and the re-

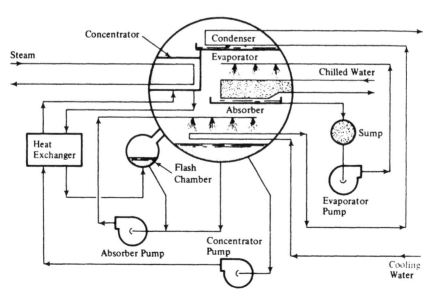

Figure 8-6. One-Shell Lithium Bromide Cycle Water Chiller
(Source: Trane Air Conditioning Manual)

frigerant on the outside of the tubes.

Absorber—The refrigerant vapor produced in the evaporator migrates to the bottom half of the shell where it is absorbed by a lithium bromide solution. Lithium bromide is basically a salt solution which exerts a strong attractive force on the molecules of refrigerant (water) vapor. The lithium bromide is sprayed into the absorber to speed up the condensing process. The mixture of lithium bromide and the refrigerant vapor collects in the bottom of the shell; this mixture is referred to as the dilute solution.

Concentrator—The dilute solution is then pumped through a heat exchanger where it is preheated by a hot solution leaving the concentrator. The heat exchanger improves the efficiency of the cycle by reducing the amount of steam or hot water required to heat the dilute solution in the concentrator. The dilute solution enters the upper shell containing the concentrator. Steam coils supply heat to boil away the refrigerant from the solution. The absorbent left in the bottom of the concentrator has a higher percentage of absorbent than it does refrigerant; thus it is referred to as concentrated.

Condenser—The refrigerant vapor boiling from the solution in the concentrator flows upward to the condenser and is condensed. The condensed refrigerant vapor drops to the bottom of the condenser and from there flows to the evaporator through a regulating orifice. This completes the refrigerant cycle.

The single-stage absorption unit consumes approximately 18.7 pounds of steam per ton of capacity (steam consumption at full load based on typical manufacturer's data). For a single-state absorption unit

$$COP = \frac{1 \text{ ton} \times 12{,}000 \text{ Btu/ton}}{18.7 \text{ lb} \times 955 \text{ Btu/lb}} = 0.67$$

The single-stage absorption unit is not as efficient as the mechanical chiller. It is usually justified based on availability of low pressure steam, equipment considerations or use with solar collector systems.

Example Problem 8-1

Compute the energy wasted when 15 psig steam is condensed prior to its return to the power plant. Comment on using the 15 psig steam directly for refrigeration.

Answer

From Steam Table 15-14 for 30 psia steam, hfg is 945 Btu per pound of steam; thus, 945 Btu per pound of steam is wasted. In this case where *excess low pressure* steam cannot be used, absorption units should be considered in place of their electrical-mechanical refrigeration counterparts.

Example Problem 8-2

2000 lb/hr of 15 psig steam is being wasted. Calculate the yearly (8000 hr/yr) energy savings if a portion of the centrifugal refrigeration system is replaced with single-stage absorption. Assume 20 kW additional energy is required for the pumping and cooling tower cost associated with the single-stage absorption unit. Energy rate is $.09 kWh and the absorption unit consumes 18.7 lb of steam per ton of capacity.

The centrifugal chiller system consumes 0.8 kWh per ton of refrigeration.

Answer

Tons of mechanical chiller capacity replaced = 2000/18.7 = 106.95 tons. Yearly energy savings = 2000/18.7 × 8000 × 0.8 × $.09 = $61,602.

Two-Stage Absorption Unit

The two-stage absorption refrigeration unit (Figure 8-7) uses steam at 125 to 150 psig as the driving force. In situations where excess medium pressure steam exists, this unit is extremely desirable. The unit is similar to the single-stage absorption unit. The two-stage absorption unit operates as follows:

Medium pressure steam is introduced into the first-stage concentrator. This provides the heat required to boil out refrigerant from the dilute solution of water and lithium bromide salt. The liberated refrigerant vapor passes into the tubes of the second-stage concentrator, where its temperature is utilized to again boil a lithium bromide solution, which in turn further concentrates the solution and liberates additional refrigerant. In effect, the concentrator frees an increased amount of refrigerant from solution with each unit of input energy.

The condensing refrigerant in the second-stage concentrator is piped directly into the condenser section. The effect of this is to reduce the cooling water load. A reduced cooling water load decreases the size of the cooling tower which is used to cool the water. The remaining portions of the system are basically the same as the single-stage unit.

The two-stage absorption unit consumes approximately 12.2 pounds of steam per ton of capacity; thus, it is more efficient than its single-stage counterpart. The associated COP is

$$COP = \frac{1 \text{ ton} \times 12{,}000 \text{ Btu/ton}}{12.2 \text{ lb} \times 860 \text{ Btu/lb -}} = 1.14$$

Either type of absorption unit can be used in conjunction with centrifugal chillers when it is desirable to reduce the peak electrical demand of the plant or to provide for a solar collector addition at a later date.

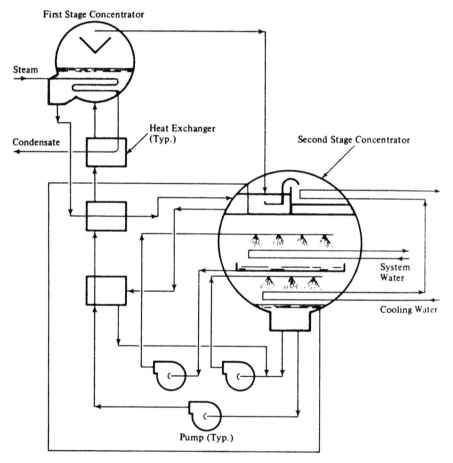

Figure 8-7. Two-stage Absorption Unit

9

Cogeneration: Theory and Practice

Because of its enormous potential, it is important to understand and apply cogeneration theory. In the overall context of energy management theory, cogeneration is just another form of the conservation process. However, because of its potential for practical application to new or existing systems, it has carved a niche that may be second to no other conservation technology.

This chapter is dedicated to development of a sound basis of current theory and practice of cogeneration technology. It is the blend of theory and practice, or praxis of cogeneration, that will form the basis of the most workable conservation technology currently available.

DEFINITION OF "COGENERATION"

Cogeneration is the sequential production of thermal and electric energy from a single fuel source. In the cogeneration process, heat is recovered that would normally be lost in the production of one form of energy. That heat is then used to generate the second form of energy. For example, take a situation in which an engine drives a generator that produces electricity: with cogeneration, heat would be recovered from the engine exhaust and/or coolant, and that heat would be used to produce, say, hot water.

Making use of waste heat is what differentiates cogeneration facilities from central station electric power generation. The overall fuel utilization efficiency of cogeneration plants is typically 70-80% versus 35-40% for utility power plants.

This means that in cogeneration systems, rather than using energy in the fuel for a single function, as typically occurs, the available energy is cascaded through at least two useful cycles.

To put it in simpler terms: cogeneration is a very efficient method of making use of all the available energy expended during any process generating electricity (or shaft horsepower) and then utilizing the waste heat.

A more subjective definition of cogeneration calls upon current practical applications of power generation and process needs. Nowhere more than in the United States is an overall system efficiency of only 30% tolerated as "standard design." In the name of limited initial capital expenditure, all of the waste heat from most processes is rejected to the atmosphere.

In short, present design practices dictate that of the useful energy in one gallon of fuel, only 30% of that fuel is put to useful work. The remaining 70% is rejected randomly.

If one gallon of fuel goes into a process, the designer may ask, "How much of that raw energy can I make use of within the constraints of the overall process?"

In this way, cogeneration may be taken as a way to use a maximum amount of available energy from any raw fuel process. Thus, cogeneration maybe thought of as *just good design.*

COMPONENTS OF A
COGENERATION SYSTEM

The basic components of any cogeneration plant are:

- A prime mover
- A generator
- A waste heat recovery system
- Operating control systems

The prime mover is an engine or turbine which, through combustion, produces mechanical energy. The generator converts the mechanical energy to electrical energy. The waste heat recovery system is one or more heat exchangers that capture exhaust heat or engine coolant heat and convert that heat to a useful form. The operating control systems insure that the individual system components function together.

The prime mover is the heart of the cogeneration system. The three basic types are steam turbines, combustion gas turbines and internal combustion engines. Each has advantages and disadvantages, as explained below.

Steam Turbines

Steam turbine systems consist of a boiler and turbine. The boiler can be fired by a variety of fuels such as oil, natural gas, coal and wood. In many installations, industrial by-products or municipal wastes are used as fuel. Steam turbine cogeneration plants produce high-pressure steam that is expanded through a turbine to produce mechanical energy which, in turn, drives a device such as an electric generator. Thermal energy is recovered in several different ways, which are discussed in Chapter 5. Steam turbine systems generally have a high fuel utilization efficiency.

Combustion Gas Turbines

Combustion gas turbine systems are made up of one or more gas turbines and a waste heat recovery unit. These systems are fueled by natural gas or light petroleum products. The products of combustion drive a turbine which generates mechanical energy. The mechanical energy can be used directly or converted to electricity with a generator. The hot exhaust gases of the gas turbine can be used directly in process heating applications, or they can be used indirectly, with a heat exchanger, to produce process steam or hot water.

A variation on the combustion gas turbine system is one that uses high-pressure steam to drive a steam turbine in conjunction with the cogeneration process. This is referred to as a combined cycle.

Internal Combustion Engines

Internal combustion engine systems utilize one or more reciprocating engines together with a waste heat recovery device. These are fueled by natural gas or distillate oils. Electric power is produced by a generator which is driven by the engine shaft. Thermal energy can be recovered from either exhaust gases or engine coolant. The engine exhaust gases can be used for process heating or to generate low-pressure steam. Waste heat is recovered from the engine cooling jacket in the form of hot water.

Cogeneration plants that use the internal combustion engine generate the greatest amounts of electricity for the amount of heat produced. Of the three types of prime movers, however, the fuel utilization efficiency is the lowest, and the maximum steam pressure that can be produced is limited.

AN OVERVIEW OF
COGENERATION THEORY

As discussed in the introduction, and as may be seen from Figure 9-1, standard design practices make use of, at best, 30% of available energy from the raw fuel source (gas, oil, coal).

Of the remaining 70% of the available energy, approximately 30% of the heat is rejected to the atmosphere through a condenser (or similar) process. An additional 30% of the energy is lost directly to the atmosphere through the stack, and finally, approximately 7% of the available energy is radiated to the atmosphere because of the high relative temperature of the process system.

With heat recovery, however, potential useful application of available energy more than doubles. Although in a "low quality" form, *all* of the condenser-related heat may be used, and 40% of the stack heat may be recovered. This optimized process is depicted (in theory) as Figure 9-2.

Thus, it may be seen that effective use of all available energy may more than double the "worth" of the raw fuel. System efficiency is increased from 30% to 75%.

This higher efficiency allows the designer to use low grade energy for various cogeneration cycles.

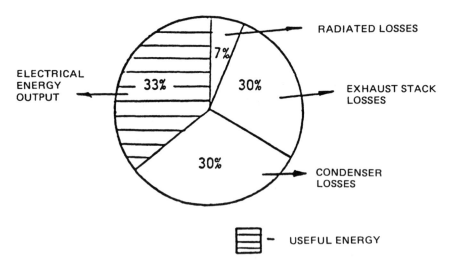

Figure 9-1. Energy Balance without Heat Recovery

Example Problem 9-1

A cogeneration system vendor recommends the installation of a 20-megawatt cogeneration system for a college campus. Determine the approximate *range* of useful *thermal* energy. Use the energy balance of Figure 9-2.

Analysis

Step 1: "Range" is defined by the best and worst operating *times* of the installed system:

At best, system will operate 365 days per year, 24 hours per day.

At worst, system will operate 5 days per week, 10 hours per day.

Step 2: Perform heat balance. See diagram.

$$Q = E_{\text{electricity}} + E_{\text{condenser}} + E_{\text{stack}} + E_{\text{radiated}} \qquad \textit{Formula (9-1)}$$

Step 3: Calculate *available thermal* energy.

From Figure 9-2, available energy equals condenser energy and 40% stack energy.

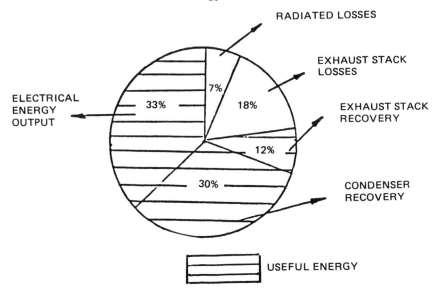

Figure 9-2. Energy Balance with Heat Recovery

$$E_{\text{available}} = E_{\text{condenser}} + E_{\text{stack}} \qquad\qquad \textit{Formula (9-2)}$$

$$E_{\text{condenser}} = .3Q \qquad\qquad \textit{Formula (9-3)}$$

$$E_{\text{stack}} = 0.12Q$$

$$E_{\text{available}} = 0.30Q + 0.12Q \qquad\qquad \textit{Formula (9-4)}$$

$$E_{\text{electrical}} = .33Q \qquad\qquad \textit{Formula (9-5)}$$

$$Q = \frac{E_{\text{electrical}}}{.33} \qquad\qquad \textit{Formula (9-6)}$$

By substitution $E_{\text{available}} = 0.3 \dfrac{E_{\text{electrical}}}{0.33} + 0.12 \dfrac{E_{\text{electrical}}}{0.33}$

$$E_{\text{available}} = 1.272 \text{ Electrical} \qquad\qquad \textit{Formula (9-7)}$$

Calculate available energy

$$Q = Q_1 \times t = K_1 \, E_{\text{available}} \times K_2 \times t$$

K_1 = 3413 Btu/kWh
K_2 = 1000 Kilowatts per megawatt
t = equipment hours of operation
Q_1 = 3413 Btu/kWh × (1.272 × 20 megawatt) × 1000 $\dfrac{\text{kW}}{\text{MWatt}}$

Q_1 = 86.82 × 10^6 Btu/hr

Worst Case: System operates
5 days/wk × 52 wks/yr × 10 hrs/day = 2600 hrs/yr

Best Case: System operates
365 days/yr × 24 hrs/day = 8670 hrs/yr

Thus range is

Q = 86.82 × 10^6 × 2600 = 225.7 × 10^9 Btu/yr
Q = 86.82 × 10^6 × 8760 = 760.5 × 10^9 Btu/yr

APPLICATION OF THE CO GENERATION CONSTANT

The *cogeneration constant* may be used as a fast check on any proposed cogeneration installation. Notice from the sample problem which follows, the ease with which a thermal vs. electrical comparison of end needs may be made.

Example Problem 9-2

A cogeneration system vendor recommends a 20 megawatt installation. Determine the approximate rate of *useful* thermal energy.

Analysis

$$E_{available} = E_{electrical} \times K_c \qquad \textit{Formula (9-8)}$$

E is the cogeneration system electrical rated capacity
K_c is the cogeneration constant
$E_{available}$ = 20 MW × 1.272
= 25.4 MW of *useful* heat
or
$$= 25.4 \text{MW} \times 1000 \ \frac{\text{kW}}{\text{MW}} \ \times \ \frac{3413 \text{ Btu}}{\text{kWh}} \ \times \ \frac{\text{Therm}}{100{,}000 \text{ Btu}}$$

$E_{available}$ = 866.9 therms/hour

APPLICABLE SYSTEMS

To ease the complication of matching power generation to load, and because of newly established laws, it is most advantageous that the generator operate in parallel with the utility grid which thereby "absorbs" all generated electricity. The requirement for "qualify facility (QF) status," and the consequent utility rate advantages which are available when "paralleling the grid," is that a significant portion of the thermal energy produced in the generation process must be recovered. Specifically, Formula 9-10 must be satisfied:

$$\frac{\text{Power Output} + 1/2 \text{ Useful Thermal Output}}{\text{Energy Input}} \geq 42.5\% \quad \text{(for any calendar year)}$$

$$\textit{Formula (9-9)}$$

Note that careful application of the *cogeneration constant* will generally assure that the qualifying facility status is met.

BASIC THERMODYNAMIC CYCLES

Bottoming and Topping Cycles

Cogeneration systems can be divided into "bottoming cycles" and "topping cycles."

Bottoming Cycles

In a bottoming cycle system, thermal energy is produced directly from the combustion of fuel. This energy usually takes the form of steam that supplies process heating loads. Waste heat from the process is recovered and used as an energy source to produce electric or mechanical power.

Bottoming cycle cogeneration systems are most commonly found in industrial plants that have equipment with high-temperature heat requirements such as steel reheat furnaces, clay and glass kilns and aluminum remelt furnaces. Some bottoming cycle plants operating in Georgia are Georgia Kraft Company, Brunswick Pulp and Paper Company and Burlington Industries.

Topping Cycles

Topping cycle cogeneration systems reverse the order of bottoming cycle systems: electricity or mechanical power is produced first; then heat is recovered to meet the thermal loads of the facility. Topping cycle systems are generally found in facilities which do not have extremely high process temperature requirements.

Figure 9-3 on the following page shows schematic examples of these two cogeneration operating cycles.

A sound understanding of basic cogeneration principles dictates that the energy manager should be familiar with two standard thermodynamic cycles. These cycles are

1.   Brayton cycle
2.   Rankine cycle

The Brayton cycle is the basic thermodynamic cycle for the simple gas turbine power plant. The Rankine cycle is the basic cycle for a vapor-

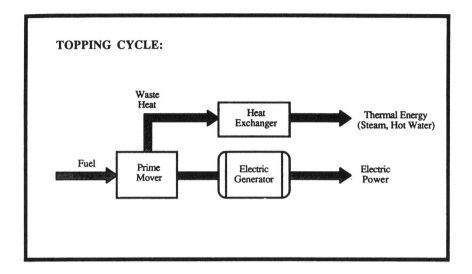

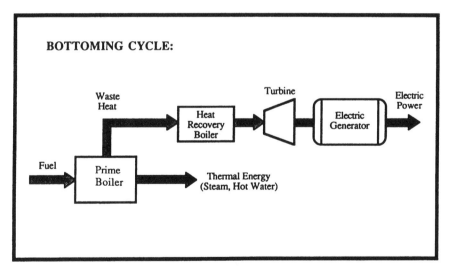

Figure 9-3. Cogeneration Operating Cycles

liquid system typical of steam power plants. An excellent theoretical discussion of these two cycles appears in Reference 10.

The Brayton Cycle

In the open Brayton cycle plant, energy input comes from the fuel that is injected into the combustion chamber. The gas/air mixture drives the turbine with high-temperature waste gases exiting to the atmosphere. (See Figure 9-4.)

The basic Brayton cycle, as applied to cogeneration, consists of a gas turbine, waste heat boiler and a process or "district" heating load. This cycle is a full-load cycle. At part loads, the efficiency of the gas turbine goes down dramatically. A simple process diagram is illustrated in Figure 9-5.

Because the heat rate of this arrangement is superior to that of all other arrangements at full load, this simple, standard Brayton cycle merits consideration under all circumstances. Note that to complete the loop in an efficient manner, a deaerator and feedwater pump are added.

The Rankine Cycle

The Rankine cycle is illustrated in the simplified process diagram, Figure 9-6. Note that this is the standard boiler/steam turbine arrangement found in many power plants and central facility plants throughout the world.

The Rankine cycle, or steam turbine, provides a real-world outlet for waste heat recovered from any process or generation situation. Hence, it is the steam turbine which is generally referred to as the topping cycle.

Combined Cycles

Of major interest and importance for the serious central plant designer is the combined cycle. This cycle forms a hybrid which includes the Brayton cycle on the "bottoming" portion and a standard Rankine cycle on the "topping" portion of the combination. A process diagram with standard components is illustrated in Figure 9-7.

The combined cycle, then, greatly approximates the cogeneration Brayton cycle but makes use of a knowledge of the plant requirements and

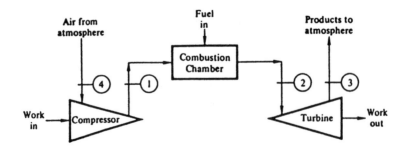

Figure 9-4. Open Gas-Turbine Power Plant (Brayton Cycle)

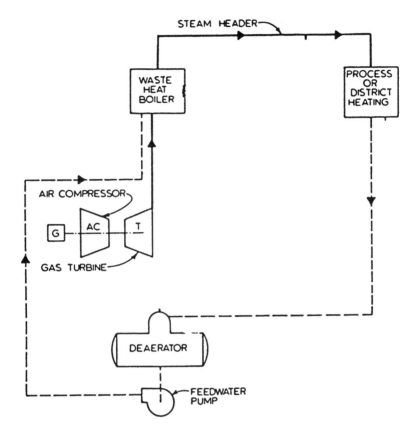

Figure 9-5. Process Diagram—Brayton Cycle

an understanding of Rankine cycle theory. Note also that the ideal mix of power delivered from the Brayton and Rankine portions of the combined cycle is 70% and 30%, respectively.

Even within the seemingly limited set of situations defined as "combined cycle," many variations and options become available. These options are as much dependent on any local plant requirements and conditions as they are on available equipment.

Some examples of combined cycle variations are:

1. Gas turbine exhaust used to produce 15 psi steam for Rankine cycle turbine with no additional fuel burned. This situation is shown in Figure 9-7.

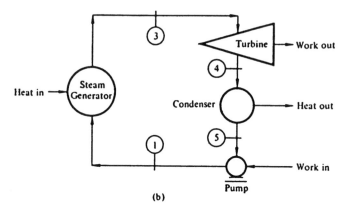

Figure 9-6. Simple Rankine Cycle

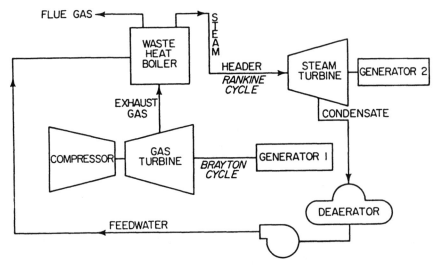

Figure 9-7. Combined Cycle Operation

2.   Gas turbine exhaust fired in the duct with additional fuel. This pro-
      vides a much greater amount of produced power with a correspond-
      ingly greater amount of fuel consumption. This situation generally
      occurs with a steam turbine pressure range of 900-1259 psig.

3.   Gas turbine exhaust fired directly and used directly as combustion
      air for a conventional power boiler. Note here that the boiler pres-
      sure range may vary between 200 and 2600 psig. (See Figure 9-8.)

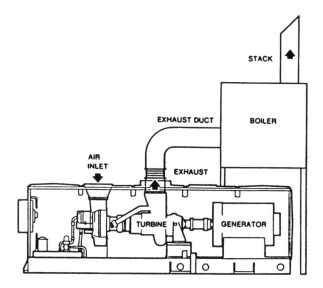

Figure 9-8. Gas Turbine Used As Combustion Air

One other note to keep in mind is that in any combined cycle case, the primary or secondary turbine may supply direct mechanical energy to a refrigerant compressor. As discussed, the variations are endless. However, a thorough understanding of the end process generally will result in a final, and best, cogeneration system selection.

DETAILED FEASIBILITY EVALUATION*

This section will introduce the parameters affecting the evaluation, selection, sizing and operation of a cogeneration plant and shows a means of evaluating those parameters where it counts—on the bottom line.

This section will enable you to answer two basic questions:

1.   Is cogeneration technically feasible for us, given our situation?

2.   If it is technically feasible, is it also economically feasible, considering estimated costs, energy savings, current utility rates and regulatory conditions in Georgia?

Georgia Cogeneration Handbook, Governor's Office of Energy Resources, August 1988.

Keep in mind that this chapter is not designed to take the place of a full-scale feasibility study. It is designed to help you and your engineering staff decide—in-house and at minimal expense—whether such a study is warranted. You will reach one of the following conclusions:

• No, cogeneration is not feasible in our case.
• Yes, cogeneration looks promising and a feasibility study is warranted.
• Maybe, but I have some questions.

If your conclusion is "no," you will have saved yourself the expense of a feasibility study. If your conclusion is "yes," you will be assured that the expense of employing outside consultants to perform the study is justified by the potential benefits. If your conclusion is "maybe," you may wish to discuss your situation with qualified engineers before deciding whether to proceed.

Utility Data Analysis
The first step in the feasibility evaluation is the gathering and analysis of utility data. These data are necessary to determine the limits of the thermal and electric energy consumed at your facility. They are also necessary for determining the timing of maximum use of each of these energy sources. From this information, you will also determine the thermal and electric load factors for your facility.

Load Factor Defined
A load factor is defined as the average energy consumption rate for a facility divided by the peak energy consumption rate over a given period of time, and it is an important simplification of energy use data. These values are usually expressed as either a decimal number or a percent. To calculate a load factor you need two pieces of data: the total energy consumption for a given time period and the maximum energy demand observed during that time period.

In the electric power industry, load factors serve as a measure of the utilization of generation equipment. A utility load factor of 0.90, or 90%, indicates very good utilization of power generation facilities, while a load factor of 0.30 or 30% indicates poor utilization. The lower the load factor, the larger and more expensive your generating equipment must be merely to meet peak power demands for short time periods.

The thermal and electric load factors of your facility are of great importance in sizing a cogeneration plant. In fact, they are even more important to a cogeneration plant than to an electric utility because the cogeneration plant does not have the advantage of an electric power grid to diversify the variations in energy use.

A high thermal or electrical load factor generally indicates that a cogeneration plant would be utilized a major portion of the time and therefore would provide a favorable return on your investment. An ideal situation exists when both load factors are high, indicating that a properly sized cogeneration plant would efficiently utilize most of its output. If both the thermal and electrical load factors are small, you may not be a practical candidate for cogeneration.

In the analysis of your electric and thermal energy consumption data, you will calculate the load factors for your facility. In this evaluation, we will concentrate on annual load factors. It is important to note, however, that monthly and daily load factors may be important in your final analysis.

Electric Energy Consumption Analysis

In order to obtain the necessary energy-use information, you must refer to your monthly electric utility bills for one complete year. From those bills, calculate the following information:

- Determine your annual kWh consumption by tabulating and adding the monthly kWh consumption.

- Tabulate the actual kW demand metered for each month. Be sure to use the actual metered demand and not the billed demand, as the billed demand may be based on time of year and a percent ratchet of the previous year's peak demand.

- Identify the maximum monthly kW demand value as the annual peak kW.

- Determine the annual average kW demand by dividing the annual kWh consumption by 8,760 hours per year.

$$\text{Annual Average kW} = \frac{\text{Annual kWh}}{8,760 \text{ hours}/\text{year}}$$

- Determine the annual electric load factor by dividing the annual average kW demand by the annual peak kW demand.

$$\text{Annual Electric Load Factor} = \frac{\text{Annual Average kW}}{\text{Annual Peak kW}}$$

Thermal Energy Consumption Analysis

If your boiler plant has an accurate steam or Btu metering system, the following data can be gathered directly from your metered output data. If you do not have this degree of metering, your boiler plant fuel bills for one complete year will be required to obtain the necessary data. Monthly natural gas bills can be analyzed with the same method used for analyzing electrical consumption. If you use fuel oil, propane or coal, make sure that accurate measurements of the reserve supply were taken at the beginning and end of the year.

- Determine your annual fuel consumption. From Figure 9-9, select the Btu value per unit of fuel for the type of fuel you use. To obtain the annual fuel Btu Input, multiply the annual fuel consumption by this value.

- Determine your boiler's fuel-to-steam-efficiency. An estimated fuel-to-steam efficiency of 78% may be used for a well-maintained boiler plant that operates only enough boilers to keep them well loaded. A poorly maintained boiler plant with some leaks, missing insulation, and oversized boilers that cycle frequently may have a fuel-to-steam-efficiency of 60% or lower.

- Determine your annual Btu output by multiplying your annual fuel Btu input by your fuel-to-steam-efficiency.

$$\text{Annual Btu Output} = \frac{\text{Annual Fuel}}{\text{Btu Input}} \times \frac{\text{Fuel-to-Steam}}{\text{Efficiency}}$$

- Determine the annual average Btu/hour demand by dividing the annual Btu output by 8,760 hours per year.

$$\text{Annual Average Btu/Hour Demand} = \frac{\text{Annual Btu Output}}{8,760 \text{ hours/year}}$$

Fuel Type	HHV Higher Heating Value (Approximate)	LHV/HHV Lower/Higher Heating Value (Approximate)
Natural Gas (Dry)	1,000	0.90
Butane	3,200	0.92
Propane	2,500	0.92
Sewage Gas	300-600	0.90
Landfill Gas	300-600	0.90
No. 2 Oil	139,000 Btu/gal	0.93
No. 6 Oil	154,000 Btu/gal	0.96
Coal, Bituminous	14,100 Btu/lb	

Figure 9-9. Typical Fuel Caloric Values (Btu/CF)

If you have metered steam production data, select the maximum value for Btu/hour output that occurred last year as the annual peak Btu/hour demand. If these data are not available, estimate your annual peak Btu/hour demand as a percentage of the maximum possible output of your boiler plant.

NOTE: Boilers that operate at approximately 150 psig and below use the terms "lbs per hour" and "mbh output" interchangeably to represent 1,000 Btu/hour. We will use the unit "mbh" to simplify notation when referring to 1,000 Btu/hour units.

- Determine the annual thermal load factor by dividing the annual average Btu/hour demand by the annual peak Btu/hour demand.

$$\frac{\text{Annual Thermal}}{\text{Load Factor}} = \frac{\text{Annual Average Btu/hour Demand}}{\text{Annual Peak Btu/hour Demand}}$$

If your annual minimum Btu/hour demand is extremely low or your boiler plant actually shuts down for several months during the summer, your annual thermal load factor is not an accurate indicator of your cogeneration potential. It is probably too large, and a closer examination of the number of hours at the annual minimum Btu/hour demand may be required to assess feasibility. This can be done using monthly and daily load profiles.

If your annual minimum Btu/hour demand is zero for a large number of hours per year (for example, if your boiler plant shuts down during the summer months), you may not be a good candidate for cogeneration.

Thermal/Electric Load Ratio

For cogeneration to be feasible, the demands for thermal and electric energy must overlap much of the time. Therefore, once the thermal and electric demands of your facility are known, you must determine the ratio of heat demand to electric demand that may be expected to occur together. This is done by using the thermal/electric (T/E) load ratio. The T/E load ratio is defined as the quantity of heat energy that is coincident with a quantity of electrical energy. In making these calculations, you will attempt to find an optimum match between your facility's T/E load where

$$\text{Thermal/Electric Load Ratio} \quad = \quad \frac{\text{Thermal Demand}}{\text{kW Demand}}$$

If the thermal and electric load factors calculated earlier are high, there is a good possibility that your facility's demand for thermal energy occurs at about the same time as your demand for electric energy. In that case, your facility's annual average thermal/electric load ratio is approximately equal to the annual average Btu/hour demand divided by the annual average kW demand. For convenience, we will use the term "mbh" to represent 1,000 Btu/hr.

$$\frac{\text{Annual Average Thermal/}}{\text{Electric Load Ratio}} \quad = \quad \frac{\text{Annual Average mbh Demand}}{\text{Annual Average kW Demand}}$$

If either the thermal or the electric load factor is small, then a worst-case assumption should be made. This assumption is called the minimum demand thermal/electric load ratio and is calculated as follows:

$$\frac{\text{Minimum Demand Thermal/}}{\text{Electric Load Ratio}} \quad = \quad \frac{\text{Annual Minimum mbh Demand}}{\text{Annual Minimum kW Demand}}$$

The use of either of the thermal/electric load ratios above must be tempered with the knowledge that these are only approximations of the load correlations and are useful only for a preliminary evaluation. Most facilities will require a more detailed examination of thermal and electric load profiles to obtain the number of hours per year that loads overlap and can be served with a cogeneration plant.

Generally, cogeneration opportunities are good for facilities with T/E load ratios above 5 and are best for ratios above 10, with average annual Btu/hour demand above 10,000,000 Btu/hour, or 10,000 mbh.

If your T/E load ratio is less than two, it is reasonable to assume that you are not a good candidate for cogeneration. This would certainly be the case if your thermal demand is extremely small during the summer months and is only significant during a few winter months. However, if your annual electric load factor is small, you may consider using electric power generation only, in a strategy known as peak shaving. This is discussed in more detail under operating strategies in this chapter.

Equipment Sizing Considerations

For efficient cogeneration, the T/E load ratio in your facility must correspond to the T/E output ratio of the cogeneration systems. T/E output ratios vary for different prime movers and cogeneration plant configurations.

In an analysis of cogeneration options at your facility, you must decide whether to size the cogeneration plant to match your peak electric load, which will produce some waste heat and lower overall plant efficiency, or to size the cogeneration plant to match your heat load and supplement your electric needs with more expensive purchased power. If your load factor is small for either thermal or electric demand, you should size toward the minimum value of that demand.

Operating Strategies

Electric Dispatch

One method of operating a cogeneration plant is to supply, your facility's total electrical requirements as a first priority and generate steam as a second priority. This mode of operation is referred to as "electric dispatch."

Under the electric dispatch mode, the cogeneration plant is sized for annual peak kW demand, with some additional capacity for growth. A cogeneration plant sized in this manner can operate totally independently of the electrical utility company.

Standby electric service from your local utility is required for scheduled and unscheduled equipment shutdown. The cost of this standby power can be very expensive, especially if demand is high or if the utility company must provide additional generation or transmission equipment to serve your facility. Total on-site backup using standby emergency power generating equipment is rarely economical.

Electric dispatch requires that you operate your cogeneration plant to meet your electric load demand requirement. Operation of the major electric loads of your facility should be coordinated with the cogeneration plant to prevent sudden load spikes or irregular, repetitive load spikes that could exceed the capacity of your generating plant.

Also, if your annual electric load factor is low, your cogeneration equipment will be oversized for your facility load most of the time. This will reduce efficiency, increase maintenance costs and lower the return on investment.

Peak Shaving

Peak shaving is the practice of selectively dropping electric loads or generating on-site electricity during periods of peak electric demand. This procedure is not cogeneration; normally, no heat recovery equipment is installed and no heat is recovered from the generating process. Peak shaving is most commonly used to reduce annual electric utility costs. Peak shaving with an electric generator has a lower initial cost than a true cogeneration system since no heat exchangers or associated recovery equipment is installed.

Thermal Dispatch

Thermal dispatch operation is the complement to electric dispatch. In this mode of operation, the cogeneration plant is sized to meet your facility's annual peak Btu/hour (or mbh) demand. You will then purchase a part of your electric power from the local utility and cogenerate the rest. For this mode of operation to be successful, thermal load requirements should parallel each other in the same way required for electric dispatch.

For maximum savings on electric energy, electric generation should be near its maximum capacity during the summer peak electric demand period. This means that the ideal candidate for thermal dispatch operation would have maximum thermal loads occurring during summertime electric peak periods. If this is not the case, you will save less on electric energy.

Hybrid Strategy

A "hybrid strategy" utilizes the best features of the electrical and thermal dispatch strategies. As the name suggests, this operating strategy is a hybrid of the electric dispatch and thermal dispatch operating strategies periodically adjusted to minimize operating costs and maximize return on investment.

With the hybrid strategy, the cogeneration plant is sized smaller than for electrical or thermal dispatch. The hybrid strategy calls for the plant to be operated to achieve maximum savings during electric peak demand periods. At other times, it would be operated in a thermal load following mode. To satisfy the total thermal load requirements, it may be necessary to maintain existing boilers or install additional ones. As in the thermal dispatch strategy, your facility still must purchase a portion of its electric energy from the utility.

Prime Mover Selection

Choosing the most appropriate prime mover for a cogeneration project involves evaluation of many different criteria, including:

- Hours of operation.
- Maintenance requirements.
- Fuel requirements.
- Capacity limits.

Following are some general guidelines for evaluating each of the different prime movers with respect to these criteria.

Hours of Operation

Most cogeneration plants are designed for continuous operation, with the exception of some small reciprocating engine packages. Large plants are not economical if they cycle on and off on an hourly or even daily basis.

Because cogeneration equipment is relatively expensive as compared with boilers of similar thermal capacity, the equipment must be operated as much as possible to achieve an acceptable return on investment.

If an analysis of your facility shows a low electric load factor, indicating relatively few hours of high electric demand, cogeneration is probably not a cost-effective option. Instead, you may wish to examine the potential of peak shaving with reciprocating engine generators.

Maintenance Requirements

All cogeneration equipment requires some type of periodic maintenance. The frequency and amount of maintenance varies considerably between prime movers.

Reciprocating engine generators have the highest maintenance requirements. Like automotive engines, diesel engines require routine maintenance such as oil and filter changes as often as once a week. However, natural gas fired reciprocating engines require more overhauls of the head and block, and they have a shorter life expectancy.

Gas turbine engines are mechanically simpler and require less frequent maintenance. They can operate for longer periods between major maintenance intervals, and the major maintenance is usually simpler, such as bearing inspection and replacement.

Steam turbines require even less mechanical maintenance than gas turbines since they have no combustion equipment attached. Occasional bearing inspection and replacement is generally the extent of steam turbine maintenance. The cost is a function of the size and number of turbines.

Fuel Requirements

Reciprocating engines are not flexible with respect to their fuel requirements. Generally, you buy either a diesel engine or a natural gas fired engine. Special design adaptations can produce engines that burn other fuels, such as low Btu gas or heavier oils. Occasionally, natural gas engines can be adapted to switch between natural gas and propane as an alternate fuel. Diesel engines are more efficient in their fuel-to-energy conversion. However, air pollution regulations may impose greater constraints on these engines than gas turbines, usually requiring catalytic converters and carburetion limits.

Gas turbines can be switched from natural gas to diesel fuel to take advantage of price fluctuations and backup fuel requirements. In addition, they can be adjusted for other fuel oils, low Btu gas, and, occasionally, organic by-products of industrial processes such as "black liquor" from pulp and paper mills.

Steam turbines are limited only by the fuel for their steam source. In addition to the above fuels, coal, wood, waste, peanut shells, any suitable biomass, and incinerated municipal waste can be used to generate steam for steam turbines.

Capacity Limits

The three prime movers discussed here have separate and distinct capacity ranges. Reciprocating engine generators range in size from about

40 kW to over 3,000 kW. Generally, small electric demand plants with still smaller heat requirements can be satisfied with reciprocating engines. The quality of the heat recovered from these engines can be a limitation. Only about 35% of the recoverable heat is available as 125 psig steam; the rest of it is available only as 180°F hot water.

Almost all of the gas turbine heat is recoverable as 125 psig steam. The lower limit of gas turbine equipment size is about 480 kW, and the upper limit is over 30,000 kW. Combined-cycle gas and steam turbine plants can produce over 100,000 kW.

Steam turbines are the most limited with respect to power generation. Their practical lower limit is about 1,000 kW. Their conversion efficiency for power generation is below 15% if practical upper limits of 200 psig of superheated inlet steam and 100 psig of saturated outlet steam are maintained. This efficiency can be improved by increasing the inlet pressure and temperature, which dramatically increases the cost of the steam production plant. Another alternative for improving electric generating efficiency is to use a condensing turbine, but this lowers the temperature of the outlet steam to the point that it cannot be used for much more than low-temperature hot water production.

Figure 9-10 on the following page shows the thermal and electric energy output typically available from several different cogeneration system configurations.

Energy Savings Analysis

To arrive at an estimate of the energy savings potential of cogeneration at your facility, you will need to make two simplifying assumptions. For this analysis we will assume that the proposed cogeneration system is sized such that 100 percent of its thermal and electric output is utilized. We will also assume that the plant will operate at or near full output capacity at all times. These assumptions represent a "best case" operating scenario for your cogeneration system. In reality, your plant probably will not operate in this manner; however, these assumptions are necessary here for the purpose of this evaluation.

In addition to the above assumptions, you will need to obtain the following information on your proposed cogeneration plant:

- The thermal and electric output capacities of the cogeneration equipment.
- The fuel consumption rate of the cogeneration equipment.

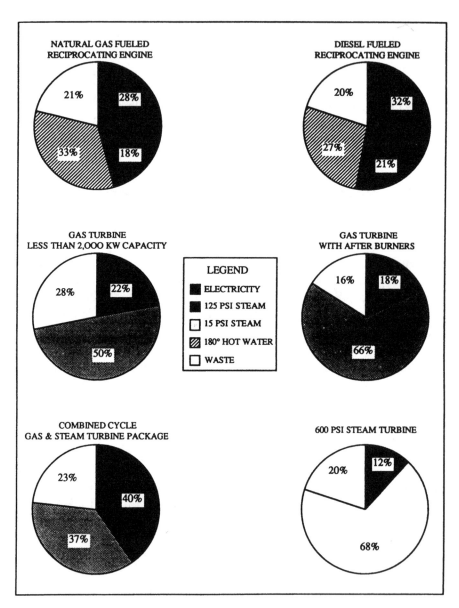

Figure 9-10. Energy Production in Cogeneration Systems

- The number of hours per year that you plan to operate the cogeneration system.

This information can be obtained from equipment data available from the manufacturer. You will need to determine the annual operating hours based on knowledge of your facility operating characteristics.

Energy Production

The thermal and electric energy produced by your cogeneration plant will be used at your facility to offset thermal energy produced by your boiler plant and electricity purchased from your local utility. The next step is to determine the annual thermal and electric energy production of the proposed cogeneration equipment.

The annual electric energy production is calculated as follows:

$$\text{Annual Electric Energy Production} = \text{Electric Output Capacity} \times \text{Annual Operating Hours}$$

The annual thermal energy production is calculated as follows:

$$\text{Annual Thermal Energy Production} = \text{Thermal Output Capacity} \times \text{Annual Operating Hours}$$

Fuel Consumption

To determine the fuel use requirements of your proposed cogeneration plant, you will need to know the rate of fuel consumption for the equipment being evaluated here. This information can be obtained from the equipment manufacturer. Your cogeneration plant fuel consumption is calculated as follows:

$$\text{Cogeneration Fuel Consumption} = \text{Equipment Fuel Consumption Rate} \times \text{Annual Operating Hours}$$

Since you have assumed that all of the output will be used by your facility, you will now calculate the value of the cogenerated energy based on your current utility costs.

Current Utility Costs

To determine your annual average cost per kWh, first tabulate electric energy costs from the bills that you used for gathering the electric demand and consumption data. (Generally, a more detailed analysis of your electric rate structure is needed to develop an accurate unit cost of the purchased power you will offset with cogeneration. For this evaluation, we will assume the unit cost of electricity is the same as the annual average unit cost of electricity.) Next, divide your annual electric cost by your annual electric consumption.

$$\text{Annual Average Cost/kWh} = \frac{\text{Annual Electric Cost}}{\text{Annual Electric Consumption}}$$

You can use the following method to determine your thermal energy cost:

• Tabulate monthly fuel costs from the bills used to obtain the fuel consumption data earlier.

• Determine your average fuel unit cost by dividing the annual fuel cost by your annual fuel consumption.

$$\text{Fuel Unit Cost} = \frac{\text{Annual Fuel Cost}}{\text{Annual Fuel Consumption}}$$

• Determine your current cost per thermal MMBtu of steam or hot water. To do this, first multiply the fuel unit cost by the Btu/unit value from Figure 9-9, then divide by your boiler's fuel-to-steam-efficiency. Divide the result by 1,000,000 to convert to MMBtu units.

$$\frac{\text{Current Cost}}{\text{Thermal MMBtu}} = \frac{\text{Fuel Unit Cost} \times \text{Btu/Fuel Unit}}{(\text{Fuel-to-steam Efficiency}) \, (1,000,000)}$$

Annual Savings

The annual energy savings of your cogeneration plant is calculated as the savings from electric and thermal energy production less the cost of operating the plant. The annual electric energy savings can be determined from the following equation:

$$\frac{\text{Annual Electric}}{\text{Energy Savings}} = \frac{\text{Annual Electric}}{\text{Energy Production}} \times \frac{\text{Annual Average}}{\text{Cost/kWh}}$$

The annual thermal energy savings is given by the following:

$$\frac{\text{Annual Thermal}}{\text{Energy Savings}} = \frac{\text{Annual Thermal}}{\text{Energy Production}} \times \frac{\text{Current Cost}}{\text{Thermal MMBtu}}$$

Operating and Maintenance Costs

Operating and maintenance costs are, to a large degree, dependent on plant operating hours and, therefore, proportional to fuel consumption. For the purpose of this evaluation, maintenance costs can be approximated as 15% of fuel costs for reciprocating engines and 7% of fuel costs for gas or steam turbines. Use a lower figure for steam turbine maintenance costs if some boiler maintenance is already included in your operating expenses.

The cost of the fuel to operate your cogeneration plant can be determined using the following equation:

$$\frac{\text{Cogeneration}}{\text{Fuel Cost}} = \frac{\text{Cogeneration Fuel}}{\text{Consumption}} \times \frac{\text{Fuel}}{\text{Unit Cost}}$$

You now have all the information necessary to determine your annual savings from the operation of a cogeneration plant. This is calculated as follows:

$$\frac{\text{Annual}}{\text{Savings}} = \frac{\text{Annual Electric}}{\text{Energy Savings}} + \frac{\text{Annual Thermal}}{\text{Energy Savings}} -$$

$$\frac{\text{Cogeneration}}{\text{Fuel Cost}} - \frac{\text{Operating and}}{\text{Maintenance Cost}}$$

At this point, it is important to note that the annual savings derived from this analysis is only a preliminary estimate of the savings potential at your facility and is based on simplifying assumptions you have made.

Economic Analysis

Initial Cost

To develop the total initial cost of the cogeneration plant you are evaluating, first refer to Figure 9-11 below for approximate equipment cost ranges per kW of installed cogeneration electric capacity.

The cost figures in Figure 9-11 do not include the cost of new buildings to house the equipment, nor do they include allowances for the electrical wiring necessary to connect your facility to your utilities. If additional space must be constructed or major work is required to connect

Reciprocating Engine Packages (High Speed)
Large 900 to 3,000 kW packages........................ $600 to $400/kW
Medium 400 to 800 kW packages..................... $500 to $700/kW
Small 45 to 300 kW packages............................ $700 to $1200/kW

Gas Fired Turbine Engine Packages
Large 4,000 to 10,000 kW .... $800 to $1,000/kW
Medium 500 to 4,000 kW................................. $1,200 to $1,800/kW

*Steam Powered Turbine Engines**
Less than 125 psig inlet turbines..............................$100 to $130/kW
Less than 250 psig inlet turbines................................$90 to $120/kW
*Note initial costs for steam powered turbines are highly dependent on the entering and leaving pressures of the turbine. The best efficiencies are for 1,000 to 3,000 psi superheated steam. The cost of small-scale steam boilers in this range is prohibitive, and the cost of condensing steam turbines is not listed since considerable extra equipment is required. The costs as given do not include the cost of heat recovery equipment.

Figure 9-11.
Approximate Cost Per kW of Cogeneration Plant Capacity

the utilities, these expenses must be added to the equipment cost estimates to determine your total initial cost. Also, these costs were assembled at an earlier time and may need to be revalidated for your current analysis.

Investment Analysis

The final decision to build a cogeneration plant is usually based on investment analysis. Broadly defined, this is an evaluation of costs versus savings. The costs for cogeneration include the initial capital cost for the equipment or the cost of operating and maintaining that equipment, costs; fuel costs, finance charges, tax liabilities, and other system costs, that may be specific to your application. Savings include offset electrical power costs; offset fuel costs; revenues from excess power sales, if any; tax benefits; and any other applicable savings that are offset by the cogeneration plant, such as planned replacement of equipment.

A number of economic analysis techniques are available for comparing investment alternatives. The simple payback period is the most commonly used and the least complex, and is usually adequate for a go, no-go decision at this stage of project development. It is the period of time required to recover the initial investment cost through savings associated with the project. Simple payback period is calculated as follows:

$$\text{Simple Payback Period} = \frac{\text{Initial Cost, \$}}{\text{Annual Savings, \$/Yr}}$$

The simple payback period will give you an indication of the attractiveness of cogeneration at your facility and will help you decide whether a full-scale feasibility study is warranted.

Figure 9-12 is a summary of the information used in this evaluation of cogeneration at your facility.

It is important to note that many of the variables used in this evaluation are assumed values based on your knowledge of the operating characteristics of your facility and the simplifying assumptions made for this preliminary analysis. In a full-scale cogeneration feasibility study, you will want to evaluate several equipment types and sizes in conjunction with different operating strategies for each system configuration. This type of evaluation should be performed for you by an experienced consulting engineer.

UTILITY DATA ANALYSIS

    Electric Energy Consumption   _____   kWh/year

    Annual Peak Electric Demand   _____   kW

    Electric Energy Cost $_____   /year

    Thermal Energy Consumption   _____   Btu/year

    Annual Peak Thermal Demand   _____   Btu/hour

    Annual Fuel Cost $_____   /year

    Annual Electric Load Factor   _____   %

    Annual Thermal Load Factor   _____   %

    Thermal/Electric Load Ratio   _____

ENERGY ANALYSIS

    Cogeneration Plant Capacity

        Electric Output   _____   kW

        Thermal Output   _____   Btu/hour

    Annual Operating Hours   _____   hours/year

    Annual Electric Energy Production   _____   kWh/year

    Annual Thermal Energy Production   _____   Btu/year

    Cogeneration Plant Fuel Consumption   _____   Btu/year

    Current Utility Costs

        Average Annual Electric Cost $_____   /kWh

        Average Thermal Energy Cost $_____   /MMBtu

    Annual Savings

        Electric Energy Savings $_____   /year

        Thermal Energy Savings $_____   /year

    Cogeneration Fuel Cost $_____   /year

    Operating and Maintenance Costs $_____   /year

ECONOMIC ANALYSIS

    Annual Savings $_____   /year

    Initial Cogeneration Equipment Cost $_____

    Simple Payback Period   _____   years

Figure 9-12. Cogeneration Feasibility Evaluation Summary

10

Control Systems

This chapter will focus on the following subjects:

INTRODUCTION

Energy engineers have been and will continue to make significant contributions for the accomplishment of energy-efficient processes, buildings and transportation systems. In the hands of the energy engineer, control systems are an important tool to make these systems energy efficient;

the majority of the presently existing systems were designed primarily as labor- and materials-saving (low first costs) rather than as energy-saving (low operating costs) equipment.

For example, the practice in the design and selection of heating, ventilating and air-conditioning (HVAC) systems has been based on maximum load requirements. Usually, HVAC systems operate at less than design load conditions and require controllers to regulate operations during partial load conditions. Design optimization of the configuration and selection of HVAC systems and controllers is essential to meet the required environmental conditions with minimum energy consumption. Energy-related characteristics of the HVAC equipment and processes have been discussed in Chapters 7 and 8. The purpose of the present chapter is to describe the controls needed for an energy-efficient design and operation of building systems.

BASIC CONCEPTS

A *control system* is a means by which some quantity of interest in a machine, mechanism or other equipment is maintained or altered, in accordance with a desired manner. A control system consists essentially of five parts:

1. Process (system).
2. Measuring means (sensor).
3. Error detector.
4. Controller.
5. Final control element.

These five elements can be interconnected either in an *open-loop* format or in a *closed-loop* format.

An open-loop system is a system in which the output has no effect upon the input signal. A clothes dryer may be considered as an open-loop control system in which input is the status of the electric switch (on/off) and the output is the dryness (moisture content) of the clothes. When the dryer is turned on, it runs for a pre-set time and then automatically turns off. The actual moisture content of the clothes has no effect on the on-time.

The dryer could be put in a closed-loop system by installing a moisture-sensing probe which will continuously measure the moisture content of the clothes being dried and compare this value with a desired value.

The difference between the two values could be used by a controller to control the on/off switch to the dryer.

Similarly, if a home thermostat were installed outside the house, the output (furnace heat) would have no effect on the input to the thermostat (in this case, outdoor temperature). If we closed this loop by bringing the thermostat indoors, the output (furnace heat) would now affect the input (indoor temperature). This closed-loop thermostat would continue to operate until the indoor temperature reached the desired value (setpoint). In this closed-loop control system, the output is fed back and compared with the desired value, and the difference between the two is used as an actuating signal to actuate the control device (gas valve) for a gas furnace.

Thus a closed-loop (feedback) control system is a system in which the output has an effect upon the input quantity to maintain the desired output value. A block diagram of a feedback control system showing all basic elements is shown in Figure 10-1.

If Figure 10-1 were to represent a residential heating system using a natural gas furnace, different elements would be as shown in Table 10-1.

CONTROL MODES

Figure 10-1 may be condensed to Figure 10-2. In this figure, for the residential heating system, the manipulated variable m(t), gas flow rate at any instant in time, could have one of the two possible values, namely zero or one, depending upon the thermostat status e(t) which could be on or off. Such a control action is called on/off or two-position control mode.

Thus, the two-position control is a type of control action in which the manipulated variable is quickly changed to either a maximum or minimum value depending upon whether the controlled variable is greater or less than the setpoint. The minimum value of the manipulated variable is usually zero (off).

The equations for two-position control are

$m = M_1$ when $e > 0$

$m = M_0$ when $e < 0$ *Formula (10-1)*

Where

M_1 = maximum value of manipulated variable.

M_0 = minimum value of manipulated variable.

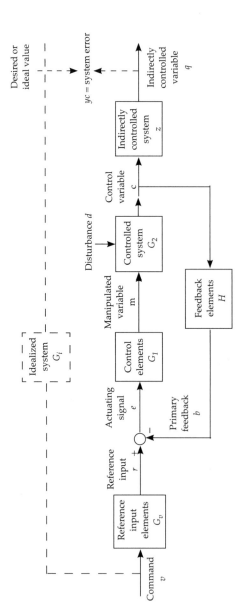

Figure 10-1. Block Diagram of a Feedback Control System Containing All Basic Elements

Table 10-1. Basic Elements of a Residential Heating System

Element	Example in a residential heating system
Command v	Desire to set the indoor temperature
Reference input element G_v	Thermostat knob
Reference input r	Position of thermostatic switch
Primary feedback b	Position of bimetallic strip
Actuating signal e	On/off status of switch
Control elements G_1	Relay-controlled gas valve
Manipulated variable m	Gas flow rate (on or off)
Controlled system G_2	Furnace burner, ducts, house, etc.
Controlled variable c	Indoor air dry-bulb temperature
Disturbance d	Heat gains, losses, infiltration, etc.
Feedback element H	Bimetallic strip in the thermostat
Indirectly controlled system z	Occupant
Indirectly controlled variable q	Thermal comfort level
Idealized system G_i	Imaginary system which would result in an indoor environment where level of occupant's satisfaction is 100 percent.
System error	Thermal discomfort level

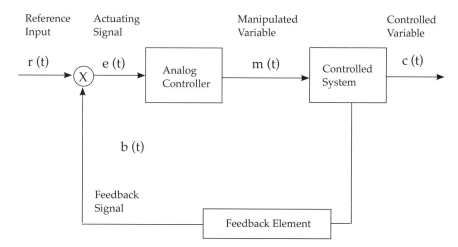

Figure 10-2. Generalized Block Diagram for a Closed-loop Control System

A differential or dead-band in two-position control causes the manipulated variable to maintain its previous value until the controlled variable has moved beyond the setpoint by a predetermined amount. In actual operation, this action may be compared to hysteresis as shown in Figure 10-3.

A differential may be intentional, as is common in domestic thermostats when employed for the purpose of preventing rapid operation of switches and solenoid valves and to enhance the life of the system.

Two-position control is simple and inexpensive, but it suffers from inherent drifts. Rapid changes of the controlled variable are possible with this type of control. Compared with other types of control actions, two-position control can be more energy-intensive. Some examples of applications of two-position control include residential heating/cooling systems and rooftop units in commercial buildings.

Proportional Control

Proportional control is a type of control in which there is a continuous linear relation between values of the actuating signal and the manipulated variable. For purposes of flexibility, an adjustment of the control action is provided and is termed proportional sensitivity. Proportional control may be described as

$$m = K_c e + M$$

Formula (10-2)

Where

K_c = proportional sensitivity

M = a constant and other terms as defined previously.

The proportional sensitivity, K_c, is the change of output variable caused by a unit

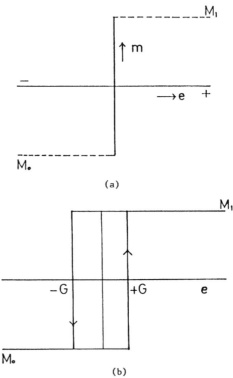

Figure 10-3 (a) Two-position Control Action; (b) Two-position Control Action with Differential Gap

change of input variable.

The constant M in Formula (10-2) may be termed as the calibration constant because the selection of a value for M determines the normal (zero actuating signal) value of the manipulated variable. The operation of proportional control action is illustrated in Figure 10-4.

For a unit step change in actuating signal
$$e = 0 \qquad t < 0$$
$$e = E \qquad t \geq 0 \qquad\qquad\qquad \textit{Formula (10-3)}$$

where E is a constant; substituting in (10-2)
$$m - M = K_c E \qquad\qquad\qquad \textit{Formula (10-4)}$$

The change in manipulated variable corresponds exactly to the change in deviation with a degree of amplification depending upon the setting of proportional sensitivity K_c. Thus, a proportional controller is simply an amplifier with adjustable gain.

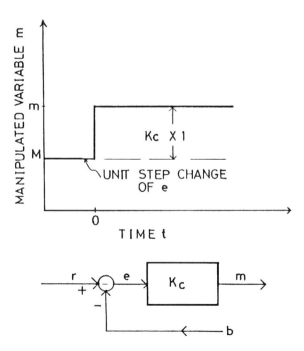

Figure 10-4. Proportional Control Action

Proportional control is more sensitive to the error signals than is two-position control. It is the least expensive of continuous-type controls but is more expensive compared to the two-position control. Calibration procedures are more difficult for proportional control than for the two-position control. Proportional control is used in air distribution systems, hydronic systems and in central systems for commercial buildings.

Integral Control

Integral action is a type of control action in which the value of the manipulated variable m is changed at a rate proportional to the actuating signal. Thus, if the actuating signal is doubled over a previous value, the final control element is moved twice as fast. When the controlled variable is at the set point (zero actuating signal), the final control element remains stationary.

Mathematically, integral control may be expressed as

$$\dot{m} = \frac{1}{t_{int}} \; e \hspace{3cm} \textit{Formula (10-5)}$$

or, in integrated form

$$m = \frac{1}{t_{int}} \int e \, dt + M \hspace{2cm} \textit{Formula (10-6)}$$

Where
$$M = \text{constant of integration}$$
$$t_{int} = \text{integral time (defined as the time of change of manipulated variable caused by a unit step change of e)}$$

The operational form of the equation is

$$m(s) = \frac{1}{t_{int}s} \; e(s) \hspace{2.5cm} \textit{Formula (10-7)}$$

and is shown in Figure 10-5.

For a step change of actuating signal

$$e = 0 \hspace{2cm} t < 0$$
$$e = E \hspace{2cm} t \geq 0 \hspace{2cm} \textit{Formula (10-8)}$$

where E is a constant. Substituting in Formula (10-6) and integrating

$$m - M = \frac{1}{t_{int}} \; Et \qquad\qquad \textit{Formula (10-9)}$$

Thus, the manipulated variable changes linearly with time and "integrates" the area under the actuating signal function. For a unit step change, if actuating signal (E = 1.0), the slope of the line is inverse of integral time (Figure 10-5).

Integral control has the advantage over proportional control in that it tends to zero the offset, but it requires more expensive calibration procedures. Its maintenance is more difficult. Some applications of integral control include control of boilers, solar storage systems and meat-processing plants.

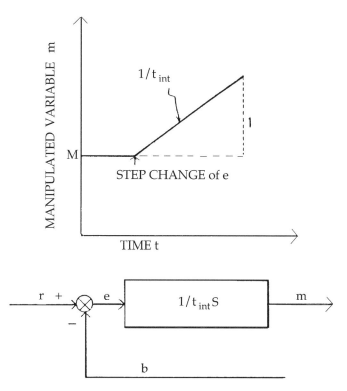

Figure 10-5. Operation of Integral Control Action

Proportional-derivative Control

Derivative control action may be defined as a control action in which the magnitude of the manipulated variable is proportional to the rate of change of actuating signal. This control mode has many synonyms, such as "pre-set," "rate," "booster" and "anticipatory control" action. Derivative control response is always used in conjunction with the proportional mode. It is not satisfactory to use this response alone because of its inability to recognize a steady-state actuating signal.

Mathematically, a proportional-derivative (PD) control action is defined by

$$m = K_c e \mid + K_c t_d \; \dot{e} \mid + \quad M \hspace{3cm} \text{Formula (10-10)}$$
$$[\text{proportional}] \; [\text{derivative}]$$

where t_d = derivative time and other variables as described previously. PD is the simple addition of proportional control and rate control action as shown by the operation formula

$$m(s) = K_c \, (1 + t_d \, s) \, e(s) \hspace{3cm} \text{Formula (10-11)}$$

Proportional-derivative action is not adequately described by applying a step change of actuating signal because the time derivative of a step change is infinite at the time of change. Consequently, a linear (ramp) change of actuating signal must be used:

$$e = Et \hspace{3cm} \text{Formula (10-12)}$$

Where
    E = a constant
    t = time

Substituting Formula (10-12) and its first time derivative into Formula (10-10).

$$m - M = K_c E \, (t + t_d) \hspace{3cm} \text{Formula (10-13)}$$

The actuating signal is defined at time t, whereas the manipulated variable is defined at $(t + t_d)$. The net effect is to shift the manipulated variable ahead by time t_d, the derivative time. As shown in Figure 10-6, the controller response leads the time change of actuating signal. Derivative time is defined as the amount of lead, expressed in units of time, that the control action is given. In other words, derivative time is the time interval by which the rate action advances the effect of proportional control action.

Proportional-derivative control has the advantage of a rapid response to the magnitude and to the rate of change in loads. However, PD control can become unstable easily because it has no zeroing capability. This control is useful for controlling environments in buildings with large variations in occupancy.

Proportional-integral Control

Integral control action is often combined additively with proportional control action. The combination is termed proportional-integral or proportional plus reset control and is used to obtain advantages of both con-

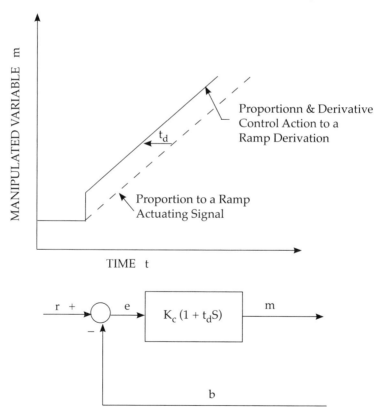

Figure 10-6. Proportional-derivative Control Action

trol actions.

Proportional-integral control action is defined by the following differential formula:

$$\frac{dm}{dt} = \frac{K_c}{t_{int}} \quad e \mid \quad + K_c \quad \frac{de}{dt} \mid \qquad \textit{Formula (10-14)}$$
$$\text{[integral]} \qquad\qquad \text{[proportional]}$$

or, in integrated form

$$m = \frac{[K_c \int e \, dt]}{t_{int}} \quad + [K_c \, e + M] \qquad \textit{Formula (10-15)}$$
$$\text{[integral]} \qquad \text{[proportional]}$$

where terms are as previously described. These formulas illustrate the simple addition of proportional and integral control actions. In operational form:

$$m(s) = K_c \left(\frac{1}{t_{int}s} + 1\right) e(s) \qquad \textit{Formula (10-16)}$$

where the system function $K_c/(t_{int}s)$ identifies the integral action and the system function K_c identifies the proportional action.

Proportional-integral (PI) control action has two adjustment parameters, the proportional sensitivity K_c and integral time t_{int}. The proportional sensitivity is defined the same way as for the proportional control action. With the integral response turned off ($t_{int} \to \infty$), the proportional sensitivity is the number of units change in manipulated variable in per-unit change of actuating signal e. As clear from Formula (10-16), the proportional sensitivity K_c affects both the proportional and integral parts of the action.

The integral action adjustment is achieved through integral time. For a step change of actuating signal e, the integral time, t_{int}, is the time required to add an increment of response equal to original step change of response as shown in Figure 10-7. Another term used with this type of control is reset rate, defined as the number of times per minute that the proportional part of response is replicated. Reset rate is therefore called "repeats per minute" and is the inverse of integral time.

For a step change of deviation
$$e = o \qquad t < 0$$
$$e = E \qquad t > 0 \qquad\qquad\qquad \textit{Formula (10-17)}$$

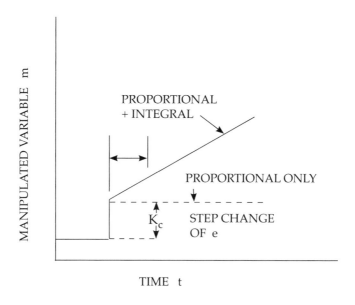

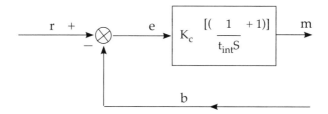

Figure 10-7. Operation of Proportional-integral Control

Substituting in Formula (10-14),

$$m - M = K_c E \left(\frac{t}{t_{int}} + 1 \right)$$ *Formula (10-18)*

This is the formula for a straight line. The first term, t/t_{int}, is the integral response, and the second term is the proportional response. The latter is indicated by the dotted line of Figure 10-7.

PI control has the advantage of compensating for changes in input in addition to compensating the deviations in the controlled variable. High cost of maintenance is a disadvantage with this type of control.

Proportional-integral-derivative Control
      The additive combination of proportional action, integral action and derivative action is termed proportional-integral-derivative action. It is described by the differential formula

$$\dot{m} = \frac{K_c e}{t_{int}} \qquad +K_c e \qquad + K_c t_d \dot{e}$$

$$\quad\;\; [\text{integral}] \qquad [\text{proportional}] \quad [\text{derivative}]$$

Formula (10-19)

or

$$m = \frac{K_c \int e\, dt}{t_{int}} \qquad +K_c e + M \qquad + K_c t_d \dot{e}$$

$$\quad\;\; [\text{integral}] \qquad\quad [\text{proportional}] \quad [\text{derivative}]$$

Formula (10-20)

$$m(s) = K_c \left(\frac{1}{s t_{int}} + 1 + t_d s \right) e(s)$$

Formula (10-21)

Proportional-integral-derivative (PID) control action is illustrated in Figure 10-8 in which the change of manipulated variable is shown for a ramp function of the actuating signal.

$$e = Et$$

Formula (10-22)

Substituting this ramp function and its time derivative into Formula (10-20):

$$m - M = K_c E \left(\frac{1}{t_{int}} \int t\, dt + t + t_d \right)$$

Formula (10-23)

Integrating the first term

$$m - M = K_c E \left(\frac{t^2}{2t_{int}} + t + t_d \right)$$

Formula (10-24)

The proportional part of the control action repeats the change of actuating signal (lower straight line) in Figure 10-8. The derivative part of the control action adds an increment of manipulated variable proportional to the area under the deviation line and, as Figure 10-8 shows, the increment

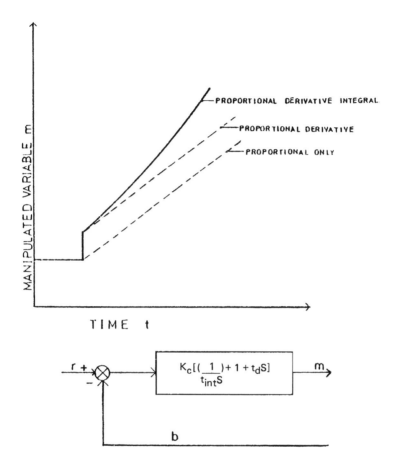

Figure 10-8. Operation of Proportional-integral-derivative Control

increases because the area increases at an increasing rate. The combination of proportional, integral and derivative actions may be made in any sequence, because these actions are described by linear differential equations.

PID control action has many advantages. It can compensate for magnitude and rate of change in input. It zeroes the deviation in controlled variable and is the most energy-efficient. However, it is the most expensive of all the controls and is difficult to calibrate. It is very hard to keep PID control stable; it requires frequent calibration checks. PID control is applied in environmental control chambers for scientific research and in variable air volume HVAC systems.

TYPES OF CONTROLLERS

Controllers for an energy (conversion and/or utilization) system may be one of the following types:

1.   Electric or electronic controllers.
2.   Pneumatic controllers.
3.   Computer-based controllers.

Electric or Electronic Controllers

Electric switches operated manually, or with relays and timers, are an important class of controllers for implementing several control strategies like start-stop of equipment, duty cycling, night setback, control of lighting, etc. Single-pole, double-throw (SPDT) switching circuits are used to control three-wire uni-directional motors. SPDT circuits are also used for heating-cooling applications.

Manually controlled electric switches used to "turn it off when it's not needed" offer great opportunities for energy savings. For example, a manufacturing company happened to hire an energy engineer whose first assignment was to identify and implement two energy-saving strategies within the first month of his starting the job. The condition was that the first cost of these projects should be zero and the savings should at least be equal to his salary. Manually operated electric switches were the answer for the engineer. He identified that the company had a policy of leaving all its office lights on for the janitor crew at night. He implemented a policy that all the employees were required to turn off their office lights manually, while leaving at 5:00 p.m. Further, the janitors would turn on the lights in an office only when they are cleaning that office. It turned out that 10,000 fluorescent tubes (40 watts each) could be turned off for 4 hours/day-5 days/week-52 weeks/year. The company was paying $0.07/kWh and $8.00/kW. Thus, savings from this strategy alone were

$$\frac{10,000 \times 40 \times 4}{1000} \ (\text{kWh}/\text{day}) \times 5 \ \text{days}/\text{week} \times$$
$$52 \ \text{weeks}/\text{year} \times \$0.07/\text{kWh} = \$29,120/\text{year}.$$

Secondly, he identified that an exhaust fan in the grinding shop was kept running all the time, even though the shop operated for 16 hours/day and 5 days/week. He implemented a policy that the workers at the end of the second shift would *manually turn off* this fan and the workers at the start of the first shift would manually start this fan. The specifications of the fan and heating equipment were as follows:

Horsepower of the fan = 10 hp
Air exhausted = 20,000 ft^3/minute
Cost of electricity = $0.07/kWh
Efficiency of the gas heater = 0.70
Cost of gas = $0.60/therm
Heating season = 24 weeks/year
Setpoint = 70°F
Avg. outdoor temperature = 35°F

Calculations:
a. *Gas savings*

$$\frac{(20{,}000\ \text{ft}^3)}{\text{minute}} \cdot \frac{(60\ \text{minutes})}{\text{hour}} \cdot \left[\frac{(5\ \text{days})}{\text{week}} \frac{(8\ \text{hours})}{\text{day}} + \frac{(2\ \text{days})}{\text{week}} \frac{(24\ \text{hours})}{\text{day}}\right] \cdot$$

$$\frac{(24\ \text{weeks})}{\text{year}} \cdot \frac{(0.077\ \text{lbm})}{\text{ft}^3} \frac{(0.248\ \text{Btu})}{\text{lbm }°\text{F}} \cdot$$

$$(70°\text{F} - 35°\text{F}) \cdot \frac{(1\ \text{Therm})}{10^5 \text{Btu}} \frac{(1)}{0.70} \frac{(\$0.60)}{\text{Therm}} = \$14{,}518/\text{year}.$$

b. *Electricity savings*

$$10\ \text{hp} \cdot \frac{0.746\ \text{kW}}{\text{hp}} \left[\frac{(5\ \text{days})}{\text{week}} \cdot \frac{(8\ \text{hours})}{\text{day}} + \frac{(2\ \text{days})}{\text{week}} \cdot \frac{(24\ \text{hours})}{\text{day}}\right] \cdot$$

$$\frac{(52\ \text{weeks})}{\text{year}} \cdot \frac{\$0.07}{\text{kWh}} = \$2{,}389/\text{year}.$$

Total savings from manually operated switches were $46,027/year, which covered the starting salary of the energy engineer, which was $40,000/year.

The example illustrates the usefulness of manually controlled switches.

TIMERS

Either single-pole, single-throw (SPST) or SPDT circuits can be interfaced to timers. These timers can range from very simple clocks to sophisticated central-time clocks with multiple-channel capability for controlling equipment on different time schedules.

Timers can be used to implement control strategies like night setback, duty cycling and automatic start-stop of equipment.

Night Setback

Energy expended to heat unoccupied buildings up to comfort conditions is wasted, and most buildings are indeed unoccupied at night-time. Energy is saved by setting back the temperature levels at these times.

Night setback offers opportunities for significant energy savings with little or no capital expenditure in most of the residential, institutional, commercial and industrial buildings. However, one should be careful that the night setback does not:

(i)   Effect process conditions, e.g., tolerances on the manufactured goods.

(ii) Cause the air conditioners to turn on to meet the setback temperatures.

Duty Cycling

In duty cycling, loads are cycled on and off by a timing device. Duty cycling has been used to duty cycle the operation of exhaust fans and space heating/cooling equipment with some possible savings. Savings are only a possibility because some experts believe that duty cycling is not an economical way of achieving savings in the long run. The idea that duty cycling can cause detrimental damage to equipment must be considered. If, for example, a company can save $10,000 per year on its electricity bill but has to spend about the same amount in maintenance and repair, then the use of duty cycling is not feasible.

In some instances, the breakdown of equipment is not acceptable at all. This is the case in a hospital. If an unforeseen problem occurs in the operating room, no financial savings could pay for the problems that might result.

Duty cycling has been with us for years. Ever since the energy crisis, duty cycling has become a broader area. Since air-conditioning systems are usually based on the most extreme operating conditions, the basic theory behind duty cycling is that the mechanical equipment is oversized during a large part of the operating life.

The standard thermostat has been used to cycle air conditioners for decades. Thermostats essentially cycle the air-conditioning equipment on and off to maintain comfortable environmental conditions. Consequently, mechanical equipment controlled by thermostats is designed to tolerate cycling for normal operation. Problems may be caused by the addition of the cycle-timing devices in addition to the thermostat controls. Therefore, the problem of accelerated deterioration of mechanical equipment has been attributed to duty cyclers.

The idea that the age of the equipment may be altered by duty cycling has to be considered first. There have been ideas on both sides. Some feel that the life will be extended and some feel otherwise.

In one case, it was reported that duty cycling control towers caused the breaking up of gear boxes. When discussed at a forum on duty cycling, it was found that, when the fan changed speed from high to low, the gears acted as a brake for slowing the fan. The added stress on the gears was not taken into consideration when the towers were designed. This problem was easily fixed by a) shutting off the fan for approximately 45 seconds so the fan speed decreased enough to avoid any more damage to the gears or b) a more sophisticated way to use decelerating relays.

A serviceman mentioned trouble he had had with air-handling equipment in a television studio. The cycling pattern was 27 minutes off and 21 minutes on. Within one year, 25% of the fan drives had failed. The problem was either a motor-drive end bearing had failed or the fan-drive end bearing had failed. The solution to this problem was simple. There was over-tension in the belts. This was initially done to keep the air-conditioning equipment from squeaking. After the belts were loosened to the proper specification, there were no problems reported within the next year.

The same serviceman also duty cycled air-handling equipment at a hospital. The off time was never less than 5 minutes. There were no problems reported there.

Another concern has been the decrease in compressor or starter motor life. It is estimated that the starter motor life would be reduced by one half for every 10-degree Celsius increase in winding temperature. For example, a motor with a 15-year normal life that is cycled once a day with an increase in winding temperature of 20°C will last only about 4 years. In other words, if the same motor with a 30°C increase in temperature were cycled four to five times a day, it would last only 2 years.

In addition, the bearings in the compressor and starter motor may get damaged by duty cycling. Moreover, the contacts and control equipment could also have a shorter life. This decrease in life would be dependent upon the number of cycles.

The problem with the contacts wearing out prematurely occurs because the original equipment is not made for the excessive use that occurs during duty cycling. All that has to be done is to put in heavy-duty contacts.

When it comes to the problem with bearings, an HVAC dealer in Peoria stated that, if the roller bearings are greased properly, there will be no problem. However, he did not recommend duty cycling sleeve bearings. Heavy-duty or over-specified sleeve bearings must be used. For collar-mounted bearings (either screw-locking or eccentric), the collars may work loose if duty cycled; so these must be watched very carefully.

As for the life of the compressor and compressor motor, this is a more complex problem. Some motors cannot be duty cycled. This will depend upon several parameters. Problems could occur because the motor was not properly maintained during its years of operation. Such a motor should not be cycled. Even if the older motor was properly maintained, it should not be aggressively cycled (cycled too often). Improper cycling can inhibit oil return to the compressor and cause it to cease. Most older motors were not manufactured for duty cycling; that is why special care should be taken.

Currently, larger motors are not used as often. A 100-hp motor will consist of four 25-hp motors. The amperage of the in-rush at start is much less for these smaller motors, and the chance of damage actually occurring is significantly less. Duty cycling of these smaller motors is much more efficient also. They are either on or off and run at optimum efficiency at all times.

Generally, heat is the worst enemy of motors. If a motor is duty cycled, it runs cooler. Even though the highest temperature occurs during start-up and there may be very many start-ups when duty cycling is used, the manufacturer takes this into consideration. This is done when the minimum on and minimum off times are specified. The average temperatures at which the starter motors and compressors run are determined. The average duration without duty cycling will not be exceeded, when duty cycling, if the minimum on and off times are followed. Thus, if a compressor is cycled off 10% of the time, the number of run hours will not be decreased. In this case, the number of days or years of the compressor life will be increased by 10%.

Publicity concerning the problems with duty cyclers led mechanical equipment manufacturers to issue stipulations in the warranties of air-conditioning equipment in the past. One such stipulation was that, if the air conditioner was duty cycled at all, the warranty was voided.

Now, the warranties are straightforward. For example, the warranty for a compressor is for 5 years. This is straight across the board (no stipulations against duty cycling). For a top-of-the-line compressor, the warranty could be for 15 years because of the additional safety equipment that is included. The only guidelines specify that the minimum on and off times stated by the manufacturer must be followed to protect the equipment against short cycling.

While some people consider duty cycling is bad for the mechanical equipment, many manufacturers have introduced duty-cycle timers to their lines and will give the buyer information on cycling on and off times. Why so? Can duty cycling actually reduce energy usage without the added expense of premature equipment failure? The answer is yes. Duty cycling can reduce energy costs and, if the cycle rates are not excessive, duty cycling will not cause the premature failure of well-designed mechanical equipment.

Determining On Time and Off Time

The National Electrical Manufacturers Association (NEMA) has published a standard (NEMA Standard MG 10) in which the problems of duty cycling of electrical motors are examined. The standard was prepared as a guide for selecting the optimal on and off times of a given electrical motor, taking into account motor type, horsepower rating, speed, starting frequency, restrictions on in-rush current, power demand charges and the extra winding stress imposed by repeated accelerations. A portion of the table outlining NEMA recommendations is shown in Table 10-2.

The parameters in Table 10-2 represent the following:

hp = NEMA horsepower rating of the motor
A = Maximum number of starts per hour
B = Maximum product of A times load inertia (lb × ft^2)
C = Minimum reset time in seconds (convert to minutes)

Table 10-2.* Allowable Number of Starts and Minimum Time Between Starts for Design A and Design B Motors

HP	2-Pole A	2-Pole B	2-Pole C	4-Pole A	4-Pole B	4-Pole C	6-Pole A	6-Pole B	6-Pole C
2.0	11.5	2.4	77	23	11	39	26.1	30	35
5.0	8.1	5.7	83	16.3	27	42	18.4	71	37
7.5	7.0	8.3	88	13.9	39	44	15.8	104	39
15.0	5.4	16	100	10.7	75	50	12.1	200	44
20.0	4.8	21	110	9.6	99	55	10.9	262	48
25.0	4.4	26	115	8.8	122	58	10.0	324	51
30.0	4.1	31	120	8.2	144	60	9.3	382	53
40.0	3.7	40	130	7.4	189	65	8.4	503	57
50.0	3.4	49	145	6.8	232	72	7.7	620	64

*Adapted from NEMA Standard MG10.

The table lists the minimum off time and maximum number of starts permitted per hour. What is actually needed is the information for determining the duty period (on time plus off time). Using the numbers in the table, Item A (maximum starts per hour), the shortest duty period permissible for a 2-pole, 25-hp motor, taking into consideration that the motor and load are sized correctly, can be calculated as follows:

$$\frac{60 \text{ min/hr}}{4.4 \text{ starts/hr}} = 60 \text{ min/4.4 starts} = 13.53 \text{ min/start}$$

Therefore, the duty period = 14 minutes.
The minimum off time is 115 seconds (approx. 2 minutes).
    On Time + Off Time = Duty Period
    On Time + 2 Min. = 14 Minutes
    On Time = 12 Minutes

To verify the above data, exchange the 14-minute duty period for the 4.4 starts per hour in the above equation. The number of starts per hour is 4.28.

Since 4.28 is less than 4.4, which was specified as the maximum number of required starts per hour, the duty period of 12 minutes on and 2 minutes off will meet the requirements of the NEMA Standard MG 10.

Increasing either the off time and/or the on time from the minimums listed above will decrease the number of starts per hour and save wear and tear on the mechanical equipment.

The NEMA standards recommend the minimum on and off time for cycling motor loads. It is also imperative to contact the compressor manufacturer for their recommendation as a minimum off time to allow the compressor to pump down and not start against a high-pressure head. Remember that duty cycling must be done properly in order to get the maximum savings.

Energy-saving Cycle Strategy

Now that the minimum on and off times have been calculated to protect the equipment from short cycling and premature failures, an energy-saving duty cycling strategy must be determined. The measure of the effectiveness of a duty cycle program is to determine that the loads are cycled in a manner to save electrical demand (to be discussed in programmable controllers) as well as consumption (kWh). Straight-time duty cycling is accomplished by starting the duty cycler at one point in time (for example, at 8:00 a.m.), Monday through Friday for a cycle of 15 minutes on time and 15 minutes off time, then stopping the duty cycle (say at 5:00 p.m.).

Whenever more than one load is to be duty cycled, there are two methods to provide the cycling program. The first method is to start all the loads at the same time and have them cycle on and off at the same time. This method is called parallel duty cycling. The second method to provide duty cycling to more than one load is to stagger the on and off times to avoid the in-rush current while evenly spreading the distribution time to each load. The decision now is which method to choose and what times to use.

The demand period is the time interval that the utility company checks the power consumption of a user to determine the maximum number of kW used during this interval. In most cases this period is 15 minutes.

The duty period is equal to the on time plus the off time. So, if the duty period is less than the demand period interval, the parallel duty cycling can be used. During the demand interval, the utility can detect if all of the loads are running at the same time. If the duty cycle period exceeds the demand time, then the staggered duty cycling is necessary if savings in kW demand are to be achieved.

Mechanical equipment manufacturers have specified that minimum duty cycle periods must be, in most cases, greater than the demand periods to ensure their equipment's life span. So, staggering load starts are usually a must to save demand charges. The mechanical equipment manufacturers are primarily concerned about short-cycle protection for the air-conditioner compressors. This allows the compressor to pump down before trying to start against a high-pressure head and thus burn out the motor.

Stagger Cycling and Savings

To show an example of the stagger method of duty cycling, assume that the manufacturer was contacted and he recommends a duty cycle of 20 minutes on and 10 minutes off as the minimum duty cycle. Anything greater in the on or off time would be acceptable. The duty period is 30 minutes. Say three compressors, each rated at 10 kW (approx. 3 tons), have to be cycled and the demand interval is 15 minutes.

$$\frac{\text{ON PERIOD} + \text{OFF PERIOD}}{\text{number of loads} = 3 \text{ loads}} = (20 + 10) \text{ min.}$$

The stagger rate equals 10 minutes per load.

Figure 10-9 shows the relationship between the stagger rate of 10 minutes per load, demand interval and the kW savings possible using the stagger function. Figure 10-10 has been generated from Figure 10-9. The demand for the three loads would be the highest demand recorded over any 15-minute interval. Since the highest demand recorded was 19.99 kW and the maximum potential demand would have been 30 kW, the savings are:

$$\frac{30 \text{ kW} - 19.90 \text{ kW}}{30 \text{ kW}} = 0.2933 = 2933\%$$

If duty cycling were done using the parallel method, there would be no savings of demand because all three loads would have been on for 20 minutes. This peak would show up during the first demand period and the second demand period.

Regardless of the cycling method, the actual energy consumption would be the same because the on time would, of course, be the same.

Percent consumption savings:

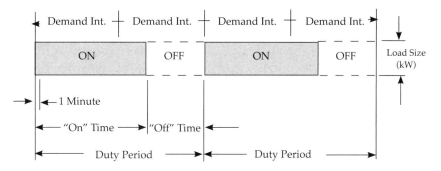

Figure 10-9. Stagger Rate

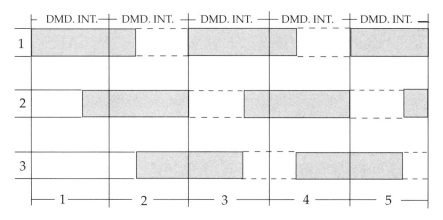

Figure 10-10. Load Stagger Duty Cycle

OFF TIME × 100/DUTY PERIOD
10 MIN OFF × 100/30 MIN = 33.33%

Duty cycling can save electrical expenses when tied to air-conditioning equipment, provided that the proper installation is done and proper maintenance followed. (For example, the mechanical equipment is not short-cycled—the correct on and off times are calculated for the equipment). Also, maintenance routines including proper lubrication, tightened belts, and clean filters should be regularly performed.

Thermostats Considered
The duty cycle example discussed did not take into account the cycle rate of the thermostat in the estimates of potential savings. To calculate the

actual savings, the thermostatic cycle rate must be known for a given condition of cooling requirements. For example, if the thermostat cycle rate were 30%, the savings in both demand and consumption would be divided by .30 to calculate the additional savings caused by cycling the load in addition to having the thermostat wired in series with the duty cycler.

The thermostat is a temperature-compensating duty cycler. It is essential in many applications where environmental temperature conditions along with energy savings are paramount.

Designers using duty cycles traditionally must balance energy savings against space comfort. With the newer compensated duty cycling, savings plus comfort can be attained.

Duty cycling has been done for years, but a duty cycler was not accomplishing the duty cycling-a thermostat was cycling the load on and off to maintain the set temperature conditions. Duty cyclers and compensated duty cycling programs based on time rather than set specific temperature have replaced the standard thermostat. The time element must be taken into account, particularly since the utility company is charging for electricity based on time.

Pneumatic Controllers

Pneumatic controllers are usually combined with sensing elements with a force or position output to obtain a variable air pressure output. The control mode is usually proportional, but other modes such as proportional-integral can be used. These controllers are generally classified as non-relay, relay, direct- or reverse-acting type.

The *non-relay* pneumatic controller uses a restrictor in the air supply and a bleed nozzle. The sensing element positions an air exhaust flapper that varies the nozzle opening, causing a variable air-pressure output applied to the controlled device.

A *relay-type* pneumatic controller, directly or indirectly through a restrictor, nozzle and flapper, actuates a relay device that amplifies the air volume available for control.

Direct-acting controllers increase the output signal as the controlled variable increases. For example, a direct-acting pneumatic thermostat increases output pressure when the sensing element detects a temperature rise.

Reverse-acting controllers increase the output signal as the controlled variable decreases. A reverse-acting pneumatic thermostat increases output pressure when the temperature drops.

As pointed out earlier, pneumatic controllers can be operated in either proportional or proportional-integral mode. System characteristics in this control mode having influence on energy efficiency are proportional band, reset time, over-capacity of the heating/cooling equipment and part-load operation. Proportional band is the ratio of the percent change in the actuating signal to the percent change in the manipulated variable. Reset time is defined as the time required by the reset to repeat the proportional action.

Programmable Controllers

A programmable controller is a control device that has logic potential but is not powerful enough to be called a computer-interfaced controller. This type of control fills a need for applications requiring more than a timer but not a computer. A programmable controller can do all that timers can do and much more but at a cost considerably less than that of a computer.

Because of the logic capability, programmable controllers are very useful in implementing control strategies like:

a) Demand control.
b) Economizer control.
c) Enthalpy control.
d) Combustion air control.
e) Temperature reset.

Demand Control

Demand charge is that portion of the electric bill which reflects the utility's capital requirements in generating plant, transmission lines, transformers, etc., to provide power to a facility. The demand load is the maximum concurrent load to occur during any 15- or 30-minute interval during the billing month, and it is measured in kilowatts. It is in the interest of the customer and the utility to keep the peak demand as low as possible. While this does not directly save energy, it definitely saves money. The following methods can be used to reduce peak demand:

(i)   Off-peak schedule.
(ii)   Thermal storage (see Chapter 12).
(iii) Demand-limiting controls.
(iv)   Programmable controllers.

In demand control by a programmable controller, when the controller senses that the electrical demand is approaching a critical (preset) level, it shuts off equipment on a programmed priority basis to restrict the demand from passing the critical level. Since demand charge generally amounts to approximately 25%-35% of the total electric bill, savings from controlling demand can be significant.

Economizer Cycle

An economizer cycle is the adaptation to the fresh-air intake which permits the use of outside air for cooling when temperatures are sufficiently low. When nighttime outdoor temperatures are below indoor setpoint by 5°F or more, the controller shuts off the refrigeration system and the return air dampers and opens the outdoor air dampers fully. Using outdoor air for night cooling will save energy in most areas.

Enthalpy Control

During occupied periods, the opportunities to use outdoor air for cooling will depend on the outdoor wet-bulb (WB) temperature as well as dry-bulb (DB) temperature. If the outdoor air is brought into the building above 60°F WB, the cooling load is increased. The wet-bulb temperature is a measure of the total energy content (enthalpy) of the air.

In enthalpy control, the controller is interfaced to DB sensors and to WB sensors for both the outdoor as well as the indoor air. The controller is programmed to calculate the enthalpies; if the enthalpy of the outdoor air is less than the enthalpy of the indoor air, 100 percent of the outdoor air is used for cooling, resulting in energy savings. (See Chapter 7 for Enthalpy Control Savings Calculations.)

Combustion Air Control

The programmable controllers can be used to control the excess air in a combustion process. By sensing CO_2 or O_2 or CO levels in exhaust, the controller can adjust the combustion air intake for optimal combustion efficiency. Continuous control through the use of a programmable controller allows the air intake to be adjusted in accordance with the demand on the unit.

Temperature Reset

The logic capability of the programmable controllers can be used to reset the setpoints on the chillers and the boilers. The controller can be interfaced to DB, WB, wind velocity and solar radiation sensors. The arithmetic logic of the controller can be used to calculate the optimal setpoints and reset accomplished which can result in significant energy savings and can enhance the comfort levels in the buildings.

COMPUTER CONTROL

The control of physical systems with a digital computer is becoming more and more common. Aircraft autopilots, mass transit vehicles, oil refineries, paper-making machines and countless electromechanical servomechanisms are among the many existing examples. Furthermore, many new digital control applications are being simulated by microprocessor technology, including on-line computer controllers in automobiles and household appliances. Among the advantages of digital logic for control are the increased flexibility of the control programs and the decision-making or logic capability of digital systems.

One desirable characteristic of a control system is to have a satisfactory response. "Satisfactory response" means that the output, c(t), is to be forced to follow or track the reference input, r(t), (See Figure 10-2) despite the presence of disturbance inputs to the plant (system) and despite errors on the sensor. It is also essential that the tracking succeeds even if the dynamics of the plant should change somewhat during the operation. The process of holding c(t) to r(t) is called *regulation.* A system which has good regulation in the face of changes in the system dynamics is said to have low *sensitivity* to system's parameters. A system which has both good disturbance rejection and low sensitivity is called *robust.* Use of digital computers as controllers can help create robust systems.

Figure 10-11 shows the use of a digital computer as a control element. Note that the notation on this figure has been changed compared with Figure 10-2 to incorporate digital computers as controllers.

In Figure 10-11, consider first the action of the analog-to-digital (A/D) converter on a signal. This device acts on a physical variable, most commonly an electrical voltage, and converts it into a stream of numbers (pulses). In Figure 10-11, the A/D converter acts on the indicated error signal, ê, and applies the numbers to the digital computer. It is also common for the sensor output, y, to be sampled and have the error formed in the computer. We need to know the times at which these numbers arrive if we are to analyze the dynamics of the system.

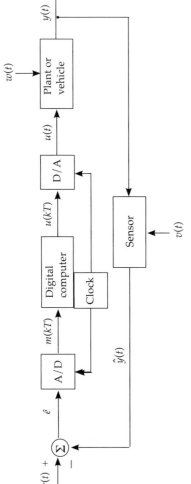

Figure 10-11. Digital Control System Block

Notation:

r = reference or command inputs
u = control or actuator input signal
y = controlled or output signal
$\hat{y}$ = instrument or sensor output, usually an approximation to or estimate of y. (For any variable, say θ, the notation $\hat{\theta}$ is now commonly taken from statistics to mean an estimate of (θ)).
$\hat{e}$ = $r - y$ = indicated error
e = $r - y$ = system error
w = disturbance input to the plant
v = disturbance or noise in the sensor
A/D = analog-to-digital converter
D/A = digital-to-analog converter

We may assume that all the numbers (pulses) arrive with the same fixed period T, called the *sampling period*. In practice, digital control systems sometimes have varying sample periods and/or different periods in different feedback paths. Typically, there is a clock as part of the computer logic which supplies a pulse every T seconds (See Figure 10-12), and the A/D converter sends a pulse (number) to the computer each time the pulse from the clock arrives. Thus, in Figure 10-11 we identify the sequence of numbers as m(kT). We conclude from the periodic sampling action of the A/D converter that some of the signals in the digital control system, like m(kT), are variable only at discrete times. We call these variables *discrete signals* to distinguish them from variables like ê and y which change continuously in time. A system containing only discrete variables is called a *discrete-time* system. A system having both discrete and continuous signals is called a *sampled-data* system.

In addition to generating a discrete signal, however, the A/D converter also provides a *quantized* signal. By this we mean that the output

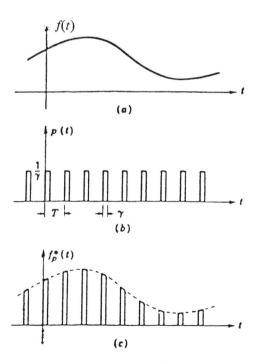

Figure 10-12. (*a*) **Continuous Function** *f*(*t*); (*b*) **Sampling Pulse Train** *p*(*t*); (*c*) **Sampled Function** $f_p^*(t)$.

of the A/D converter must be stored in digital logic composed of a finite number of digits. Most commonly, of course, the logic is based on binary digits (i.e., bits) composed of 0's and 1's, but the essential feature is that the representation has a finite number of digits. A common situation is that the conversion of ê to m is done so that m may be thought of as a number with a fixed number of places of accuracy. If we plot ê versus m, we may get a plot as shown in Figure 10-13.

In Figure 10-13, we would say that m has been truncated to one decimal place, or that m is *quantized* with q of 0.1 since m changes only in fixed quanta of, in this case, 0.1 units. A signal which is both discrete and quantized is called a *digital signal.*

In a real sense, the problems of analysis and design of *digital controls* are concerned with accounting for the effects of the sampling period T and the quantization size q. If both T and q are small, digital signals are nearly continuous, and continuous methods of analysis can be used.

Design of Digital Systems

As mentioned previously, we can place the systems of interest in three categories according to the nature of the signals present. These are:

1.   Discrete systems.
2.   Sampled-data systems.
3.   Digital systems.

In *discrete systems,* all signals vary at discrete times only. For the analysis of discrete systems, one must learn z-transforms of discrete signals

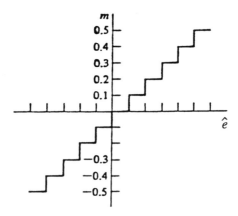

Figure 10-13. Plot of Output vs. Input Characteristics of A/D Device

and "pulse"-transfer functions for linear constant discrete systems. Of special interest is the characterization of the dynamic response of discrete systems. As a special case of discrete systems, we consider discrete equivalents of continuous systems, which is one aspect of the current field of *digital filters* (Kalman filter).

If quantization effects are ignored, digital filters are discrete systems which are designed to process discrete signals in such a fashion that the digital device (a digital computer program, for example), can be used to replace a continuous filter.

A *sampled-data* system has both discrete and continuous signals. In these systems, we are concerned with the question of data extrapolation to convert discrete signals (as they emerge from a digital computer) into the continuous signals necessary for providing the input to one of the plants. This action typically occurs in conjunction with the D/A conversion.

Digital controllers for the building systems may be one of the following types:

a. Direct digital control (DDC) or distributed control or remote control or on-site control.
b. Supervising control or central control.
c. Hierarchy control (DDC plus supervisory control).

A central control system provides the energy engineer with a means of constant surveillance of the building and helps in making efficient and effective use of physical plant systems and personnel. Proprietary central computerized control systems are marketed by each of the major temperature control manufacturers. These systems have common features and can accomplish a similar range of tasks, but each manufacturer uses coding and computer languages which are unique to the system.

Each manufacturer's system is made up of standard hardware, but the application is always tailored to the specific project. Basically, any system is composed of the following four major components:

• Interface panels which are located at strategic points throughout the building.

• The transmission system between the central console and all interface panels. This system can be single-core cable for digital transmission or multi-core cable for multiple transmissions.

- A central control console and associated computer hardware located in a control room. The console, computer and associated hardware form the point at which the operator enters all instructions and retrieves all data.

- Software programs. Programs for common applications such as start-stop action, load shedding, reset, alarm, optimization, etc. are available from the major control system suppliers.

Experiences with central control systems reported thus far in literature have not been very positive. Some of the reasons cited for ineffective use of the central control systems are inadequate training facilities for the operators, inadequate buyer commitment, poor vendor assessment and ineffective interface with the building system. In spite of these difficulties, if the energy engineer is involved with larger, more complex buildings (or groups of buildings), possible use of central computer-based control systems should be fully explored.

References
1. D'Azzo, J.J.; and Houpis, C.R. *Feedback Control Systems: Analysis and Synthesis.* New York. McGraw Hill Book Co., 1983.
2. Mehta, D. Paul. Dynamic Thermal Responses of Buildings and Systems. Ph.D. Dissertation. Iowa State University, Ames, IA, 1979.
3. *HVAC Systems and Applications Handbook.* ASHRAE, Atlanta, GA, 1987.
4. Smith, C.B. *Energy Management Principles.* Pergamon Press, 1981.
5. Kennedy W.J.; and Turner, W.C. *Energy Management.* Prentice-Hall Inc.; Englewood Cliffs, NJ, 1984.
6. Dubin, F.S.; and Long, C.G. *Energy Conservation Standards.* New York. McGraw Hill Book Co., 1978.

11

Energy Management

Energy management is the judicious and effective use of energy to maximize profits (minimize costs) and to enhance competitive positions. A successful energy management program is more than conservation. It is a total program that involves every area of a business. A comprehensive energy management program is not purely technical. It takes into account planning and communication as well as salesmanship and marketing. It affects the bottom line profits of every business; thus the individual who is assigned the role of "energy manager" has high visibility within the organization.

Energy management includes energy productivity, which is defined as reducing the amount of energy needed to produce one unit of output. Energy management includes energy awareness, which is essential in motivating employees to save energy. Probably the highest initial rate of return will occur through the establishment of a good maintenance management program.

This chapter reviews the basics of maintenance management and energy management organization.

MAINTENANCE MANAGEMENT

There are obvious losses from poor maintenance such as steam and air leaks and uninsulated steam lines. There are also losses from less obvious areas. For example Tables 11-1 and 11-2 illustrate the hidden effect of dirty evaporators and condensers on equipment performance. These losses go undetected and result in decreased capacity of equipment and an increase in energy usage. In Table 11-1, for a reciprocating compressor under the dirty condenser and evaporator conditions, capacity is reduced 25.4% and the increase in brake horsepower per ton is 39%. In Table 11-2, for an absorption chiller under similar conditions, capacity is reduced 23.8% and power requirements are increased 7.5%.

**Table 11-1. The Effects of Poor Maintenance on the Efficiency of a
Reciprocating Compressor, Nominal 15-ton Capacity**

Conditions	(1) °F	(2) °F	(3) Tons	(4) %	(5) HP	(6) HP/T	(7) %
Normal	45	105	17.0	—	15.9	0.93	
Dirty Condenser	45	115	15.6	8.2	17.5	1.12	20
Dirty Evaporator	35	105	13.8	18.9	15.3	1.10	18
Dirty Condenser and Evaporator	35	115	12.7	25.4	16.4	1.29	39

(1) Suction Temp, °F
(2) Condensing Temp, °F
(3) Tons of refrigerant
(4) Reduction in capacity %
(5) Brake horsepower
(6) Brake horsepower per ton
(7) Percent increase in compressor bh per/ton

**Table 11-2. The Effects of Poor Maintenance on the Efficiency of
an Absorption Chiller, 520-ton Capacity**

Conditions	Chilled Water °F	Tower Water °F	Tons	Reduction in Capacity %	Steam lb/ton/H	Per-cent
Normal	44	85	520	—	18.7	—
Dirty Condenser	44	90	457	12	19.3	3
Dirty Evaporator	40	85	468	10	19.2	2.5
Dirty Condenser and Evaporator	40	90	396	23.8	20.1	7.5

A third major loss is in missing the opportunity to upgrade the facilities within the spare parts program. For example, when a motor burns out it is usually replaced with the same model or sent to a shop for rewinding. If the motor is sent to a rewind shop, it will usually have poorer efficiency and power factor characteristics than before. Thus the energy manager should consider upgrading the replacement with either a high-efficiency motor or requiring a higher specification from the shop rewinding the motor.

The following summarizes key elements of the maintenance management program.

Preventive Maintenance Survey

This survey is made to establish a list of all equipment on the property that requires periodic maintenance and the maintenance that is required. The survey should list all items of equipment according to physical location. The survey sheet should list the following columns:

1. Item.
2. Location of item.
3. Frequency of maintenance.
4. Estimated time required for maintenance.
5. Time of day maintenance should be done.
6. Brief description of maintenance to be done.

Preventive Maintenance Schedule

The preventive maintenance schedule is prepared from the information gathered during the survey. Items are to be arranged on schedule sheets according to physical location. The schedule sheet should list the following columns:

1. Item.
2. Location of item.
3. Time of day maintenance should be done.
4. Weekly schedule with double columns for each day of the week (one column for "scheduled" and one for "completed").
5. Brief description of maintenance to be done.
6. Maintenance mechanic assigned to do the work.

Use of Preventive Maintenance (PM) Schedule—At some time before the beginning of the week, the supervisor of maintenance will take a copy of the schedule. The copy the supervisor prepares should be available in a three-ring notebook. He will go over the assignments in person with each mechanic.

After completing the work, the mechanic will note this on the schedule by placing a check under the "completed" column for that day and the index card system for cross-checking the PM program.

The supervisor of maintenance or the mechanic will check the schedule daily to determine that all work is being completed according to the plan. At the end of the week, the schedule will be removed from the book and checked to be sure that all work was completed. It will then be filed.

Preventive Maintenance Training

The supervisor of maintenance or a mechanic is responsible for assisting department heads in the training of employees in handling, daily care and the use of equipment. When equipment is mishandled, he must take an active part in correcting this through training.

Spare Parts

All too often equipment is replaced with the exact model as presently installed. Excellent energy conservation opportunities exist in upgrading a plant by installing more efficient replacement parts. Consideration should be given to the following:

*   Efficient line motor to replace standard motors.
*   Efficient model burners to replace obsolete burners.
*   Upgrading lighting systems.

Leaks-steam, Water and Air

The importance of leakage cannot be understated. If a plant has many leaks, this may be indicative of a low standard of operation involving the loss not only of steam but also of water, condensate, compressed air, etc.

If, for example, a valve spindle is worn, or badly packed, giving a clearance of 0.010 inch between the spindle, for a spindle of 3/4-inch diameter the area of leakage will be equal to a 3/32-inch diameter hole. Table 11-3 illustrates fluid loss through small holes.

Although the plant may not be in full production for every hour of the entire year (i.e., 8760 hours), the boiler plant water systems and compressed air could be operable. Losses through leakage are usually, therefore, of a continuous nature.

Thermal Insulation

Whatever the pipework system, there is one fundamental: It should be adequately insulated. Table 11-4 gives a guide to the degree of insulation required. Obviously, there are a number of types of insulating materials with different properties and at different costs, each one of which will give a variance return on capital. Table 11-4 is based on magnesia insulation, but most manufacturers have cataloged data indicating various benefits and savings that can be achieved with their particular product.

Table 11-3. Fluid Loss Through Small Holes

| Diameter of Hole | Steam- lb/hour | | Water-gals/hour | | Air SCFM |
	100 psig	300 psig	20 psig	100 psig	80 psig
1/16"	14	33	20	45	4
1/8"	56	132	80	180	26
3/16"	126	297	180	405	36
1/4"	224	528	320	720	64

Table 11-4. Pipe Heat Losses

| Pipe Dia Inches | Surface Temp °F | Insulation Thickness Inches | Heat Loss (Btu/Ft/Hr) | | Insulation Efficiency |
			Uninsulated	Insulated	
4	200	1-1/2	300	70	76.7
	300	2	800	120	85.0
	400	2-1/2	1500	150	90.0
6	200	1-1/2	425	95	78.7
	300	2	1300	180	85.8
	400	2-1/2	2000	195	90.25
8	200	1-1/2	550	115	79.1
	300	2	1500	200	86.7
	400	2-1/2	2750	250	91.0

Steam Traps

The method of removing condensate is through steam trapping equipment. Most plants will have effective trapping systems. Others may have problems with both the type of traps and the effectiveness of the system.

The problems can vary from the wrong type of trap being installed to air locking or steam locking. A well-maintained trap system can be a great steam saver. A bad system can be a notorious steam waster, particularly where traps have to be bypassed or are leaking.

Therefore, the key to efficient trapping of most systems is good installation and maintenance. To facilitate the condensate removal, the pipes should slope in the direction of steam flow. This has two obvious advantages in relationship to the removal of condensate: One is the action of gravity, and the other the pushing action of the steam flow. Under these circumstances the strategic siting of the traps and drainage points is greatly simplified.

One common fault that often occurs at the outset is installing the wrong size traps. Traps are very often ordered by the size of the pipe connection. Unfortunately, the pipe connection size has nothing whatsoever to do with the capacity of the trap. The discharge capacity of the trap depends upon the area of the valve, the pressure drop across it and the temperature of the condensate.

It is therefore worth recapping exactly what a steam trap is. It is a device that distinguishes between steam and water and automatically opens a valve to allow the water to pass through but not the steam. There are numerous types of traps with various characteristics. Even within the same category of traps, e.g., ball floats or thermo-expansion traps, there are numerous designs, and the following guide is given for selection purposes:

1.   Where a small amount of condensate is to be removed, an expansion or thermostatic trap is preferred.

2.   Where intermittent discharge is acceptable and air is not a large problem, inverted bucket traps will adequately suffice.

3.   Where condensate must be continuously removed at steam temperatures, float traps must be used.

4.   When large amounts of condensate have to be removed, relay traps must be used. However, this type of steam trap is unlikely to be required for use in the food industry.

To insure that a steam trap is not stuck open, a weekly inspection should be made and corrective action taken. Steam trap testing can utilize several methods to insure proper operation:

•   Install heat sensing tape on trap discharge. The color indicates proper operation.

- Place a screwdriver to the ear lobe with the other end on the trap. If the trap is a bucket-type, listen for the click of the trap operating.

- Use acoustical or infrared instruments to check operation.

ENERGY MANAGEMENT ORGANIZATION

A common problem facing energy managers is that they have too much responsibility and very little authority to get the job done. A second problem is the lack of definition of the job. Probably the biggest problem is the lack of true commitment from top management. As pointed out in the text, an overall energy utilization program requires an "investment" in order to get the return desired.

The first phase of the program should start with top management establishing the organization, defining the goals and providing the resources for doing an effective job. Tables 11-5 and 11-6 illustrate a checklist for top management and typical energy conservation goals.

A key element in any energy management program is "flexibility." For example, in one year natural gas prices may accelerate steeply. Unless an organization reacts quickly, it soon may be out of business. Every energy management program should have a contingency or backup plan. Periodically, the risk associated with each scenario should be reviewed. Risk assessment can help the energy manager determine the resources required to meet various emergency scenarios. A method of evaluating the cost to business for each scenario is illustrated in Formula 11-1.

$$C = P \times C_1 \qquad \qquad \textit{Formula (11-1)}$$

Where

C = cost as a result of emergency where no contingency plans are in effect.

P = the probability or likelihood the emergency will occur.

C_1 = the loss in dollars as a result of an emergency where no plan is in effect.

Implementing the Energy Management Program

The first phase of the energy management program involves the accumulation of data. Table 11-7 illustrates the minimum information required to evaluate the various energy utilization opportunities. The energy manager must also review government regulations which will affect

Table 11-5. Checklist for Top Management

A. Inform line supervisors of:
  1. The economic reasons for the need to conserve energy.
  2. Their responsibility for implementing energy saving actions in the areas of their accountability.
B. Establish a team having the responsibility for formulating and conducting an energy conservation program and consisting of:
  1. Representatives from each department in the plant.
  2. A coordinator appointed by and reporting to management.
  NOTE: in smaller organizations, the manager and his staff may conduct energy conservation activities as part of their management duties.
C. Provide the team with guidelines as to what is expected of them:
  1. Plan and participate in energy saving surveys.
  2. Develop uniform record keeping, reporting, and energy accounting.
  3. Research and develop ideas on ways to save energy.
  4. Communicate these ideas and suggestions.
  5. Suggest tough, but achievable, goals for energy saving.
  6. Develop ideas end plans for enlisting employee support and participation.
  7. Plan and conduct a continuing program of activities to stimulate interest in energy conservation efforts.
D. Set goals in energy saving:
  1. A preliminary goal at the start of the program.
  2. Later, a revised goal based on savings potential estimates from results of surveys.
E. Employ external assistance in surveying the plant and making recommendations, if necessary.
F. Communicate periodically to employees regarding management's emphasis on energy conservation action and report on progress.

Adapted from *NBS Handbook, 115.*

the program.

All too often the energy manager is the last to find out information after the fact. For example, the procurement of motors may be the electrical department's responsibility. The energy manager must get his or her input into the evaluation process prior to purchase. One way is to establish a review process where important documents are initialed by the energy manager before they are issued. Key documents include:

- Bid summaries for major equipment.
- One-line diagrams.
- Process flow diagrams.

Table 11-6. Typical Energy Conservation Goals

1. Overall energy reduction goals
   (a) Reduce yearly electrical bills by _____%.
   (b) Reduce steam usage by _____%.
   (c) Reduce natural gas usage by _____%.
   (d) Reduce fuel oil usage by _____%.
   (e) Reduce compressed air usage by _____%.
2. Return on investment goals for individual projects
   (a) Minimum rate of return on investment before taxes is _____.
   (b) Minimum payout period is _____.

   (c) Minimum ratio of $\dfrac{\text{Btu/year savings}}{\text{capital cost}}$ is _____.

   (d) Minimum rate of return on investment after taxes is _____.

Table 11-7. Information Required to Set Energy Projects on the Same Base

Fuel	Cost At Present	Estimated Cost Escalation Per Year	Energy Equivalent
1. Energy equivalents and costs for plant utilities.			
Natural gas	$_____/1000 ft^3	$_____/1000 ft^3	_____Btu/ft^3
Fuel oil	$_____/gal	$_____/gal	_____Btu/gal
Coal	$_____/ton	$_____/ton	_____Btu/lb
Electric power	$_____/kWh	$_____/kWh	_____Btu/kWh
Steam			
_____psig	$_____/1000 lb	$_____/1000 lb	_____Btu/1000 lb
_____psig	$_____/1000 lb	$_____/1000 lb	_____Btu/1000 lb
_____psig	$_____/1000 lb	$_____/1000 lb	_____Btu/1000 lb
Compressed air	$_____/1000 ft^3	$_____/1000 ft^3	_____Btu/1000 ft^3
Water	$_____/1000 lb	$_____/1000 lb	_____Btu/1000 lb
Boiler make-up water	$_____/1000 lb	$_____/1000 lb	_____Btu/1000 lb

2. Life Cycle Costing Equivalents

After tax computations required	_____
Depreciation method	_____
Income tax bracket	_____
Minimum rate of return	_____
Economic life	_____
Tax credit	_____
Method of life cycle costing (annual cost method, payout period, etc.)	_____

- Heat and material balances.
- Plot plan.
- Piping and instrument diagrams.

It is much easier to add local instrumentation to the piping and instrument diagram prior to construction. Another way of insuring that the energy manager gets timely input is to incorporate these activities into the overall planning document, such as the critical path or PERT schedules. Figure 11-1 illustrates a planning schedule incorporating input from the energy manager. Figure 11-2 illustrates how energy management activities interact with various departments.

Richard L. Aspenson, director of Facilities Engineering and Real Estate at 3M, summarized* the areas a good energy manager must master as follows:

Technical Expertise

Energy management normally begins with a solid technical background—preferably in mechanical, electrical or plant engineering. Managers will need a good grasp of both design theory and the nuts-and-bolts details of conservation programs. This includes a thorough understanding of the company's processes, products, maintenance procedures and facilities.

Communication

In the course of a single week, energy managers might find themselves dealing with lawyers, engineers, accountants, financial planners, public relations specialists, government officials, and even journalists and legislators. A good energy manager has to be able to communicate clearly and persuasively with all of these people—in *their* language. Above all, energy managers must be able to sell the benefits of their programs to top management.

Financial Understanding

To enlist the support of top management, energy managers will have to develop and present their programs as investments, with predictable returns, instead of as unrecoverable costs. They will have to demonstrate

---

*"The Skills of the Energy Manager," *Energy Economics, Policy and Management*, Vol. 2, No. 2, 1982.

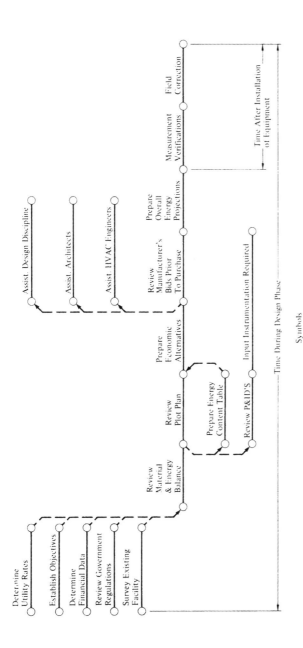

Figure 11-1. Schedule of Energy Management Activities

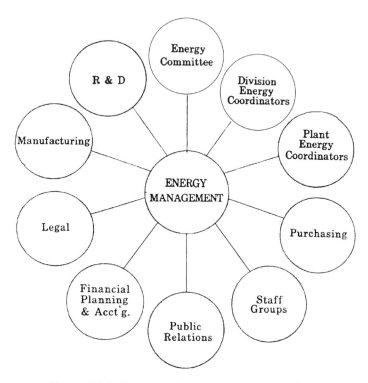

Figure 11-2. Energy Management Interactions

what kind of returns—in both energy and cost savings—can be expected from each project, and over what period of time. This means first of all developing some credible way to measure returns-a method that will be understood and accepted by the financial officers of the company.

Planning and Strategy

A basic part of energy management is forecasting future energy supplies and costs with reasonable accuracy. This means coming to grips with the complexities of worldwide supply, market trends, demand projections, and the international political climate. There is no way, of course, to predict all of these things with certainty, but every business-especially energy—intensive ones—will need some kind of reliable forecasting from now on.

Community Relations and Public Policy

Energy managers have some responsibility to go outside their companies to share their ideas and experience with a variety of publics. Trade and professional associations can become clearing houses for new ideas. Legislators and government agencies need, and often welcome, expert help in setting standards and policies. And the public needs help in understanding what is at stake in learning to use limited supplies of energy wisely and efficiently.

Energy management can be an exciting, challenging and rewarding job if you put your skills to work.

We as energy managers are a special breed of people who have unlimited opportunities if we develop the skills and implement the programs that will improve the profitability of our respective businesses. We can and must influence public policy decision by using our technical knowledge in a constructive and objective way.

Pointers for the Energy Manager
1. Be aggressive. Learn how to say *no* in a diplomatic way.
2. Energy is not a soft sell. Stick your neck out.
3. Be sure you anticipate questions before you request approval on energy expenditures. Prepare factual data—think ahead.
4. Be creative. Identify needs. Prioritize your action plan tasks.
5. Be a positive person—one with perpetual motion-a catalyst for action.
6. Establish a free-thinking environment. Be a good listener.
7. Develop a five-year plan. Update yearly. Include a strategy for implementation—action plans.
8. Establish credibility through an accurate energy accountability system.
9. Efficient use of energy is an evolutionary process. Be patient but demanding. Follow through on commitments.
10. Give individual recognition for achievements.

Energy managers can turn the problems of a changing energy situation into opportunities. At the same time, we can grow to become better corporate citizens of our society.

12

Compressed Air System Optimization

INTRODUCTION

Compressed air is often referred to as the fourth utility, along with electricity, oil/gas, and water. The cost of compressed air like all utilities is not free and must be managed. Industry sources have estimated that the total connected horsepower of factory compressed air systems in the U.S. exceeds 17 million. This represents a worthy target for the application of energy efficient technologies because many energy-conscious engineers exposed to factory environments believe that from 10% to 35% of this could be saved. Using a national energy cost per kilowatt-hour of 8 cents, this translates to billions of dollars in operating cost reductions for the manufacturers. Hence, understanding of the strategies for optimizing the compressed air systems is very important for energy engineers.

COMPONENTS OF A COMPRESSED AIR SYSTEM

Compressed air systems consist of a supply side, which includes compressors and air treatment, and a demand side, which includes distribution and storage systems and enduse equipment. A properly managed supply side will result in clean, dry, stable air being delivered at the appropriate pressure in a dependable, cost-effective manner. A properly managed demand side minimizes wasted air and uses compressed air for appropriate applications. Improving and maintaining peak compressed air system performance requires addressing both the supply and demand sides of the system and how the two interact.

A typical modern industrial compressed air system is composed of several major subsystems and many sub-components. Major subsystems include the compressor, prime mover, controls, treatment equipment and accessories, and the distribution system. The compressor is the mechanical device that takes in ambient air and increases its pressure. The prime mover powers the compressor. Controls serve to regulate the amount of compressed air being produced. The treatment equipment removes contaminants from the compressed air and accessories keep the system operating properly. Distribution systems are analogous to wiring in the electrical world-they transport compressed air to where it is needed. Compressed air storage can also serve to improve system performance and efficiency. Figure 12-1 shows a representative industrial compressed air system and its components.

TYPES OF AIR COMPRESSORS

Several types of air compressors are available in the market today. Each type is specially designed to best operate under some pre-set operating conditions. As shown in Figure 12-2, there are two basic compressor types: positive-displacement and dynamic. In the positive-displacement type, a given quantity of air or gas is trapped in a compression chamber and the volume which it occupies is mechanically reduced, causing a corresponding rise in pressure prior to discharge. Dynamic compressors impart velocity energy to continuously flowing air or gas by means of impellers rotating at very high speeds. The velocity energy is changed into pressure energy both by the impellers and the discharge volutes or diffusers.

The two main types of compressors that are commonly used are the reciprocating and the screw compressors. Both of these compressors are classified as positive displacement compressors.

The reciprocating compressors compress air with the use of a piston/cylinder assembly. Consecutive quantities of air are trapped in the cylinder. The piston (driven by the electric motor) reduces the volume of the air causing its pressure to rise. When the pressure reaches a preset value, it is automatically discharged out of the cylinder by the opening

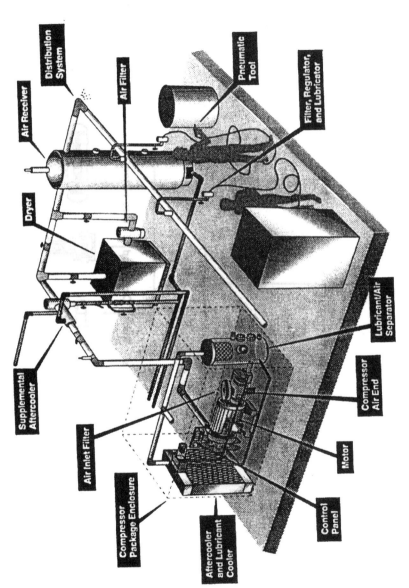

Figure 12-1. Components of an Industrial Compressed Air System

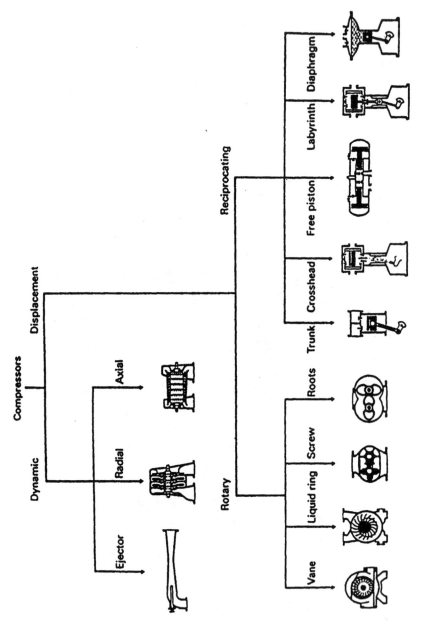

Figure 12-2. Air Compressor Types

of a valve. Many types of reciprocating compressors are available in the market. They could be classified as single or multiple stage, single or multiple port, single or double acting, or any combination of these. The most common type being is the two- port two-stage compressor.

With screw air compressors, the air is compressed between the intermeshing helical lobes and the rotor housing as it moves axially towards the discharge end. Screw compressors are classified as single or multiple stage. Reciprocating compressors are characterized by the vibrations they cause while in operation. The vibrations are caused by the movement of the pistons. Screw compressors are characterized by the noise they produce.

Most air compressors are controlled in a similar manner. When the demand for compressed air is high, the compressor would operate in a loaded state where it consumes power closely matching its rated nameplate horsepower. When the demand for compressed air is low, the compressor stops pumping air and it would then run in an unloaded state. In that case, the compressor would consume power between 20 and 70% of its rated power as is the case with screw compressors or it could simply shut off, as it is usually the case for reciprocating compressors.

Reciprocating compressors are best used in applications requiring high pressure delivery (above 140 psig), and when the demand for compressed air varies significantly throughout the working day. Most common sizes are under 30 hp. Screw compressors are selected for a low pressure requirement (under 125 psig) and in cases where constant demand for air is needed. Most common sizes start at 25 hp.

Some manufacturers classify air compressors by their isentropic efficiency, which is the efficiency of the compression cycle. For reciprocating compressors, the average isentropic efficiency is estimated to be about 75%. For screw compressors, the isentropic efficiency is estimated to be about 82%. In addition to the isentropic efficiency, the efficiency of the electric motor driving the compressor should be considered. For our study, we should also consider the operating efficiency, which is based on the selection of an air compressor to fit certain operating conditions. As mentioned previously, reciprocating compressors are selected for applications of varying demands for compressed air while the screw compressors are best selected for applications when a constant demand for compressed air exists.

COOLING

In a compression cycle, only a small fraction of the work input is used to compress air and the balance is converted into heat. And thus, the compressor and the compressed air need to be cooled down to acceptable temperatures. For example, a 100 hp screw air compressor could generate well over 260,000 Btu/hr of heat. Cooling could be done by either mechanical ventilation or by the use of a liquid to liquid heat exchanger. In the later technology a cooling fluid, usually city water, is used.

In the case of mechanical ventilation, compressed air mixed with oil go through a liquid to air heat exchanger (resembles the radiator in a car). Air is blown through the heat exchanger by the use of a fan. The fan could be driven by the compressor motor or by a separate motor. The oil and air are cooled and then separated in the oil separator. In this case, waste heat could be easily captured and utilized for space heating purposes or any other purpose.

In the case of water cooling, city water runs through a liquid to liquid heat exchanger where the other liquid is the compressor's oil. The oil is cooled and then separated from the air in the oil separator. In some cases, water is circulated through a cooling tower. For either of these cases usually additional equipment is needed to reduce the consumption of water and to enable heat recovery.

COMPRESSED AIR QUALITY

Compressed air is pressurized atmospheric air. The composition of dry air is 78% Nitrogen, 21 % Oxygen, and 1 % other gases. In addition, this air is humid and carries many impurities and solid particles. On the average, ambient air, at 75% relative humidity and 75°F carries approximately 18 gallons of water per day into a compressor with a capacity of 100 CFM operating at 100 psig. If this water is not collected properly, it would be carried with the air in the lines causing malfunction of the equipment. For example, in a painting application, water in the lines could cause rejects in the paint leading to rework, loss of time, and increased energy consumption.

The six main levels of compressed air quality are listed below:

Level 1: Filtered centrifugal separator
Removes all solids three microns and larger. Removes liquids; 99% of water droplets, 40% of oil aerosol. Recommended to be used as the primary stage for all compressed air treatment levels.

Level 2: Refrigerated compressed air dryer, oil line filter
Removes moisture and all water content. Removes 79% of oil aerosols and all solid particles one micron and larger. Recommended for air tools, sand blasting, and pneumatic control systems.

Level 3: Refrigerated compressed air dryer, oil removal filter
Removes moisture and all water content. Removes 99.99% of oil aerosols and solid particles 0.025 microns and larger. Recommended for paint spraying, powder coating, and packaging machines.

Level 4: Refrigerated air dryer, oil removal filter and vapor absorber
Removes moisture and all water content. Removes 99.99% of oil aerosols and solid particles 0.025 microns and larger. In addition, removes oil vapor, oil smell and taste. Recommended for food industries, laboratories, chemical, and pharmaceutical industry.

Level 5: Air line filter, oil removal, and low dew point drier
Removes moisture producing a –40 to –150°F pressure dew point. Removes 99.99% of oil aerosols and solid particles 0.025 microns and larger. Recommended for outdoor pipelines, breweries, and electronics industries.

Level 6: Breathing air system
Removes common harmful compressed air contaminants and will produce grade D breathing air. Recommended for breathing air.

COMPRESSED AIR SYSTEM AUDITS

A compressed air system audit can highlight the true costs of compressed air and identify simple opportunities to improve efficiency and productivity.

Compressed air system users should consider using an independent auditor to assess their compressed air system. Several firms exist that specialize in compressed air system assessments. Audits are also performed by electric utilities, equipment distributors and manufacturers, energy service companies, engineering firms, and industrial assessment centers. An informed consumer should be aware that the quality and comprehensiveness of audits can vary.

A comprehensive compressed air system audit should include an examination of both air supply and usage and the interaction between the supply and demand. Auditors typically measure the output of a compressed air system, calculate energy consumption in kilowatt-hours, and determine the annual cost of operating the system. The auditor may also measure total air losses due to leaks and locate those that are significant. All components of the compressed air system are inspected individually and problem areas are identified. Losses and poor performance due to system leaks, inappropriate uses, demand events, poor system design, system misuse, and total system dynamics are calculated, and a written report with a recommended course of action is provided.

Data Collection

A data sheet of the type shown in Figure 12-3 should be completed for each one of the compressors in a facility during the audit visit. As shown in the figure, data collection includes obtaining the compressor's power draw and cycling time. The compressor's power draw is best measured using a true rms wattmeter. Such a meter would compensate for power factor and load factor influences as well as the compressor's motor efficiency. Hence, results obtained correspond directly to utility meter data. The power draw should be measured during the loaded and unloaded cycles.

The cycling time of the compressor can be obtained by timing the compressor, in seconds for instance, during the loaded and unloaded cycles. The loaded and unloaded cycles can be identified by simply hearing the compressor changing cycles (when loaded the compressor operates at a higher noise level than when it is unloaded). If that can not be done by simple hearing, observing the wattmeter should lead to the same results (when the compressor is loaded the wattmeter reads a higher power draw than when it is not).

Cycling times of the compressor should be done at different time periods during the working day. As an example, a set of five data points

Compressor Data Sheet

Compressor Brand: _____   Type: _____

Horsepower: _____   Efficiency: _____

Operating Pressure: _____   Maximum Pressure: _____

Capacity: _____   Intake Temperature _____

Power Draw (Loaded) _____   Power Draw (Unloaded): _____

Timing Table

Time	Operation	Time in seconds				
	Loaded					
	Unloaded					
	Loaded					
	Unloaded					
	Loaded					
	Unloaded					
	Loaded					
	Unloaded					
	Loaded					
	Unloaded					

Fraction Loaded: _____   Fraction Unloaded: _____

Exhauast Rate of Flow: _____   Exhaust Temperature:_____

Type of Oil Used: _____   Gearbox Capacity:_____

Water Cooled? _____   Water Flow Rate: _____

Water Inlet Temperature: _____   Water Output Temperature: _____

Date: _____   Signature: _____

Figure 12-3. Data Sheet

can be obtained once in the morning, once before lunch break, once after lunch, once after second shift starts, and once before second shift ends. This would give a total of five sets of five data points (a total of 25 data points) taken throughout the entire working day. The data points are then averaged out to obtain a fraction for the compressor when it is loaded and one when its unloaded. The more data points collected during different time periods, the better the results would be.

ENERGY CONSUMPTION IN COMPRESSED AIR SYSTEMS

The prime mover is the main power source providing energy to drive the compressor. The prime mover must provide enough power to start the compressor, accelerate it to full speed, and keep the unit operating under various design conditions. This power can be provided by any one of the following sources: electric motors, diesel or natural gas engines, or steam engines or turbines. Electric motors are by far the most common type of prime mover. Thus, to calculate the energy consumption of an air compressor, one should calculate the energy consumption of the driving motor. In some cases, the compressor is cooled by a cooling fan that could be powered by a different motor. The energy used by that motor should also be considered.

ENERGY SAVINGS IN COMPRESSED AIR SYSTEMS

Generally, an assessment of the compressed air systems can lead to one or more of the following assessment recommendations (AR's) to save energy and to reduce the operating costs.

Use Outside Air for Compressor Intakes

Whenever feasible, the intake duct for an air compressor should be run to the outside of the building, preferably on the north side or the coolest side. Since the average outdoor temperature is usually well below that in the compressor room, it is possible to reduce the energy requirement for compression by increasing the average density of the intake air. Higher is the density, lower is the work required to compress the air to a desired pressure.

Reduce Compressor Air Pressure

If air actuated equipment in a compressed air system is surveyed and it is determined that the air compressor operating pressure can be lowered a certain amount without causing any performance problems, the energy required for compression can be decreased. When operating at a lower pressure, air compressors take less time loaded to reach that pressure. Thus savings would be generated by decreasing the loaded fraction of time and consequently, increasing the unloaded time or shut off in some reciprocating applications.

A survey or a test should be conducted to estimate the pressure drop in the compressed air lines. The pressure drop found should be added to the minimum pressure required to operate the air-assisted equipment properly in order to determine the new set point. In many cases you could simply set the compressor operating pressure about 10 or 15 psig above the maximum required on the line.

Use Synthetic Lubricants

For new compressors, it is always recommended to use synthetic lubricants and grease to lubricate bearings, gearboxes, transmissions, motor drives, and other contact points that require lubrication. In older compressors such a recommendation is limited by the condition of the compressor.

The use of synthetic oils has been demonstrated to provide energy savings through reduced friction and by providing an extended oil life of about four times that of premium motor oils. Manufacturers claim energy savings of four to eight percent when synthetic lubricants replace conventional lubricants. A six-percent energy savings can be assumed to be on the conservative side. The savings would occur due to a reduction in the power draw by the motor while loaded and while unloaded.

Change Current Compressor Operations

The operation of screw air compressors can be controlled in several ways. For example, a compressor might be programmed to operate loaded all the time regardless of the air requirements. In such a case, when the air requirements are met and the pressure in the receiver tank reaches its set limit, an intake valve closes in the compressor disallowing intake air into the compressor. The power draw by the compressor in such application is almost constant (fluctuates by a magnitude of less than one horsepower) and usually equals to the nameplate horsepower. This type of operation is used in plants where a steady and constant demand of compressed air occurs. This operation is considered efficient if the demand for compressed air is very close to the compressor's capacity (within 20 cfm). In other cases, the compressor is allowed to cycle loaded and unloaded (modulate) depending on the demand for compressed air. This type of operations is considered to be the most efficient.

The two operations described above are the most common ones.

Other cases include the use of an electronic controller, which modulates the operations of the compressor. In this case, compressor would run loaded and unloaded based on a preset schedule and not based on the needs for air.

Tests should be performed before selecting a particular type of operation. The test should first show that all air driven equipment would run with no deficiencies. Also, the test should include an analysis to determine the annual operating cost (mainly electrical cost) of the compressor under each type of operation. The results can be used to select the most efficient mode of operation.

Install a Small Compressor

Depending on the demands for compressed air, some applications would be more efficient by running different size air compressors at different times. It is recommended to run a smaller air compressor to more closely match the air requirements during periods of low compressed air usage. Using a smaller compressor during these periods will reduce the energy usage and associated costs since the smaller compressor will operate at a better overall efficiency than the larger unit (during low-load periods).

Potential savings from this measure are due to the fact that compressors that operate in a loaded-unloaded fashion operate more efficiently when running loaded for a large percentage of the time, as opposed to running unloaded for much of the time.

The selection of a suitable size air compressor can be done after estimating the total cfm needed. The cfm can be estimated after identifying what machinery is going to be used during the low-load periods. Another alternative is to estimate the cfm needed from the operation of the current compressor. An average required cfm could be calculated based on the capacity and on the average loaded period of the current compressor. If the compressor delivers, let us say, 500 cfm under the current operating condition and operates loaded for about 10% of the time, average air needs are 50 cfm only.

Recover Waste Heat From An Air Cooled Air Compressor

In most compressed air applications, the heat generated by the compressor and in some cases, by the air dryer can be captured and used for space heating. If the compressor is located in a separate room, it is always recommended to install ductwork and controls to recover the waste heat generated by the compressor for space heating. The duct work should be designed so that the waste heat can be exhausted to the outside during the summer. As an alternative, the compressor's waste heat could be used to preheat combustion intake air or for drying processes.

Usually, space heating is done by natural gas fired heaters and the energy savings in such a case would be a reduction in the natural gas energy usage (therms). If the plant is heated electrically, the same principle and formula can be used to calculate the energy savings (kWh). In addition, a complete study should be conducted to estimate the demand reduction associated with this recommendation.

Recover Waste Heat From a Water Cooled Compressor

Some air compressors use water for cooling. Usually, in such a case, the compressor uses city water which would run through cooling coils inside the compressor and the intercooler and then it would be discharged into the sewer system. When water is not recycled or reused for other processes, it is recommended that a closed loop cooling system should be installed to eliminate the need for city water for cooling and to recover the waste heat generated by the compressor.

The closed loop cooling system contains a liquid to air heat exchanger that eliminates the use of water and may be used for space heating. A typical system consists of an elevated heat exchanger, a cooling fan, and a circulation pump. The system is directly connected to the existing cooling system. Usually a fluid made up of a mixture of 50% glycol and 50% water is used to transfer heat.

Sometimes it may be more economical to install water flow controls than to install a closed loop cooling system. This will be true if electric costs to operate the cooling fans and the circulating pumps are much higher than the cost savings from water conservation and heat recovery.

Reduce Compressed Air Leaks

Leaks can be a significant source of wasted energy in an industrial compressed air system, sometimes wasting 20-30% of a compressor's output. A typical plant that has not been well maintained will likely have a leak rate equal to 20% of total compressed air production capacity. On the other hand, proactive leak detection and repair can reduce leaks to less than 10% of compressor output.

In addition to being a source of wasted energy, leaks can also contribute to other operating losses. Leaks cause a drop in system pressure, which can make air tools function less efficiently, adversely affecting production. In addition, by forcing the equipment to cycle more frequently, leaks shorten the life of almost all system equipment (including the compressor package itself). Increased running time can also lead to additional maintenance requirements and increased unscheduled downtime. Finally, leaks can lead to adding unnecessary compressor capacity.

While leakage can come from any part of the system, the most common problem areas are:

- Couplings, hoses, tubes, and fittings,
- Pressure regulators,
- Open condensate traps and shut-off valves, and
- Pipe joints, disconnects, and thread sealants.

CALCULATIONS
FOR ENERGY CONSUMPTION AND SAVINGS

A summary of the mathematical models to calculate energy consumption, energy savings, and cost savings for several assessment recommendations (AR's) described above are given in this section. Applications of these models for the various AR's are illustrated in the next section with the help of numerical examples.

Nomenclature For Mathematical Models

C_1 = conversion constant, 0.746 kW/hp
C_2 = conversion constant, 12 month/yr
C_3 = conversion constant, 60 min/hr
C_4 = conversion constant, 144 in²/ft²
C_5 = conversion constant, 3.03×10^{-5} hp - min/ft-lb
C_6 = isentropic sonic volumetric flow constant, 49.02 ft/sec °R$^{0.5}$
C_7 = conversion constant, 60 sec/min
C_8 = coefficient of discharge for square edged orifice, 0.6 no units
C_9 = pythagorean constant, 3.1416, no units
Cp = specific heat of air, 0.24×10^{-6} MMBtu/lbm°F
CS = cost savings, $/yr

D   = leak diameter, inches (estimated from observation)

DEN = density of water, 8.342 lbs/gal

DF   = demand factor, estimated fraction of compressor power seen as monthly demand by the utility company, no units

DI   = demand increase in the utility bill due to compressor operation, kWm/yr

DR   = annual peak demand reduction, kWm/yr

E   = effectiveness of the heat exchanger, no units

EF   = efficiency of driving motor, no units

EFF = efficiency of space heating system, no units

El   = increase in energy, kWyr

ES   = energy savings, expressed in kWh/yr for electrical related savings and in MMBtu/yr for natural gas related savings

EWF = estimated weighted fraction of leaks eliminated, 85%

FLc = estimated fraction of the time current compressor is loaded, no units

FLp = estimated fraction of the time proposed compressor is loaded, no units

FR   = calculated ratio of proposed power consumption to current power consumption based on operating pressure, no units

FS   = estimated fractional energy savings, 6%

FTL = fraction of operating time loaded, no units

FTU = fraction of operating time unloaded, no units

FUc = estimated fraction of the time current compressor is unloaded, no units

FUp = estimated fraction of the time proposed compressor is unloaded,

H   = hours per year of compressor operations, hrs/yr

HF   = fraction of the heat that can be utilized, no units

HH   = annual hours during which waste heat can be used for heating, hrs/yr

HLc = horsepower draw by the current compressor while loaded, hp

HLp = horsepower draw by the proposed compressor while loaded, hp

HP   = horsepower of driving motor, hp

HPc = current average horsepower draw by the compressor, hp

HPL = actual horsepower draw while loaded, hp

HPU = actual horsepower draw while unloaded, hp

HT = heat transfer from compressor to water, Btu/hr

HUc = horsepower draw by the current compressor while unloaded, hp

HUp = horsepower draw by the proposed compressor while unloaded, hp

k = ratio of specific heats, (for air, k = 1.4), no units

L = power loss due to loads, hp

LF = Estimated fraction of rated load at which the equipment will operate, no units

M = number of months considered, months/yr

MPDc = monthly peak demand due to current compressor operation, kWm/yr

MPDp = monthly peak demand due to proposed compressor operation, kWm/yr

N = number of stages, no units

NL = number of leaks, no units

OPC = operating cost, $/yr

Pdc = current discharge pressure, psia

Pdp = proposed discharge pressure, psia

Pi = inlet pressure, 14.7 psia

PL = line pressure at leak in questions, psia

Q = rate of air flow from the cooling system, cfm

RHO = density of air at exhaust temperature, lbM/ft³ estimated to be equal to [(Pi) × 144)] / [(53.34 × (Te + 460)]

Ta = setpoint temperature in room to be heated, °F

Te = average exhaust air temperature, °F

TI = average temperature of inside air, °F

TL = average line temperature, °F

TO = annual average outside air temperature, °F

Vf = volumetric flow rate of free air, cfm

WFR = water flow rate, gal/hr

WH = waste heat that could be recovered, Btu/yr

WR = reduction in compressor work due to utilizing cooler intake air, no units

WS = water savings, gal/yr

Mathematical Models

Energy Consumption

$$EC_c = [(HPL \times FTL) + (HPU \times FFU)] \times C_1 \times H$$

Demand Increase

$$DI = [(HPL \times FTL) + (HPU \times FFU)] \times C_1 \times DF \times M$$

AR No. 1 - Use Outside Air For Compressor Intakes

$$ES \quad [(HPL \times FTL) + (HPU \times FTU)] \times C_1 \times H \times WR$$

$$DR \quad [(HPL \times FTL) + (HPU + FTU)] \times C_1 \times DF \times M \times WR$$

where
$$WR \quad [(TI - TO)/(TI + 460)]$$

AR No. 2 - Reduce Air Compressor Pressure

$$ES = [(HPL \times FTL) + (HPU \times FTU)] \times H \times C_1 \times (1 - FR)$$
$$DR = [(HPL \times FTL) + (HPu + FTu)] \times C_1 \times DF \times M \times (1 - FR)$$

where
$$FR = \frac{(Pdp/Pi)^{N \times (k-1)/k} - 1}{(Pdc/Pi)^{N \times (k-1)/k} - 1}$$

AR No. 3 - Use Synthetic Lubricant

$$ES = [(HPL \times FTL) + (HPU \times FTU)] \times C_1 \times H \times FS$$
$$DR = [(HPL \times FTL) + (HPU \times FFU)] \times C_1 \times DF \times M \times FS$$

AR No. 4 - Change Current Compressor Operations

$$ES = ECc - Ecp$$
and
$$DR = MPDc - MPDp$$

where

$$ECc = HPc \times C_1 \times H$$
$$ECp = [(HPL \times FTL) + (HPU \times FTU)] \times C_1 \times H$$
$$MPDc = HPc \times C_1 \times DF \times C_2$$
$$MPDp = HPL \times C_1 \times DF \times C_2 \quad \text{(if the compressor is always loaded more than 30 minutes at a time)}$$

or

$$MPDp = HPU \times C_1 \times DF \times C_2 = \text{(if the compressor is never loaded more than 30 minutes at a time)}$$

AR No. 5 - Install a Small Compressor

$$ES = [(HLc \times FLc) + (HUc \times FUc) - (HLp \times FLp) - (HUp \times FUp)] \times C_1 \times H$$

and

$$DR = (Huc - HLp) \times C_1 \times DF \times C_2 \quad \text{(if proposed compressor is expected to run loaded more than 30 minutes at a time)}$$

or

$$DR = (HUc - HUp) \times C_1 \times DF \times C_2 \quad \text{(if proposed compressor is expected to run loaded less than 30 minutes at a time)}$$

AR No. 6 - Recover Waste Heat from Air Cooled Compressor

$$ES = RHO \times Q \times C_3 \times Cp \times (Te - Ta) \times HH \times HF / EFF$$

AR No. 7 - Recover Waste Heat From Water Cooled Compressor

$$HT = WFR \times DEN \times Cp (TO - TI)$$
$$WS = WFR \times H$$
$$WH = HT \times HH \times E / EFF$$
$$EI = HP \times LF \times C_1 \times H / EF$$

$$DI = HP \times LF \times C_1 \times DF \times M/EF$$
$$OPC = EI \times (\text{effective energy cost}) +$$
$$DI \times (\text{effective demand cost})$$

AR No. 8 - Reduce Compressed Air Leaks

$$ES= (Ecb\text{-}Eca) \times EWF$$

where

$$EC = [(HPL \times FTL) + (HPU \times FTU)] \times H \times C_1$$

and

$$DR = (HPL \text{ - } HPU) \times C_1 \times DF \times C_2$$

Alternatively

$$ES = L \times H \times C_1$$
$$DR = L \times C_1 \times DF \times M$$
$$L = Pi \times C_4 \times V_f \times k/(k\text{–}1) \times N \times C_5 \times$$
$$[(Po/pi^{(k\text{–}1)/(k\times N)} - 1]/(Ea \times Em)$$
$$V_f = NL \times (Ti = 460) \times (PL/PI) \times$$
$$C_6 \times C_7 \times C_8 \times D^2/4 \times C_9/[C_4 \times (TL + 460)^{0.5}]$$

SOLVED EXAMPLES

Example 12-1: Use Outside Air For Compressor Intakes

Assuming that a facility has a 150 hp screw air compressor that draws about 145 hp while loaded and 60 hp while unloaded. After timing the compressor it was determined that the compressor runs approximately 80% of the time loaded and 20% of the time unloaded. Based upon measurements, the temperature at the intake of the compressor was estimated to be 98°F throughout the year. The year round outdoor air temperature is estimated, based from weather data, to be 52°F. The compressor runs approximately 6,240 hours per year. Thus, WR, is found to be:

$$WR = (98\text{-}52)/(98 + 460)$$
$$WR = 0.0824$$

This implies that if outside air is used as intake air, the energy consumed by the compressor would drop by 8.24% from the current requirements.

Thus, the energy savings, ES, for the 150 hp air compressor are estimated to be:

$$ES = [(145 \times 0.80) + (60 \times 0.20)] \times$$
$$(0.746) \times (6{,}240) \times (0.0824)$$
$$ES = 49{,}098 \text{ kWh/yr}$$

The demand reduction, DR, would be:

$$DR = [(145 \times 0.80) + (60 \times 0.20)] \times$$
$$(0.746) \times (1.00) \times 12 \times (0.824)$$
$$DR = 94.4 \text{ kWm/yr}$$

Estimating that each kWh costs about $0.035 and each kWm costs about $11.00, the total cost savings can be estimated to be $2,757/yr.

The most common material used for ducting outside air to the compressor intakes is plastic (PVC) pipe. One end of the pipe is attached to the air cleaner intake and the other end is routed through the wall or ceiling to the outside. For a 150 hp compressor a five-inch diameter PVC pipe is needed. The length of the pipe varies depending on the location of the compressor. Assuming that a total of $500 is needed for the implementation, thus, the payback period is about 0.18 years.

Example 12-2: Reduce Air Compressor Pressure

Consider the same 150 hp air compressor given in the previous example. The compressor draws about 145 hp while loaded and 60 hp while unloaded. Overall, it runs approximately 80% of the time loaded and 20% of the time unloaded. Assume that the compressor is operated at a pressure of 140 psig (154.7 psia) and the maximum pressure needed for the machinery is about 85 psig. After estimating the minimum operating pressure, let us say to be 100 psig (114.7 psia), FR can be calculated to be:

$$FR = \frac{(114.7/14.7)^{1 \times (1.4-1)/1.4} - 1}{(154.7/14.7)^{1 \times (1.4-1)/1.4} - 1}$$

FR = 0.8327

Thus, the current energy consumed by the compressor at the current operating pressure can be reduced by about 16.73% (1-0.8327) at the proposed pressure. The annual energy savings, ES, can be estimated to be:

$$ES = [(145 \times 0.8) + (60 \times 0.2)] \times$$
$$6{,}240 \times 0.746 \times (1 - 0.8327)$$
$$ES = 99{,}685 \text{ kWh/yr}$$

The demand reduction, DR, can be estimated as:

$$DR = [(145 \times 0.8) + (60 \times 0.2)] \times$$
$$0.746 \times 1.00 \times 12 \times (1 - 0.8327)$$
$$DR = 191.7 \text{ kWm/yr}$$

If the cost of electric energy is about $0.035/kWh, and the cost of demand is $11.00, thus the total cost savings, ECS, are then estimated to be about $5,598/yr.

The implementation cost of such a recommendation is estimated to be negligible since it takes about 10 minutes to adjust the pressure setting on a compressor.

Example 12-3: Use Synthetic Oil

Considering the 150 hp compressor discussed in the previous examples, the energy savings, ES, due to the use of synthetic oils instead of premium oils can be calculated as follows:

$$ES = [(145 \times 0.80) + (60 \times 0.20)] \times$$
$$0.746 \times 6{,}240 \times 0.06$$
$$ES = 35{,}751 \text{ kWh/yr}$$

The demand reduction, DR, for the compressor is estimated to be:
$$DR = [(145 \times 0.80) + (60 \times 0.20)] \times 0.746 \times 1.0 \times 12 \times 0.06$$
$$DR = 68.8 \text{ kWm/yr}$$

Thus, the cost savings can be found to be about $2,008 per year (considering $0.035/kWh and $11.00/kWm).

Based on manufacturers' information, the cost of synthetic oil is approximately three times that of conventional oil. Considering that the synthetic oil has a rated life four times longer than that of conventional oil, the differential cost between synthetic oil and conventional oil is insignificant. In fact, this represents a reduction in used oil disposal costs.

The implementation cost considered for this recommendation is considered to be the cost associated with the drainage of the current conventional oil and the clean up of the interior of the gearboxes of the compressor. It is estimated that the total implementation cost is about $250. Thus, the savings of $2,008/yr will pay for the implementation cost of $250 within about 0.12 years.

Example 12-4: Change Current Compressor Operations

Assume that the compressor of the previous examples was operated in a continuously loaded manner and that it draws about 150 hp constantly when operated. After conducting several tests to investigate the feasibility of operating the compressor under the loaded-unloaded mode, it was determined that all the air lines would have adequate pressure. Assuming that the same operating conditions discussed earlier exist for this case (loaded 145 hp and 80%; unloaded 60 hp and 20%), thus, the energy savings, ES, can be calculated to be:

$$
\begin{aligned}
ES &= [150 \times 0.746 \times 6{,}240] - \\
&\quad ([(145 \times 0.8) + (60 \times 0.2)] \times 0.746 \times 6{,}240 \\
ES &= (698{,}256) - (595{,}845) \\
ES &= 102{,}411 \text{ kWh/yr}
\end{aligned}
$$

Assuming that the demand factor, DF, for the compressor is 100% and it remains at the same value after the change in operation, the demand reduction, DR, can be found to be:

$$
\begin{aligned}
DR &= (150 \times 0.746 \times 1.00 \times 12) - \\
&\quad [(145 \times 0.8) + (60 \times 0.2)] \times 0.746 \times \\
&\quad 1.00 \times 121 \\
DR &= (1{,}343 - 1{,}146) \\
DR &= 197 \text{ kWm/yr}
\end{aligned}
$$

Considering the energy savings and demand reduction found

above, the cost savings are estimated to be approximately $5,751 per year.

In most cases, the implementation cost for such a measure is estimated to be negligible. It is estimated to be the cost of 20 minutes needed to reprogram the compressor. In other cases, if the compressor has no controls to allow the recommended programming, the installation cost of the controls should be included. Different compressor models and sizes require different controls and thus, the implementation cost varies. For this example, the implementation cost is estimated to be about $4,000. Thus, the payback period is about 0.70 years.

Example 12-5: Install a Small Compressor

Considering the same 150 hp considered previously. Based on tests conducted during the third shift, it was estimated that the compressor operates loaded approximately 7% and unloaded for about 93% of the time. Form the current compressor's catalogue, it was found that the compressor delivers approximately 507 cfm at the current operating conditions. Thus, based on this data, it is estimated that approximately 35 cfm of air are needed during the third shift (2,080 hours per year).

The proposed unit is selected based on cfm required at a pressure of 140 psig. Several options are available but a 15 brake-horsepower reciprocating unit is selected. At maximum load conditions, the unit delivers approximately 50 cfm at 140 psig. Thus, it is estimated that the unit would run loaded for approximately 7 1 % of the time. To be conservative, it is estimated that the unit will operate at a 100% load fraction.

The energy savings, ES, can be calculated to be:

ES= $[(140 \times 0.07) + (60 \times 0.93) - (15 \times 0.71) - (0 \times 0.29)] \times 0.746 \times 2,080$

ES= 85,265 kWh/yr

In this example, it is assumed that there will be no demand reduction since no demand charges occur during the third shift. Thus, the cost savings are estimated to be about $2,984/yr.

It is estimated that the maintenance cost associated with the operation of the proposed unit is comparable with the operating cost of the existing unit as if it would remain in operation during the third shift.

The implementation cost for this measure is estimated to be the cost to purchase and install a 15 hp reciprocating air compressor. The cost varies from a compressor distributor to another but on the average such a unit would cost approximately $2,500 installed. Thus, the payback period for this example is about 0.84 years.

Example 12-6:
Recover Waste Heat From Air Cooled Compressor
The same 150 hp is considered. The rate of airflow through the compressor cooling coil was measured to be approximately 7,000 cfm. The air is discharged into the compressor room at a temperature of 125°F. Considering an ambient pressure of 14.7 psig, the density of the exhaust air can be calculated to be about 0.06784 lbm/ft³. The temperature of the area to be heated is set at 65°F during the heating season (October through April). During that period, the compressor operates about 3,640 hours per year.

It is estimated that by ducting the exhausted air from the compressor room which is located outside the building, about 10% of that heat would be lost to the environment through the duct work. Assuming an efficiency of 75% for the natural gas fired space-heating equipment thus, the energy savings, ES, that can be realized by recovering the heat generated by the air compressor are estimated to be:

$$ES = 0.06784 \times 12,000 \times 60 \times$$
$$0.24 \times 10^{-6} \times (125 - 65) \times 3,640 \times 0.90/0.75$$
$$ES = 17,922 \text{ therms/yr}$$

Considering an average cost of $0.35/therm, the cost savings are then calculated to be $6,273/yr.

The implementation cost for such a measure depends on the location of the compressor, the size of the compressor, the areas to be heated, and on the type of controls (dampers and thermostats) needed. These costs can be estimated after identifying the materials needed for the implementation and after consulting an HVAC contractor. For our example, the implementation cost is estimated to be about $2,000. Thus, the payback period is about 0.32 years.

Example 12-7: Recover Waste Heat
From Water Cooled Compressor

Considering the same compressor, the water flow rate through the compressor was found to be approximately 30 gallons per minute (1,800 gal/hr).

The inlet and outlet water temperatures were measured to be about 64°F and 90°F, respectively. Thus, the heat transfer, HT, from the compressor to the water can be estimated to be:

$$HT = 1,800 \times 8.342 \times 1.0 \times (90\text{-}64)$$
$$HT = 390,406 \text{ Btu/hr}$$

The water savings, WS, for installing a closed loop system can be estimated to be:

$$WS = 1,800 \times 6,240$$
$$WS = 11,232,000 \text{ gal/yr}$$

The water cost savings are then calculated to be about $22,464/yr assuming that the effective water cost, including surcharges, are about $0.002/gal.

The waste heat, WH, that could be recovered is estimated to be:

$$WH = 390,406 \times 3,640 \times 0.90/0.75$$
$$WH = 1,705 \text{ MMBtu/yr}$$

Thus, at $0.35/therm ($3.5/MMBtu) the cost savings are then $5,968/yr.

Assuming that the cooling unit has a capacity of 500,000 Btu/hr thus, the estimated load fraction under which the unit would operate at, is about 78%. A unit of this size would contain three cooling fans of I hp each and one I hp water pump with an overall motor efficiency of 80%, then the annual energy increase, El, and the annual demand increase, DI, from the operation of the new system is estimated as follows:

$$EI = 4 \times 0.78 \times 0.746 \times 6,240/0.8$$
$$EI = 18,155 \text{ kWh/yr}$$

And

$$DI = 4 \times 0.78 \times 0.746 \times 1.00 \times 12/0.8$$
$$DI = 34.9 \text{ kWm}$$

Thus, the total operating cost is estimated to be about $1,109/yr. Therefore, the net cost savings, CS, associated with this measure are estimated to be:

$$CS = \$22,464/yr + \$5,968/yr - \$1,109/yr$$
$$CS = \$27,413/yr$$

The implementation of this measure is estimated to be about $20,000. Thus, the payback period is about 0.73 years.

Example 12-8:
Reduce Compressed Air Leaks

As an example, considering a 1/16 inch diameter leak, at line pressure of 139.7 psia and *average* line temperature of 70°F. The same one stage compressor with intake temperature as 98°F operating at 154.7 psia. The compressor motor efficiency is 94% and its isentropic efficiency is 82%.

Therefore, Vf could be calculated to be:

$$Vf = (1)(98 + 460)(139.7/14.7)(49.02)$$
$$(60)(0.6)(0.0625^{2/4})(3.1416)$$
$$/[(144)(70 + 460)^{0.5}]$$

$$Vf = 8.66 \text{ cfm}$$

Therefore, the power loss, L, due to the 1/16 inch leak is calculated as follows:

$$L = (14.7)(144)(8.66)(1.4/0.4)(1)(3.03 \times 10^{-5})$$
$$[(154.7/14.7)^{0.4[(1.4)(1)]} - 1]/[(0.82)(0.94)]$$
$$L = 2.42 \text{ hp}$$

Thus, the estimated savings, ES, for the leak considered can be calculated as follows:

ES = (2.42)(6,240)(0.746)
ES = 11,265 kWh/yr

The annual demand reduction, DR, can be estimated by the following:

DR = (2.42)(0.746)(1.00)(12)
DR = 29 kWm/yr

Thus the annual cost savings are estimated to be $713/yr.

Assuming that this leak is due to a broken seal on a pressure regulator, the cost to repair such a leak is conservatively estimated to be about $100. Thus, the payback is about 0.14 years.

CONCLUSION

It has been attempted in this chapter to discuss the basics of a compressed air system including its components and various types of compressors as used in the manufacturing sector. Several energy saving strategies related to compressed air systems have been presented. It is hoped that the material presented in this chapter will be found to be useful by the energy engineers when they are attempting to audit and optimize the operation of the compressed air systems in their plants.

References

1. D. Paul Mehta, "Air Is Not Free For Agile Manufacturing," Proceedings of the International Conference on Agility, ICAM, Lafayette, LA. February 23-25, 1997.
2. _____ "Improving Compressed Air System Performance: A Source-Book for Industry." *Motor Challenge*, U.S. Department of Energy. 1998.
3 Mark D. Oviatt and Richard K. Miller. *Industrial Pneumatic Systems: Noise Control and Energy Conservation*. The Fairmont Press Inc., 1991.
4. Fred L. Eargle. *Applied Pneumatics Handbook*. University of North Carolina at Raleigh, 1964.
5. Charles W. Gibbs. *Compressed Air and Gas Data*. Second edition. Ingersoll-Rand, 197 1.
6. John Rollins. *Compressed Air and Gas Handbook*. Fourth edition. Compressed Air and Gas Institute, 1973.
7. F.W. O'Neil. *Compressed Air Data*, Fifth edition. Bowling Green Building,

1939.

8. H.J. Thorkelson. *Air Compression and Transmission*. McGraw-Hill Book Co., 1913.

9. Vladimir Chlumsky. *Reciprocating and Rotary Compressors*. E & FN Spon Lts., 1965.

10. A.H. Shapiro, *The Dynamics and Thermodynamics of Compressible Fluid Flow*, Vol. 1. Ronald Press, N.Y. 1953.

13

Financing Energy Projects

INTRODUCTION

Every day spent by an organization without having installed the appropriate energy efficiency measures means lost savings and lost opportunities. Performance contracting is a sophisticated solution to this problem. As with any sophisticated system, there are elements of complexity to be managed. Therefore, communication, knowledge and experience are essential for successful project completion.

DEFINITIONS AND CLARIFICATIONS

For the purposes of this chapter, it is assumed that the energy services company (ESCO) is providing the energy conservation measures (ECMs) such as audit, design, installation, monitoring and maintenance for the customer and that a separate third party, such as a bank or investment company, is providing the capital for the project. Often, energy services companies market themselves to customers as providing financing. In many cases, there is an independent financing source involved in the background. Alternatively, an ESCO can be a utility subsidiary which uses the utility's shareholder money to finance projects. For simplicity's sake, we will treat the ultimate source of capital for projects as a separate lender with its own guidelines.

FINANCING ALTERNATIVES

There is a variety of options for financing energy conservation projects. Some of the most common are:

General Obligation Bond

Specifically applicable to municipalities, these bonds are based on the general credit of a state or local government. The process is long and complicated, but interest rates are low.

Municipal Lease

Specifically applicable to municipalities, state entities and local entities. The lessor earns tax-exempt credit and the borrowing entity pays low interest rates.

Commercial Loan

This is a loan to the customer from a conventional bank based on the customer's assets and credit quality. This form of financing is rarely offered by typical finance companies or financial institutions for energy conservation projects requiring less than $5 million of capital.

Taxable Lease

There are a number of leasing vehicles with a variety of names such as: operating lease, capital lease, guideline lease, tax-oriented lease, and non-tax-oriented lease. The tax ramifications of leasing are often not well understood. Figure 13-1 will help clarify some of those distinctions. Figure 13-2 will show the dynamics of the parties involved.

It is important to note that in all of the categories listed above, the customer is directly obligated to make payments relating to the installed energy-saving measures REGARDLESS OF PERFORMANCE of either the equipment or the ESCO. The customer may well have recourse under a separate contract to the service provider or equipment manufacturer, but will still owe under the financing instrument. The interest rate and any additional loan terms are based almost exclusively on the creditworthiness of the customer.

RELATIVE BENEFITS OF PROJECT FINANCING

Many companies and government entities are undergoing severe budget cutbacks. When these entities are approached by an ESCO offering virtually no up-front investment and a guaranteed amount of savings, new possibilities are created for energy projects to be developed and completed. If structured properly, performance contract financing has

Figure 13-1. Leasing Options

Guideline Lease/True Lease/Tax-oriented Lease

Compliance with all of the IRS guidelines including those listed below is required:

(a) The total lease term (including extensions and renewals at a predetermined, fixed rate) must not exceed 80% of the estimated useful life of the equipment at the start of the lease, i.e., at the end of the lease the equipment must have an estimated remaining useful life equal to at least 20% of its originally estimated useful life. Also, this remaining useful life must not be less than one year thereby limiting the maximum term of the lease.

(b) The equipment's estimated residual value at the expiration of the lease term must equal at least 20% of its value at the start of the lease. This requirement limits the maximum lease term and the type of equipment to be leased.

(c) No bargain purchase option.

(d) The lessee cannot make any investment in the equipment.

(e) The equipment must not be "limited use" property. Equipment is "limited use" property if no one other than the lessee or a related party has a use for it at the end of the lease.

(f) Tax-oriented leasing is 100% financing. Guideline leases (tax-oriented leases) may be either a capital lease or an operating lease for reporting purposes under Financial Accounting Standards Board Rule #13.

Capital Lease

A capital lease is one that fulfills any ONE of the following criteria:

(a) The lease transfers ownership of the property to the lessee by the end of the lease term.

(b) The lease contains a bargain purchase option (less than fair market value).

(c) The lease term is equal to 75% or more of the estimated economic life of the leased property.

(d) The present value of the minimum lease payments equals or exceeds 90% of the fair value of the leased property.

The capital lease shows up on the lessee's balance sheet as an asset and a liability.

Operating Lease

An operating lease does not meet ANY of the above criteria for a capital lease.

An operating lease is not booked on the lessee's balance sheet but is recorded as a periodic expense on the income statement.

Sources
Accounting For Leases, Financial Accounting Standards Board
Leasing and Tax Reform: A Guide Through The Maze, General Electric Credit Corporation
Handbook of Leasing, General Electric Credit Corporation

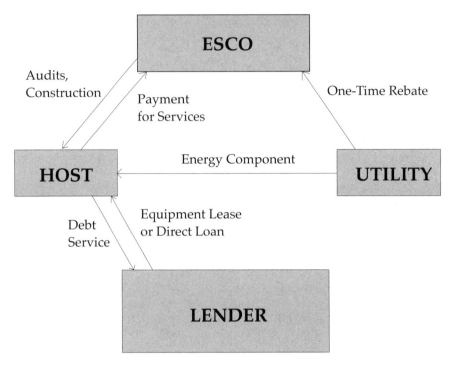

Figure 13-2. Equipment lease/direct financing.

the benefits to the customer of an operating lease, i.e., off-balance-sheet treatment, with the added benefit of being non-recourse to the customer. By maintaining this financing out of its balance sheet, the customer can retain the use of available credit for expansion, research and development, additional inventory or business emergencies.

BASIC STRUCTURE

Performance contract financing from a lender's perspective is a challenging combination of business credit analysis and project evaluation. The project financing is structured as follows:

(1)   The ESCO contracts with the customer to provide energy conservation measures (lighting, variable speed drives, etc.) as well as ongoing services which may include warranties, handling and disposal of

wastes, operation and maintenance of installed equipment, repair and replacement of measures, and measurement, monitoring and verification of savings. The Energy Services Agreement (ESA) is the foundational document upon which a lender will rely to confirm that the customer and ESCO have a clear understanding of all aspects of the ECMs being installed. The ESA incorporates a payment from the customer which is designed to cover the costs to finance the project as well as paying the ESCO for the services provided. (A sample ESA is included in Figure 13-3. Note that it is a guideline only. Be sure to have your legal counsel review any document before you authorize it.)

(2)   The lender lends directly to the ESCO. The ESCO uses the funds to recoup its project development expenses and purchase and install the equipment. Repayment to the lender is made by the ESCO out of the funds paid to it by the customer. One variation is to create a single-purpose entity exclusively to hold the assets of the project financing. Because the loan is technically made to this entity, it has the added benefit of keeping the transaction off the balance sheet of the ESCO as well as the customer (see Figure 13-4). In addition, the lender can lend up to 95% of the project amount.

PROPOSAL REVIEW

The lender will analyze a proposed project with these questions in mind:

•   Will the revenues generated by the energy savings, utility rebates and customer payments be sufficient to cover debt service?

•   Can the ESCO perform as required under its contract with the customer and/or the utility for the term contract?

•   Are the risk allocations among all parties fair from both business and legal perspectives?

•   Will the customer and/or the utility be able to meet its payment obligations for the term of the financing?

The lender will want to see three years of audited financial statements: balance sheet, income statement and cash flow statement with explanatory notes from the customer. These will be used to make trend and financial ratio analyses. The ESCO's financial statements are also important and will receive a similar review.

With utility payments and/or customer payments based on actual savings, the lender needs to make a thorough analysis of the projected savings of the energy conservation measures installed. The project figures (simple payback, types of equipment, maintenance savings, if any, construction period) will be carefully reviewed to determine how the savings will cover the loan payments and to determine the effects of factors such as energy price fluctuations and inflation.

Any information which the parties can provide early on in the proposal negotiation to make the lender feel more comfortable with these issues will save a tremendous amount of time, money and effort to all parties.

CONCLUSION

The key point here is that if an event of default occurs in the ESA between the ESCO and the customer which gives the customer the right to reduce or cease its payment obligation, it is the lender which is most at risk of suffering a loss. The lender must either be comfortable that the ESCO will cure the default or be confident that it can hire another ESCO to cure the project default and force the customer to resume payments. The lender is also at risk of the energy savings not being estimated or measured properly. Despite all of this, good projects are being funded and customers are extremely satisfied with the resulting benefits.

Figure 13-3. Sample Energy Services Agreement

Note: Consult your legal counsel before authorizing any legal agreement.

I.   TERM SHEET.
The purpose of the Agreement is the evaluation, engineering, design, procurement, installation, financing and monitoring by **ESCo** of Energy Conservation Measures ("ECMs") at Customer's facility(ies) identified in Appendix A attached hereto ("Premises").

ESCo and Customer (the "Parties") agree to the following terms pursuant to which this Agreement shall be performed:

1. EXECUTION DATE: _____

2. TERM OF AGREEMENT: _____

_____
     (years after ECM Commencement Date)

3. LOCATION: _____

4. OWNERSHIP OF PREMISES: _____

PROJECT FIGURES

	Preliminary Estimates	Final Installed Figures
5. **ESCo** PERCENTAGE OF ENERGY SAVINGS:	_____	_____
6. PROJECT TOTAL CAPITAL COST ($ 000):	_____	_____
7. ECM COMMENCEMENT DATE:	_____	_____
8. VALUE OF FIRST YEAR ENERGY SAVINGS ($):	_____	_____
9. **CUSTOMER** FIRST YEAR PAYMENT TO ESCo (line 5 × line 8) ($):	_____	_____
10. **CUSTOMER** FIRST YEAR MONTHLY PAYMENT (line 9 12) ($):	_____	_____
11. **CUSTOMER** FIXED MONTHLY PAYMENT (OPTIONAL):	_____	_____
	Please Initial:	ESCo

CUSTOMER   _____

Final Installed Figures appearing in the right-hand column above shall be completed in accordance with Section V and the Parties shall then re-execute the Agreement below and enter their initials above.

As indicated in term #11 above, upon completion of any determinations required by Section V, **Customer** shall have the option to fix its monthly payment to **ESCo**, as required under Section VI, by multiplying One Hundred and _____ Percent (1____%) of **ESCo's** Percentage of Energy Savings by the Final Installed Figure for the Value of First Year Energy Savings and paying such product to **ESCo** each year in twelve (12) monthly payments ("**Customer** Fixed Monthly Payment"). **Customer** shall indicate its exercise of such option by initialing the appropriate line below. Except as set forth in this Agreement, such monthly payments shall be due and payable each month of this Agreement from and after the month in which the Commencement Date occurs and shall be made by **Customer** without regard to the amount of Energy Savings in any such month or year. Any excess of Energy Savings over the Final Estimated Value of First Year Energy Savings shall be retained by **Customer. Customer's** Fixed Monthly Payment shall not be revised except as may be specifically required in accordance with the terms of this Agreement.

ESCo: **Customer:**

_____   _____

_____   _____
          (business organization)   (business organization)

By: _____

By: _____
                    (entity)   (entity)
By: _____

By: _____
                    (entity)   (entity)
Its: _____

Its: _____
                    (title)   (title)

_____   _____
                    (title)   (title)

ACCEPTANCE OF FINAL INSTALLED FIGURES

ESCo: **Customer:**

_____   _____

_____   _____

     (business organization)   (business organization)

By: _____

By: _____

          (entity)   (entity)

By: _____

By: _____

          (name)   (name)

Its: _____

Its: _____

          (title)   (title)

CUSTOMER ACCEPTANCE OF FIXED MONTHLY PAYMENT

Accepted   Not Accepted

_____   _____   (please initial)

II. DEFINITIONS.
When used in this Agreement, the following terms shall have the meaning specified:

2.1 **Agreement**: This Agreement between **Customer** and **ESCo**.

2.2 **ESCo's Percentage of Energy Savings**: The percentage of Energy Savings **ESCo** shall receive as compensation for its services under this Agreement, subject to Customer's option to fix monthly payments set forth in Section I, paid to **ESCo** in accordance with Section VI and Section VII.

2.3 **Current Market Value of Energy Savings**: The total market value rate expressed in r/kWh of electrical energy use and/or $/kW of electrical demand imposed by the Utility company in the current monthly period then occurring, or in any future monthly period then being considered, including applicable taxes, surcharges and franchise fees, if applicable. The **Current Market Value of Energy Savings** shall be determined in accordance with Appendix C.

2.4 **ECM Commencement Date**: The ECM Commencement Date shall be the date on which the installation of the ECMs is substantially complete. Prior thereto, **ESCo** shall give **Customer** a Notice of Substantial Completion and shall therein identify the ECM Commencement Date, which shall occur no sooner than fifteen (15) days after such notice.

2.5 **Energy Audit Report ("EAR")**: The analysis performed by **ESCo** of the electric energy use by **Customer** at the Premises, and the potential for electric energy savings. Such analysis includes, without limitation, the ECMs recommended by **ESCo** and agreed to by the Parties for installation at the Premises and the Measurement Plan for measuring the savings estimated to result from such ECMs, all as attached hereto and made a part hereof as Appendix B.

2.6 **Energy Conservation Measure ("ECM")**: The various items of equipment, devices, materials and/or software as installed by **ESCo** at the Premises, or as repaired or replaced by **Customer** hereunder, for the purpose of improving the efficiency of electric consumption, or otherwise to reduce the electric utility costs of the Premises.

2.7 **Energy Savings**: Electric energy reduction (expressed in kilowatt-hours of electric energy and/or kilowatts of electric demand and measured in accordance with the Measurement Plan) achieved through the more efficient utilization of electricity resulting from the installation of the ECMs agreed to by the Parties under this Agreement.

2.8 **Measurement Plan**: The plan for measuring Energy Savings under this Agreement, which shall be in accordance with the requirements of the Utility Agreement and shall be a part of the Energy Audit Report attached as Appendix B hereto.

2.9 **Monthly Period**: A span of time covering approximately 30 days per month, corresponding to **Customer's** billing period from its electric utility.

2.10 **Party: Customer** or **ESCo Parties** means **Customer** and **ESCo**.

2.11 **Premises**: The buildings, facilities and equipment used by **Customer**, as identified in Section I and as more fully described in the attached Appendix A, where **ESCo** shall implement the Project under this Agreement.

2.12 **Project**: The complete range of services provided by **ESCo** pursuant to this Agreement, including evaluation, engineering, procurement, installation, financing and monitoring of ECMs at the Premises.

2.13 **Uncontrollable Circumstances**: Any event or condition having a material adverse effect on the rights, duties or obligations of **ESCo**, or materially adversely affecting the Project, if such event or condition is beyond the reasonable control, and not the result of willful or negligent action or omission or a lack of reasonable diligence, of **ESCo**; provided, however, that the contesting by **ESCo** in good faith of any event or condition constituting a Change in Law shall not constitute or be construed as a willful or negligent action, or a lack of reasonable diligence. Such events or conditions may include, but shall not be limited to, circumstances of the following kind:

a. An act of God, epidemic, landslide, lightning, hurricane, earthquake, fire, explosion, storm, flood or similar occurrence, an equipment failure or outage, an interruption in supply, an act or omission by persons or entities other than a Party, an act of war, effects of nuclear radiation, blockade, insurrection, riot, civil disturbance or similar occurrences, or damage, interruption or interference to the Project caused by hazardous waste stored on or existing at the Project site;

b. strikes, lockouts, work slowdowns or stoppages, or similar labor difficulties, affecting or impacting the performance of **ESCo** or its contractors and suppliers;

c. a change in law or regulation or an act by a governmental agency or judicial authority.

2.14 Utility Agreement: The agreement entered into by **ESCo** with ____ ____, a _____ public utility company ("Utility"), pursuant to which **ESCo** is required to install certain ECMs at facilities such as **Customer's** Premises and in accordance with the terms of which **ESCo** has entered into this Agreement.

III. ECM COMMENCEMENT DATE AND TERM OF AGREEMENT.
The term of this Agreement shall commence as of the date on which this Agreement is executed and shall continue, unless sooner terminated in accordance with the terms hereof, for the period of years after the ECM Commencement Date set forth in Section I.

Upon receipt of the Notice of Substantial Completion identifying the ECM Commencement Date, **Customer** shall provide **ESCo**, within fifteen (15) days, any comments and requests for work or corrections. **ESCo** shall make all commercially reasonable efforts to respond to such comments and requests within thirty (30) days of the ECM Commencement Date, which shall occur on the identified date.

Upon the expiration or termination of this Agreement the provisions of this Agreement that may reasonably be interpreted or construed as surviving the expiration or termination of this Agreement shall survive the expiration or termination for such period as may be necessary to effect the intent of this Agreement.

At the end of the term of this Agreement, **Customer** shall purchase the ECMs and **ESCo** shall transfer title to **Customer**, free and clear of all liens and encumbrances, all as set forth in Section 6 of the General Terms and Conditions.

IV. SCOPE OF ESCO'S SERVICES.
Subject to and in accordance with the terms and conditions of this Agreement, **ESCo** shall provide the evaluation, engineering, design, procurement, installation, financing and monitoring of the ECMs set forth in the Energy Audit Report. The ECMs in the EAR may be modified pursuant to Section V. **ESCo** shall use reasonable commercial efforts to achieve the ECM Commencement Date estimated in Section I, line 7, column entitled, "Engineering Estimates."

ESCo agrees to extend its funds and install such ECMs in return for **Customer's** agreement to perform hereunder and in particular, to pay **ESCo's** Percentage of Energy Savings pursuant to Section VI. The Parties anticipate the measurement of Energy Savings, but notwithstanding this expectation or any provision of this Agreement which may suggest to the contrary, in light of the factors affecting savings which are beyond **ESCo's** reasonable control, **ESCo** assumes no obligation that any particular level of savings shall materialize due to its services hereunder.

ESCo warrants that the ECMs which have a lifetime greater than one (1) year shall be, and shall remain, free of defects for one (1) year after the ECM Commencement Date. In addition, **ESCo** agrees to assign all manufacturers warranties for such ECMs to **Customer** for the period of manufacturer warranty, subject to all exclusions and limitations as may be set forth therein.

V. DETERMINATION OF FINAL TERMS.

After the execution of this Agreement, if the Parties agree to revise the ECMs listed in the EAR, and to make all associated revisions to this Agreement, including, without limitation, to the numbers or dates, as the case may be, entered in the column entitled, "Engineering Estimates" appearing in lines 5 through 11 of Section I, the Parties shall, upon such agreement, make any associated revision and enter in the column entitled, "Final Installed Figures" the agreed-to numbers and/or dates. The Parties shall then re-execute Section I upon the entry of such numbers and/or dates and **ESCo** shall revise the Termination Values in Appendix D consistent with Appendix D and the Final Installed Project Total Capital Cost then appearing in line 6 of Section I above.

All such revisions shall be voluntary and the Parties shall not be required, absent mutual consent, to revise the ECMs listed in the EAR after the execution hereof; provided, however, **ESCo** shall not be required to install ECMs affected by Uncontrollable Circumstances. Such ECMs shall be deleted from the Project unless the Parties agree to all necessary changes to the Project numbers and/or dates required to adjust to such circumstances. Changes to Project numbers and/or dates in connection with Uncontrollable Circumstances shall be entered by **ESCo** in the column entitled, "Final Installed Figures," in lines 5 through 11 of Section I. The Parties shall then re-execute Section I and **ESCo** shall revise the Termination Values set forth in Appendix D consistent with Appendix D and the Final Installed Project Total Capital Cost then appearing in line 6 of Section I.

VI. COMPENSATION.

From and after the ECM Commencement Date, except as provided in Sections I and VII with respect to **Customer's** option to fix monthly payments, **Customer** shall pay **ESCo** an amount equal to **ESCo's** Percentage of Energy Savings, as set forth in Section I above, multiplied by the applicable Current Market Value of Energy Savings, all as determined pursuant to Appendix C.

 ESCo shall prepare and send **Customer**, and **Customer** agrees to pay, a monthly invoice calculated pursuant to Section VII below.

VII. BILLING.

ESCo will submit monthly invoices to **Customer** in amounts determined in accordance with Section VI. In the event **Customer** exercises its option under Section I to fix its monthly payments to **ESCo**, **Customer** shall pay to **ESCo** the **Customer** Fixed Monthly Payment defined in Section I. If Customer does not exercise such option, monthly payments shall be estimated as set forth in this Section VII and paid on such estimated basis, subject to reconciliation as provided hereinafter. Monthly invoices shall be paid within thirty (30) calendar days following receipt. Reconciliation payments, or refunds, as the case may be, shall be due within thirty (30) calendar days of receipt of a reconciliation invoice.

 Subject to reconciliation, invoice amounts shall be estimated for each year following the ECM Commencement Date. In the first such year, monthly payments shall be 1/12 of the product of **ESCo's** Percentage of Energy Savings and the Current Market Value of Energy Savings expected in such first year as set forth in line 8 of the column entitled, "Final Installed Figures" in Section I above. Such fixed amounts shall be reconciled and adjusted as necessary every six (6) months based on the difference between the Current Market Value of Energy Savings measured pursuant to Appendix C and the estimated value then applicable. **ESCo** shall prepare and submit to **Customer** a reconciliation invoice within thirty (30) calendar days of the expiration of each such six-month period. The estimate of the value of Energy Savings then in effect at the end of any year shall continue until replaced with reconciled amounts in the succeeding year. In the second year and following years, the estimate of the Current Market Value of Energy Savings expected in such year shall equal the expected value (in nominal dollars for the billing year in question) of the Energy Savings actually delivered in the prior year.

VIII. NOTICES.

All notices to be given by either Party to the other shall be in writing and must be delivered or mailed by registered or certified mail, return receipt requested, or sent by a courier service which renders a receipt upon delivery addressed as set forth above in Section I or such other addresses as either Party may hereinafter designate by a Notice to the other. Notices are deemed delivered or given and become effective upon mailing if mailed as aforesaid and upon actual receipt if otherwise delivered to the addresses set forth in Section I.

IX. APPLICABLE LAW.
This Agreement and the construction and enforceability thereof shall be interpreted under the laws of the state of _____.

X. FINAL AGREEMENT.
This Agreement, together with its appendices and attachments, shall constitute the full and final Agreement between the Parties, shall supersede all prior agreements, communications and understandings regarding the subject matter hereof and shall not be amended, modified or revised except in writing. **This Agreement shall bind the Parties as of the date on which it was executed.** Any inconsistency in this Agreement shall be resolved by giving priority in the following order: (a) Amendments to the Agreement, in reverse chronological order; (b) the Agreement, Sections I through XI; (c) General Terms and Conditions of the Agreement; (d) Appendices to the Agreement; and (e) Attachments to, or other documents incorporated into, the Agreement.

XI. INCORPORATION OF GENERAL TERMS AND CONDITIONS:

THIS AGREEMENT IS SUBJECT TO THE GENERAL TERMS AND CONDITIONS ATTACHED HERETO AND MADE A PART HEREOF. IN ADDITION, THE PERFORMANCE OF THIS AGREEMENT BY ESCo IS SUBJECT TO THE PROVISIONS OF THE UTILITY AGREEMENT.

IN WITNESS WHEREOF, and intending to be legally bound, the Parties hereto subscribe their names to this instrument hereinabove in Section I as of the date of execution first written above.

SUMMARY OF GENERAL TERMS AND CONDITIONS
Section 1.   Operating the Premises and Maintaining the
               Use of the ECMs.

APPENDICES:

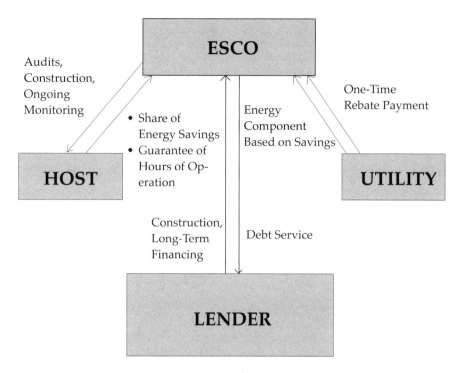

Figure 14-3. Shared savings/performance contracting.

14

Energy, Environmental, and Quality Management Standards*

ISO 50001—2011 ENERGY MANAGEMENT STANDARD

Introduction

The purpose of ISO50001 is to establish some mechanisms needed to improve energy performance. The objectives of implementing this standard are to reduce global warming and acid rain through a systematic energy management system (EnMS).

The Standard specifies the requirements for EnMS upon which an organization can develop and implement an energy policy, establish objectives and targets, and action plans which take into account legal requirements and information related to significant energy use. This Standard has the flexibility so that it can be tailored to fit the specific requirements of any organization including the complexity of the system, degree of documentation, and resources.

This International Standard is based on the Plan-Do-Check-Act (PDCA) continual improvement framework and incorporates energy management into everyday organizational practices. In the context of energy management, various steps for the PDCA approach are: a) Plan: conduct the energy review and establish the baseline, energy performance

---

indicators (EnPls), objectives, targets and action plans necessary to deliver results that will improve energy performance in accordance with the organization's energy policy; b) Do: implement the energy management action plans; c) Check: monitor and measure processes and the key characteristics of operations that determine energy performance against the energy policy and objectives, and report the results; d) Act: take actions to continually improve energy performance and the EnMS.

This International Standard can be used for certification, registration and self-declaration of an organization's EnMS and has the flexibility to be site specific. This Standard is based on the common elements of ISO management system standards, ensuring a high level of compatibility notably with ISO 14001 (Environmental Management) and ISO 9001 (Quality Management).

SCOPE

This International Standard helps an organization to follow a systematic approach in achieving continual improvement of energy performance, including energy efficiency, energy use and consumption by specifying requirements applicable to energy use and consumption, including measurement, documentation and reporting, design and procurement practices for equipment systems, process and personnel that contribute to energy performance. This International Standard is applicable to any organization wishing to ensure that it conforms to its stated energy policy and wishing to demonstrate this to others, such conformity being confirmed either by means of self-evaluation and self-declaration of conformity, or by certification of the energy management system by an external organization.

TERMS AND DEFINITIONS

The following terms and definitions apply for the interpretation and use of ISO 50001.

boundaries
physical or site limits and/or organization limits as defined by the organization, e.g. a group of processes; a site; an entire organization; or multiple sites.

continual improvement

> recurring process which results in enhancement of energy performance and the energy management system. Continual improvement achieves improvements in overall energy performance, consistent with the organization's energy policy.

correction

> action to eliminate a detected nonconformity

corrective action

> action to eliminate the cause of a detected nonconformity

energy

> electricity, fuels, steam, heat, compressed air, renewable energy and other like media.

energy baseline

> quantitative reference(s) providing a basis for comparison of energy performance. An energy baseline reflects a specified period of time. An energy baseline can be normalized using variable which affect energy use and/or consumption, e.g. production level, degree days (outdoor temperature), etc.

energy consumption

> quantity of energy applied

energy efficiency

> ratio or other quantitative relationship between an output of performance, service, goods or energy, and an input of energy and input of energy.

energy management system—EnMS

> set of interrelated or interacting elements to establish an energy policy and energy objectives, and processes and procedures to achieve those objectives.

energy management team

> person(s) responsible for effective implementation of the energy management system activities and for delivering energy performance improvements.

energy objective
>   specified outcome or achievement set to meet the organization's energy policy related to improved energy performance

energy performance
>   measurable results related to energy efficiency, energy use, and energy consumption.

energy performance indicator (EnPl)
>   Quantitative value or measure of energy performance, as defined by the organization. EnPls could be expressed as a simple metric, ratio or a more complex model.

energy policy
>   statement by the organization of its overall intentions and direction of an organization related to its energy performance, as formally expressed by top management. The energy policy provides a framework for action and for the setting of energy objectives and energy targets.

energy review
>   determination of the organization's energy performance based on data and other information, leading to identification of opportunities for improvement.

energy services
>   activities and their results related to the provision and/or use of energy

energy target
>   detailed and quantifiable energy performance requirement, applicable to the organization or parts thereof, that arises from the energy objective and that needs to be set and met in order to achieve this objective

energy use
>   manner or kind of application of energy e.g. ventilation; lighting; heating; cooling; transportation; processes; production lines etc.

interested party
> person or group concerned with, or affected by, the energy performance of the organization

internal audit
> systematic, independent and documented process for obtaining evidence and evaluation it objectively in order to determine the extent to which requirements are fulfilled

nonconformity
> non-fulfillment of a requirement

organization
> company, corporation, firm, enterprise, authority or institution, or part of combination thereof, whether incorporated or not, public or private, that has its own functions and administration and that has the authority to control its energy use and consumption

preventive action
> action to eliminate the cause of a potential nonconformity. Preventive action is taken to prevent occurrence, whereas corrective action is taken to prevent recurrence,

procedure
> specified way to carry out an activity or a process.

record
> Document stating results achieved or providing evidence of activities performed.

scope
> Extent of activities, facilities and decisions that the organization addresses through an EnMS, which can include several boundaries e.g. the scope can include energy related to transport.

significant energy use
> energy use accounting for substantial energy consumption and/or offering considerable potential for energy performance improvement.

top management
> person or group of people who directs and controls an organization at the highest level. Top management controls the organization defined within the scope and boundaries of the energy management system.

ENERGY MANAGEMENT SYSTEM REQUIREMENTS

General Requirements

The organization shall: a) establish, document, implement, maintain and improve an EnMS in accordance with the requirements of this international Standard; b) define and document the scope and boundaries of its EnMS; c) determine how it will meet the requirements of this International Standard in order to achieve continual improvement of its energy performance and of its EnMS.

Management Responsibility
Top management

Top management shall demonstrate it commitment to support the EnMS and to continually improve its effectiveness by: a) defining, establishing, implementing and maintaining an energy policy; b) appointing a management representative and approving the formation of an energy management team; c) providing the resources needed to establish, implement, maintain and improve the EnMS and the resulting energy performance; d) identifying the scope and boundaries to be addressed by the EnMS; e) communicating the importance of energy management to those in the organization; f) ensuring that energy objectives and targets are established; g) ensuring that EnPls are appropriate to the organization; h) considering energy performance in long-term planning; i) ensuring that results are measured and reported at determined intervals; j) conducting management reviews.

Management representative

Top management shall appoint a management representative(s) with appropriate skills and competence, who, irrespective of other responsibilities, has the responsibility and authority to: a) ensure the EnMS is established, implemented, maintained, and continually improved; b) identify person(s) authorized by an appropriate level of management, to work with the management representative in support of energy management activities; c) report to top management on energy performance; d) report to top management on the performance of the EnMS; e) ensure that the planning of energy management activities is designed to support the organization's energy policy; f) define and communicate responsibilities and authorities in order to facilitate effective energy management; g) determine criteria and methods needed to ensure that both the operation and control of the EnMS are effective; h) promote awareness of the energy policy and objectives at all levels of the organization.

Energy Policy

The energy policy shall state the organization's commitment to achieving energy performance improvement. Top management shall define the energy policy and ensure that it: a) is appropriate to the nature and scale of the organization's energy use and consumption; b) includes a commitment to continual improvement in energy performance; c) includes a commitment to ensure the availability of information and of necessary resources to achieve objectives and targets; d) includes a commitment to comply with applicable legal requirements and other requirements to which the organization subscribes related to its energy use, consumption and efficiency; e) provides the framework for setting and reviewing energy objectives and targets; f) supports the purchase of energy-efficient products and services, and design for energy performance improvement; g) is documented and communicated at all levels within the organization; h) is regularly reviewed, and updated as necessary.

Energy Planning

General

The organization shall conduct and document an energy planning process. Energy planning shall be consistent with the energy policy and shall lead to activities that continually improve energy performance. Energy planning shall involve a review of the organization's activities that can affect energy performance. A concept diagram illustrating energy planning is shown in Figure 14-1.

Legal requirements and other requirements

The organization shall identify, implement, and have access to the applicable legal requirements and other requirements to which the organization subscribes related to its energy use, consumption and efficiency.

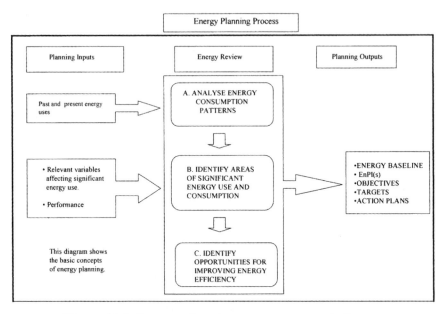

Figure 14-1. Energy planning process concept diagram

The organization shall ensure that these legal requirements and other requirements are considered in establishing, implementing and maintaining the EnMS on a continually review basis.

Energy review

The organization shall develop, record, and maintain an energy review. The methodology and criteria used to develop the energy review shall be documented. To develop the energy review, the organization shall; a) analyze energy use and consumption based on measurement and other data, i.e. (i) identify current energy sources; (ii) evaluate past and present use and consumption; b) based on the analysis of energy consumption patterns, identify the areas of significant energy use, i.e. (i) identify the energy intensive facilities, equipment, systems, processes and personnel; (ii) identify other relevant variables affecting significant energy uses; (iii) determine the current energy performance of energy intensive facilities, equipment systems and processes; (iv) estimate future energy use and consumption; c) identify, prioritize and record opportunities for improving energy performance. Opportunities can relate to potential sources of energy, use of renewable energy, or other alternative energy sources, such as waste energy.

Energy baseline

The organization shall establish an energy baseline(s) using the information in the initial energy review, considering a data period suitable to the organization's energy use and consumption. Changes in energy performance shall be measured against the energy baseline(s). Adjustments to the baseline(s) shall be made in the case of one or more of: a) EnPIs no long reflect organizational energy use and consumption, or b) major changes to processes, operations, or energy consuming equipments changes, or c) new method. The energy baseline(s) shall be maintained and recorded.

Energy performance indicators

The organization shall identify EnPIs appropriate for monitoring and measuring its energy performance. The methodology for determining and updating the EnPIs shall be recorded and regularly reviewed.

Energy objectives, energy targets and energy management action plans

The organization shall develop, implement and maintain documented energy objectives and targets at the relevant functions, levels, processes or facilities within the organization. Time frames shall be established for achievement of the objectives and targets consistent with the energy policy and targets. The organization shall establish, implement and maintain action plans for achieving its objectives and targets. The action plans shall include: a) designation of responsibility; b) the means and time frame by which individual targets are to be achieved; c) a statement of the method by which an improvement in energy performance shall be verified; d) a statement of the method of verifying the results. The action plans shall be documented, and updated at defined intervals.

Implementation and Operation
General

The organization shall use the action plans and other outputs resulting from the planning process for implementation and operation.

Competence, training and awareness

The organization shall ensure that any person(s) working for or on its behalf, related to significant energy uses are competent on the basis of appropriate education, training, skills or experience, shall identify training needs associated with the control of its significant energy uses and the operation of its EnMS, and shall provide training or take other actions to meet these needs. Appropriate records shall be maintained.

The organization shall ensure that any person(s) working for or on its behalf are aware of: a) the importance of conformity with the energy policy, procedures and the requirements of the EnMS; b) their roles, responsibilities and authorities in achieving the requirements of the EnMS; c) the benefits of improved energy performance; d) the impact, actual or potential, with respect to energy use and consumption, of their activities and how their activities and behavior contribute to the achievement of energy objectives and targets, and the potential consequences of departure from specified procedures.

Communication

The organization shall communicate internally and externally (optional) with regard to its energy performance and EnMS, and shall invite comments and suggestions for the improvement of its EnMS.

Documentation
Documentation requirements

The organization shall establish, implement and maintain information, in paper, electronic or any other medium, to describe the core elements of the EnMS and their interaction. The EnMS documentation shall include: a) the scope and boundaries of the EnMS; b) the energy policy; c) the energy objectives, targets, and action plants; d) the documents, including records, required by this International Standard; e) and other documents determined by the organization to be necessary.

Control documents

Documents required by this International Standard and the EnMS shall be controlled. The organization shall establish, implement and maintain procedure(s) to: a) approve documents for adequacy prior to issue; b) periodically review and update documents as necessary; c) ensure that changes and the current revision status of documents are identified; d) ensure that relevant versions of applicable documents are available at points of use; e) ensure that documents remain legible and readily identifiable; f) ensure documents of external origin determined by the organization to be necessary for the planning and operation of the EnMS are identified and their distribution controlled; g) prevent the unintended use of obsolete documents, and suitably identify those to be retained for any purpose.

Operational control

The organization shall identify and plan those operations and maintenance activities which are related to its significant energy uses and that are consistent with its energy policy, objectives, targets and action plans, in order to ensure that they are carried out under specified conditions, by means of a) establishing and setting criteria for the effective operation and maintenance of significant energy uses, where their absence could lead to a significant deviation from effective energy performance; b) operating and maintaining facilities, processes, systems and equipment, in accordance with operational criteria; c) appropriate communication of the operational controls to personnel working for, or on behalf of, the organization.

Design

The organization shall consider energy performance improvement opportunities and operational control in the design of new, modified and renovated facilities, equipment, systems and processes that can have a significant impact on its energy performance.

Procurement of energy services, products, equipment and energy

When procuring energy services, products and equipment that have, or can have, an impact on significant energy use, the organization shall inform suppliers that procurement is partly evaluated on the basis of energy performance. The organization shall define, document and implement energy purchasing criteria and specifications, as applicable, for effective energy use.

Checking
Monitoring, measurement and analysis

The organization shall ensure that the key characteristics of its operations that determine energy performance are monitored, measured and analyzed at planned intervals. Key characteristics shall include at a minimum: a) significant energy uses and other outputs of the energy review; b) the relevant variables related to significant energy uses; c) EnPIs; d) the effectiveness of the action plans in achieving objectives and targets; e) evaluation of actual versus expected energy consumption. The results from monitoring and measurement of the key characteristics shall be recorded. It is up to the organization to determine the means and methods of measurement. Records of calibration and other means of establishing accuracy and repeatability of the measuring instruments/equipment shall be maintained. The organization shall investigate and respond to significant deviations in energy performance. Results of these activities shall be maintained.

Evaluation of compliance with legal requirements and other requirements

At planned intervals, the organization shall evaluate compliance with legal requirements and other requirements to which it subscribes related to its energy use and consumption. Records of the results of the evaluations of compliance shall be maintained.

Internal audit of the EnMS

The organization shall conduct internal audits at planned intervals to ensure that the EnMS: a) conforms to planned arrangements for energy management including the requirements of this International Standard; b) conforms with the energy objectives and targets established; c) is effectively implemented and maintained, and improves energy performance.

Nonconformities, correction, corrective action and preventive action

The organization shall address actual and potential nonconformities by making corrections, and by taking corrective action and preventive action, including: a) reviewing nonconformities or potential nonconformities; b) determining the causes of nonconformities or potential nonconformities; c) evaluating the need for action to ensure that nonconformities do not occur or recur; d) determining and implementing the appropriate action needed; e) maintaining records of corrective actions and preventive actions; f) reviewing the effectiveness of the corrective action or preventive action taken. The organization shall ensure that any necessary changes are made to the EnMS.

Control of records

The organization shall establish and maintain records, as necessary, to demonstrate conformity to the requirements of its EnMS and of this International Standard, and the energy performance results achieved. The organization shall define and implement controls for the identification, retrieval and retention of records. Records shall be and shall remain legible, identifiable and traceable to the relevant activity.

Management review
General

At planned intervals, top management shall review the organization's EnMS to ensure its continuing suitability, adequacy and effectiveness. Records of management review shall be maintained.

Input to management review

Inputs to the management review shall include: a) follow-up actions from previous management reviews; b) review of the energy policy; c) review of energy performance and related EnPls; d) results of the evaluation of compliance with legal requirements and changes in legal requirements and other requirements to which the organization subscribes; e) the extent to which the energy objectives and targets have been met; f) EnMS audit results; g) the status of corrective actions and preventive actions; h) projected energy performance for the following period; i) recommendations for improvement.

Output from management review

Outputs from the management review shall include any decisions or actions related to: a) changes in the energy performance of the organization; b) changes to the energy policy; c) changes to the EnPls; d) changes to objectives, targets or other elements of the EnMS, consistent with the organization's commitment to continual improvement; e) changes to allocation of resources.

CORRELATING ENERGY, ENVIRONMENTAL, AND QUALITY MANAGEMENT STANDARDS

The oil embargo of 1973, was the starting point for "Energy Problem" or for an opportunity for "Energy Efficiency." The need for energy efficiency captured the attention of all sectors of our society since then and continued through for the rest of 1970's when energy supplies dwindled and prices increased. Interest in energy efficiency continued during the 1980s and 1990s primarily due to environmental concerns and secondarily because of industrial competitiveness. U.S. Government's responses to address the environmental issues like global warming and acid was to initiate, develop, and implement more legislation. Some examples of such legislation are Toxic Substances Control Act of 1976, Resource Conservation and Recovery Act of 1984, Clean Air Act of 1990, and Pollution Prevention Act of 1990.

The International Standards Organization (ISO) on the other hand developed a series of Environmental Management Standards (EMS) ISO 14000 as a mechanism for achieving improvements in environmental performance and for supporting the trade prospects of "clean" firms. Theses standards were built on an existing Quality Management System ISO 9000 series.

The EMS is becoming more and more a matter of interest outside the management of the enterprise—to workers, regulators, local residents, commercial partners, bankers, insurers, and the general public. The ISO 14000 series of standards has become the best known common framework for EMS. This series is based on the overall approach and broad success of the Quality Management Standards (QMS) prepared and issued as the ISO 9000 series.

Within the ISO system, ISO 14001 sets out the basic structure for an EMS, while 14004 provides guidance. The crucial feature of ISO 14001 is that it identifies the elements of a system which can be independently audited and certified. Certification means that a qualified body (an "accredited certifier") has inspected the EMS system that has been put in place and has made a formal declaration that the system is consistent with the requirements of ISO 14001.

Compliance with ISO 14001 does not by itself automatically ensure that an enterprise will actually achieve improved environmental performance. The standard requires that there be an environmental policy that "includes a commitment to continual improvement and pollution prevention" and "a commitment to comply with relevant environmental legislation and regulations." It also requires that the company establish procedures for taking corrective and preventive action in cases of nonconformance.

Basically, an EMS can be described as a program of continuous environmental improvement that follows a defined sequence of steps drawn from established project management practice and routinely applied in business management. These steps are as follows:

•   Review the environmental consequences of the operations.
•   Define a set of policies and objectives for environmental performance.
•   Establish an action plan to achieve the objectives.

- Monitor performance against these objectives.
- Report the results appropriately.
- Review the system and the outcomes and strive for continuous improvement.

From the above discussion it is clear that a common thread in ISO Standards is the Plan-Do-Check, Act (PDCA) continual improvement framework. This commonality is further illustrated with the help of Table 14-1. In this table a correlation with ISO 22000:2005 is also shown as an example of a particular industry namely "Food Safety." Table 14-1 "is reproduced from ISO 50001:2011 with permission of the American National Standards Institute (ANSI) on behalf of the International Organization for Standardization (ISO). No part of this material may be copied or reproduced in any form, electronic retrieval system or otherwise or made available on the internet, a public network, by satellite or otherwise without the prior written consent of ANSI. Copies of this standard may be purchased from ANSI, 25 West 43rd Street, New York, NY 10036, (212) 642-4900, http://webstore.ansi.org"

References
1. ISO 9000:2005, Quality management systems—Fundamentals and vocabulary
2. ISO 9001:2008, Quality management systems—Requirements
3. ISO 14001:2004, Environmental management systems—Requirements with guidance for use
4. ISO 22000:2005, Food safety management systems—Requirements for an organization in the food chain
5. Handbook of Pollution Prevention Practices—Nicholas P. Cheremisinoff—Marcel Dekker, Inc. New York 2001.

Table 14-1. Correspondence between ISO 50001:2011, ISO 9001:2008, ISO 14001:2004 and ISO 22000:2005

ISO 50001:2011		ISO 9001:2008		ISO 14001:2004		ISO 22000:2005	
Clause	Criteria	Clause	Criteria	Clause	Criteria	Clause	Criteria
—	Foreword	—	Foreword	—	Foreword	—	Foreword
—	Introduction	—	Introduction	—	Introduction	—	Introduction
1	Scope	1	Scope	1	Scope	1	Scope
2	Normative references	2	Normative references	2	Normative references	2	Normative references
3	Terms and definitions	3	Terms and definitions	3	Terms and definitions	3	Terms and definitions
4	Energy management system requirements	4	Quality management system	4	Environmental management system requirements	4	Food safety management system
4.1	General requirements	4.1	General requirements	4.1	General requirements	4.1	General requirements
4.2	Management responsibility	5	Management responsibility	—	—	5	Management responsibility
4.2.1	Top management	5.'	Management commitment	4.4.1	Resources, roles, responsibility and authority	5.1	Management commitment
4.2.2	Management representative	5.5 1	Responsibility and authority	4.4.1	Resources, roles, responsibility and authority	5.4	Responsibility and authority
		5.5 2	Management representative			5.5	Food safety team leader
4.3	Energy policy	5.3	Quality policy	4.2	Environmental policy	5.2	Food safety policy
4.4	Energy planning	5.4	Planning	4.3	Planning	5.3	Food safety management system planning
						7	Planning and realization for safe products
4.4.1	General	5.4.1	Quality objectives	4.3	Planning	5.3	Food safety management system planning
		7.2.1	Determination of requirements related to the product			7.1	General
4.4.2	Legal requirements and other requirements	7.2.1	Determination of requirements related to the product	4.3.2	Legal and other requirements	7.2.2	(no title)
		7.3.2	Design and development inputs			7.3.3	Product characteristics
4.4.3	Energy review	5.4.1	Quality objectives	4.3.1	Environmental aspects	7	Planning and realization of safe products
		7.2.1	Determination of requirements related to the product				
4.4.4	Energy baseline	--	—	—	—	7.4	Hazard analysis
4.4.5	Energy performance indicators	--	—	—	—	7.4.2	Hazard identification and determination of acceptable levels
4.4.6	Energy objectives, energy targets and energy management action plans	5.4.1	Quality objectives	4.3.3	Objectives, targets and programme(s)	7.2	Prerequisite programmes
		7.1	Planning of product realization				

Table 14-1 (cont'd). Correspondence between ISO 50001:2011, ISO 9001:2008, ISO 14001:2004 and ISO 22000:2005

ISO 50001:2011		ISO 9001:2008		ISO 14001 2004		ISO 22000:2005	
Clause	Criteria	Clause	Criteria	Clause	Criteria	Clause	Criteria
4.5	Implementation and operation	7	Product realization	4.4	Implementation and operation	7	Planning and realization of safe products
4.5.1	General	7.5.1	Control of production and service provision	4.4.6	Operational control	7.2.2	(no title)
4.5.2	Competence, training and awareness	6.2.2	Competence, training and awareness	4.4.2	Competence, training and awareness	6.2.2	Competence, training and awareness
4.5.3	Communication	5.5.3	Internal communication	4.4.3	Communication	5.6.2	Internal communication
4.5.4	Documentation	4.2	Documentation requirements	—	—	4.2	Documentation requirements
4.5.4.1	Documentation requirements	4.2.1	General	4.4.4	Documentation	4.2.1	General
4.5.4.2	Control of documents	4.2.3	Control of documents	4.4.5	Control of documents	4.2.2	Control of documents
4.5.5	Operational control	7.5.1	Control of production and service provision	4.4.6	Operational control	7.6.1	HACCP plan
4.5.6	Design	7.3	Design and development	—	—	7.3	Preliminary steps to enable hazard analysis
4.5.7	Procurement of energy services, products, equipment and energy	7.4	Purchasing	—	—	—	—
4.6	Checking	8	Measurement, analysis and improvement	4.5	Checking	8	Validation, verification and improvement of the food safety management system
4.6.1	Monitoring, measurement and analysis	8.2.3	Monitoring and measurement of process	4.5.1	Monitoring and measurement	7.6.4	System for monitoring of critical control points
		8.2.4	Monitoring and measurement of product				
		8.4	Analysis of data				
4.6.2	Evaluation of compliance with legal requirements and other requirements	7.3.4	Design and develop review	4.5.2	Evaluation of compliance	—	—
4.6.3	Internal audit of the EnMS	8.2.2	Internal audit	4.5.5	Internal audit	8.4.1	Internal audit
4.6.4	Nonconformities correction, corrective action and preventive action	8.3	Control of nonconforming product	4.5.3	Nonconformity, corrective action and preventive action	7.10	Control of nonconformity
		8.5.2	Corrective action				
		8.5.3	Preventive action				
4.6.5	Control of records	4.2.4	Control of records	4.5.4	Control of records	4.2.3	Control of records
4.7	Management review	5.6	Management review	4.6	Management review	5.8	Management review
4.7.1	General	5.6.1	General	4.6	Management review	5.8.1	General
4.7.2	Input to management review	5.6.2	Review input	4.6	Management review	5.8.2	Review input
4.7.3	Output from management review	5.6.3	Review output	4.6	Management review	5.8.3	Review output

Appendix

Table A-1. 10% Interest factors.

Period n	Single-payment compound-amount F/P	Single-payment present-worth P/F	Uniform series compound-amount F/A	Sinking-fund payment A/F	Capital recovery A/P	Uniform-series present-worth P/A
	Future value of $1 $$(1 + i)^n$$	Present value of $1 $$\frac{1}{(1+i)^n}$$	Future value of uniform series of $1 $$\frac{(1+i)^n - 1}{i}$$	Uniform series whose future value is $1 $$\frac{i}{(1+i)^n - 1}$$	Uniform series with present value of $1 $$\frac{i(1+i)^n}{(1+i)^n - 1}$$	Present value of uniform series of $1 $$\frac{(1+i)^n - 1}{i(1+i)^n}$$
1	1.100	0.9091	1.000	1.00000	1.10000	0.909
2	1.210	0.8264	2.100	0.47619	0.57619	1.736
3	1.331	0.7513	3.310	0.30211	0.40211	2.487
4	1.464	0.6830	4.641	0.21547	0.31147	3.170
5	1.611	0.6209	6.105	0.16380	0.26380	3.791
6	1.772	0.5645	7.716	0.12961	0.22961	4.355
7	1.949	0.5132	9.487	0.10541	0.20541	4.868
8	2.144	0.4665	11.436	0.08744	0.18744	5.335
9	2.358	0.4241	13.579	0.07364	0.17364	5.759
10	2.594	0.3855	15.937	0.06275	0.16275	6.144
11	2.853	0.3505	18.531	0.05396	0.15396	6.495
12	3.138	0.3186	21.384	0.04676	0.14676	6.814
13	3.452	0.2897	24.523	0.04078	0.14078	7.103
14	3.797	0.2633	27.975	0.03575	0.13575	7.367
15	4.177	0.2394	31.772	0.03147	0.13147	7.606
16	4.595	0.2176	35.950	0.02782	0.12782	7.824
17	5.054	0.1978	40.545	0.02466	0.12466	8.022
18	5.560	0.1799	45.599	0.02193	0.12193	8.201
19	6.116	0.1635	51.159	0.01955	0.11955	8.365
20	6.727	0.1486	57.275	0.01746	0.11746	8.514
21	7.400	0.1351	64.002	0.01562	0.11562	8.649
22	8.140	0.1228	71.403	0.01401	0.11401	8.772
23	8.954	0.1117	79.543	0.01257	0.11257	8.883
24	9.850	0.1015	88.497	0.01130	0.11130	8.985
25	10.835	0.0923	98.347	0.01017	0.11017	9.077
26	11.918	0.0839	109.182	0.00916	0.10916	9.161
27	13.110	0.0763	121.100	0.00826	0.10826	9.237
28	14.421	0.0693	134.210	0.00745	0.10745	9.307
29	15.863	0.0630	148.631	0.00673	0.10673	9.370
30	17.449	0.0673	164.494	0.00608	0.10608	9.427
35	28.102	0.0356	271.024	0.00369	0.10369	9.644
40	45.259	0.0221	442.593	0.00226	0.10226	9.779
45	72.890	0.0137	718.905	0.00139	0.10139	9.863
50	117.391	0.0085	1163.909	0.00086	0.10086	9.915
55	189.059	0.0053	1880.591	0.00053	0.10053	9.947
60	304.482	0.0033	3034.816	0.00033	0.10033	9.967
65	490.371	0.0020	4893.707	0.00020	0.10020	9.980
70	789.747	0.0013	7887.470	0.00013	0.10013	9.987
75	1271.895	0.0008	12708.954	0.00008	0.10008	9.992
80	2048.400	0.0005	20474.002	0.00005	0.10005	9.995
85	3298.969	0.0003	32979.690	0.00003	0.10003	9.997
90	5313.023	0.0002	53120.226	0.00002	0.10002	9.998
95	8556.676	0.0001	85556.760	0.00001	0.10001	9.999

Table A-2. 12% Interest factors.

Period n	Single-payment compound-amount F/P	Single-payment present-worth P/F	Uniform series compound-amount F/A	Sinking-fund payment A/F	Capital recovery A/P	Uniform-series present-worth P/A
	Future value of $1	Present value of $1	Future value of uniform series of $1	Uniform series whose future value is $1	Uniform series with present value of $1	Present value of uniform series of $1
	$(1+i)^n$	$\dfrac{1}{(1+i)^n}$	$\dfrac{(1+i)^n-1}{i}$	$\dfrac{i}{(1+i)^n-1}$	$\dfrac{i(1+i)^n}{(1+i)^n-1}$	$\dfrac{(1+i)^n-1}{i(1+i)^n}$
1	1.120	0.8929	1.000	1.00000	1.12000	0.893
2	1.254	0.7972	2.120	0.47170	0.59170	1.690
3	1.405	0.7118	3.374	0.29635	0.41635	2.402
4	1.574	0.6355	4.779	0.20923	0.32923	3.037
5	1.762	0.5674	6.353	0.15741	0.27741	3.605
6	1.974	0.5066	8.115	0.12323	0.24323	4.111
7	2.211	0.4523	10.089	0.09912	0.21912	4.564
8	2.476	0.4039	12.300	0.08130	0.20130	4.968
9	2.773	0.3606	14.776	0.06768	0.18768	5.328
10	3.106	0.3220	17.549	0.05698	0.17698	5.650
11	3.479	0.2875	20.655	0.04842	0.16842	5.938
12	3.896	0.2567	24.133	0.04144	0.16144	6.194
13	4.363	0.2292	28.029	0.03568	0.15568	6.424
14	4.887	0.2046	32.393	0.03087	0.15087	6.628
15	5.474	0.1827	37.280	0.02682	0.14682	6.811
16	6.130	0.1631	42.753	0.02339	0.14339	6.974
17	6.866	0.1456	48.884	0.02046	0.14046	7.120
18	7.690	0.1300	55.750	0.01794	0.13794	7.250
19	8.613	0.1161	63.440	0.01576	0.13576	7.366
20	9.646	0.1037	72.052	0.01388	0.13388	7.469
21	10.804	0.0926	81.699	0.01224	0.13224	7.562
22	12.100	0.0826	92.503	0.01081	0.13081	7.645
23	13.552	0.0738	104.603	0.00956	0.12956	7.718
24	15.179	0.0659	118.155	0.00846	0.12846	7.784
25	17.000	0.0588	133.334	0.00750	0.12750	7.843
26	19.040	0.0525	150.334	0.00665	0.12665	7.896
27	21.325	0.0469	169.374	0.00590	0.12590	7.943
28	23.884	0.0419	190.699	0.00524	0.12524	7.984
29	26.750	0.0374	214.583	0.00466	0.12466	8.022
30	29.960	0.0334	241.333	0.00414	0.12414	8.055
35	52.800	0.0189	431.663	0.00232	0.12232	8.176
40	93.051	0.0107	767.091	0.00130	0.12130	8.244
45	163.988	0.0061	1358.230	0.00074	0.12074	8.283
50	289.002	0.0035	2400.018	0.00042	0.12042	8.304
55	509.321	0.0020	4236.005	0.00024	0.12024	8.317
60	897.597	0.0011	7471.641	0.00013	0.12013	8.324
65	1581.872	0.0006	13173.937	0.00008	0.12008	8.328
70	2787.800	0.0004	23223.332	0.00004	0.12004	8.330
75	4913.056	0.0002	40933.799	0.00002	0.12002	8.332
80	8658.483	0.0001	72145.692	0.00001	0.12001	8.332

Table A-3. 15% Interest factors.

Period n	Single-payment compound-amount F/P	Single-payment present-worth P/F	Uniform series compound-amount F/A	Sinking-fund payment A/F	Capital recovery A/P	Uniform-series present-worth P/A
	Future value of $1	Present value of $1	Future value of uniform series of $1	Uniform series whose future value is $1	Uniform series with present value of $1	Present value of uniform series of $1
	$(1+i)^n$	$\dfrac{1}{(1+i)^n}$	$\dfrac{(1+i)^n - 1}{i}$	$\dfrac{i}{(1+i)^n - 1}$	$\dfrac{i(1+i)^n}{(1+i)^n - 1}$	$\dfrac{(1+i)^n - 1}{i(1+i)^n}$
1	1.150	0.8696	1.000	1.00000	1.15000	0.870
2	1.322	0.7561	2.150	0.46512	0.61512	1.626
3	1.521	0.6575	3.472	0.28798	0.43798	2.283
4	1.749	0.5718	4.993	0.20027	0.35027	2.855
5	2.011	0.4972	6.742	0.14832	0.29832	3.352
6	2.313	0.4323	8.754	0.11424	0.26424	3.784
7	2.660	0.3759	11.067	0.09036	0.24036	4.160
8	3.059	0.3269	13.727	0.07285	0.22285	4.487
9	3.518	0.2843	16.786	0.05957	0.20957	4.772
10	4.046	0.2472	20.304	0.04925	0.19925	5.019
11	4.652	0.2149	24.349	0.04107	0.19107	5.234
12	5.350	0.1869	29.002	0.03448	0.18448	5.421
13	6.153	0.1625	34.352	0.02911	0.17911	5.583
14	7.076	0.1413	40.505	0.02469	0.17469	5.724
15	8.137	0.1229	47.580	0.02102	0.17102	5.847
16	9.358	0.1069	55.717	0.01795	0.16795	5.954
17	10.761	0.0929	65.075	0.01537	0.16537	6.047
18	12.375	0.0808	75.836	0.01319	0.16319	6.128
19	14.232	0.0703	88.212	0.01134	0.16134	6.198
20	16.367	0.0611	102.444	0.00976	0.15976	6.259
21	18.822	0.0531	118.810	0.00842	0.15842	6.312
22	21.645	0.0462	137.632	0.00727	0.15727	6.359
23	24.891	0.0402	159.276	0.00628	0.15628	6.399
24	28.625	0.0349	194.168	0.00543	0.15543	6.434
25	32.919	0.0304	212.793	0.00470	0.15470	6.464
26	37.857	0.0264	245.712	0.00407	0.15407	6.491
27	43.535	0.0230	283.569	0.00353	0.15353	6.514
28	50.066	0.0200	327.104	0.00306	0.15306	6.534
29	57.575	0.0174	377.170	0.00265	0.15265	6.551
30	66.212	0.0151	434.745	0.00230	0.15230	6.566
35	133.176	0.0075	881.170	0.00113	0.15113	6.617
40	267.864	0.0037	1779.090	0.00056	0.15056	6.642
45	538.769	0.0019	3585.128	0.00028	0.15028	6.654
50	1083.657	0.0009	7217.716	0.00014	0.15014	6.661
55	2179.622	0.0005	14524.148	0.00007	0.15007	6.664
60	4383.999	0.0002	29219.992	0.00003	0.15003	6.665
65	8817.787	0.0001	58778.583	0.00002	0.15002	6.666

Table A-4. 20% Interest factors.

Period n	Single-payment compound-amount F/P	Single-payment present-worth P/F	Uniform series compound-amount F/A	Sinking-fund payment A/F	Capital recovery A/P	Uniform-series present-worth P/A
	Future value of $1	Present value of $1	Future value of uniform series of $1	Uniform series whose future value is $1	Uniform series with present value of $1	Present value of uniform series of $1
	$(1 + i)^n$	$\dfrac{1}{(1 + i)^n}$	$\dfrac{(1 + i)^n - 1}{i}$	$\dfrac{i}{(1 + i)^n - 1}$	$\dfrac{i(1 + i)^n}{(1 + i)^n - 1}$	$\dfrac{(1 + i)^n - 1}{i(1 + i)^n}$
1	1.200	0.8333	1.000	1.00000	1.20000	0.833
2	1.440	0.6944	2.200	0.45455	0.65455	1.528
3	1.728	0.5787	3.640	0.27473	0.47473	2.106
4	2.074	0.4823	5.368	0.18629	0.38629	2.589
5	2.488	0.4019	7.442	0.13438	0.33438	2.991
6	2.986	0.3349	9.930	0.10071	0.30071	3.326
7	3.583	0.2791	12.916	0.07742	0.27742	3.605
8	4.300	0.2326	16.499	0.06061	0.26061	3.837
9	5.160	0.1938	20.799	0.04808	0.24808	4.031
10	6.192	0.1615	25.959	0.03852	0.23852	4.192
11	7.430	0.1346	32.150	0.03110	0.23110	4.327
12	8.916	0.1122	39.581	0.02526	0.22526	4.439
13	10.699	0.0935	48.497	0.02062	0.22062	4.533
14	12.839	0.0779	59.196	0.01689	0.21689	4.611
15	15.407	0.0649	72.035	0.01388	0.21388	4.675
16	18.488	0.0541	87.442	0.01144	0.21144	4.730
17	22.186	0.0451	105.931	0.00944	0.20944	4.775
18	26.623	0.0376	128.117	0.00781	0.20781	4.812
19	31.948	0.0313	154.740	0.00646	0.20646	4.843
20	38.338	0.0261	186.688	0.00536	0.20536	4.870
21	46.005	0.0217	225.026	0.00444	0.20444	4.891
22	55.206	0.0181	271.031	0.00369	0.20369	4.909
23	66.247	0.0151	326.237	0.00307	0.20307	4.925
24	79.497	0.0126	392.484	0.00255	0.20255	4.937
25	95.396	0.0105	471.981	0.00212	0.20212	4.948
26	114.475	0.0087	567.377	0.00176	0.20176	4.956
27	137.371	0.0073	681.853	0.00147	0.20147	4.964
28	164.845	0.0061	819.223	0.00122	0.20122	4.970
29	197.814	0.0051	984.068	0.00102	0.20102	4.975
30	237.376	0.0042	1181.882	0.00085	0.20085	4.979
35	590.668	0.0017	2948.341	0.00034	0.20034	4.992
40	1469.772	0.0007	7343.858	0.00014	0.20014	4.997
45	3657.262	0.0003	18281.310	0.00005	0.20005	4.999
50	9100.438	0.0001	45497.191	0.00002	0.20002	4.999

Table A-5. 25% Interest factors.

Period n	Single-payment compound-amount F/P	Single-payment present-worth P/F	Uniform series compound-amount F/A	Sinking-fund payment A/F	Capital recovery A/P	Uniform-series present-worth P/A
	Future value of $1	Present value of $1	Future value of uniform series of $1	Uniform series whose future value is $1	Uniform series with present value of $1	Present value of uniform series of $1
	$(1+i)^n$	$\dfrac{1}{(1+i)^n}$	$\dfrac{(1+i)^n-1}{i}$	$\dfrac{i}{(1+i)^n-1}$	$\dfrac{i(1+i)^n}{(1+i)^n-1}$	$\dfrac{(1+i)^n-1}{i(1+i)^n}$
1	1.250	0.8000	1.000	1.00000	1.25000	0.800
2	1.562	0.6400	2.250	0.44444	0.69444	1.440
3	1.953	0.5120	3.812	0.26230	0.51230	1.952
4	2.441	0.4096	5.766	0.17344	0.42344	2.362
5	3.052	0.3277	8.207	0.12185	0.37185	2.689
6	3.815	0.2621	11.259	0.08882	0.33882	2.951
7	4.768	0.2097	15.073	0.06634	0.31634	3.161
8	5.960	0.1678	19.842	0.05040	0.30040	3.329
9	7.451	0.1342	25.802	0.03876	0.28876	3.463
10	9.313	0.1074	33.253	0.03007	0.28007	3.571
11	11.642	0.0859	42.566	0.02349	0.27349	3.656
12	14.552	0.0687	54.208	0.01845	0.26845	3.725
13	18.190	0.0550	68.760	0.01454	0.26454	3.780
14	22.737	0.0440	86.949	0.01150	0.26150	3.824
15	28.422	0.0352	109.687	0.00912	0.25912	3.859
16	35.527	0.0281	138.109	0.00724	0.25724	3.887
17	44.409	0.0225	173.636	0.00576	0.25576	3.910
18	55.511	0.0180	218.045	0.00459	0.25459	3.928
19	69.389	0.0144	273.556	0.00366	0.25366	3.942
20	86.736	0.0115	342.945	0.00292	0.25292	3.954
21	108.420	0.0092	429.681	0.00233	0.25233	3.963
22	135.525	0.0074	538.101	0.00186	0.25186	3.970
23	169.407	0.0059	673.626	0.00148	0.25148	3.976
24	211.758	0.0047	843.033	0.00119	0.25119	3.981
25	264.698	0.0038	1054.791	0.00095	0.25095	3.985
26	330.872	0.0030	1319.489	0.00076	0.25076	3.988
27	413.590	0.0024	1650.361	0.00061	0.25061	3.990
28	516.988	0.0019	2063.952	0.00048	0.25048	3.992
29	646.235	0.0015	2580.939	0.00039	0.25039	3.994
30	807.794	0.0012	3227.174	0.00031	0.25031	3.995
35	2465.190	0.0004	9856.761	0.00010	0.25010	3.998
40	7523.164	0.0001	30088.655	0.00003	0.25003	3.999

Table A-6. 30% Interest factors.

Period n	Single-payment compound-amount F/P	Single-payment present-worth P/F	Uniform series compound-amount F/A	Sinking-fund payment A/F	Capital recovery A/P	Uniform-series present-worth P/A
	Future value of $1	Present value of $1	Future value of uniform series of $1	Uniform series whose future value is $1	Uniform series with present value of $1	Present value of uniform series of $1
	$(1+i)^n$	$\dfrac{1}{(1+i)^n}$	$\dfrac{(1+i)^n-1}{i}$	$\dfrac{i}{(1+i)^n-1}$	$\dfrac{i(1+i)^n}{(1+i)^n-1}$	$\dfrac{(1+i)^n-1}{i(1+i)^n}$
1	1.300	0.7692	1.000	1.00000	1.30000	0.769
2	1.690	0.5917	2.300	0.43478	0.73478	1.361
3	2.197	0.4552	3.990	0.25063	0.55063	1.816
4	2.856	0.3501	6.187	0.16163	0.46163	2.166
5	3.713	0.2693	9.043	0.11058	0.41058	2.436
6	4.827	0.2072	12.756	0.07839	0.37839	2.643
7	6.275	0.1594	17.583	0.05687	0.35687	2.802
8	8.157	0.1226	23.858	0.04192	0.34192	2.925
9	10.604	0.0943	32.015	0.03124	0.33124	3.019
10	13.786	0.0725	42.619	0.02346	0.32346	3.092
11	17.922	0.0558	56.405	0.01773	0.31773	3.147
12	23.298	0.0429	74.327	0.01345	0.31345	3.190
13	30.288	0.0330	97.625	0.01024	0.31024	3.223
14	39.374	0.0254	127.913	0.00782	0.30782	3.249
15	51.186	0.0195	167.286	0.00598	0.30598	3.268
16	66.542	0.0150	218.472	0.00458	0.30458	3.283
17	86.504	0.0116	285.014	0.00351	0.30351	3.295
18	112.455	0.0089	371.518	0.00269	0.30269	3.304
19	146.192	0.0068	483.973	0.00207	0.30207	3.311
20	190.050	0.0053	630.165	0.00159	0.30159	3.316
21	247.065	0.0040	820.215	0.00122	0.30122	3.320
22	321.194	0.0031	1067.280	0.00094	0.30094	3.323
23	417.539	0.0024	1388.464	0.00072	0.30072	3.325
24	542.801	0.0018	1806.003	0.00055	0.30055	3.327
25	705.641	0.0014	2348.803	0.00043	0.30043	3.329
26	917.333	0.0011	3054.444	0.00033	0.30033	3.330
27	1192.533	0.0008	3971.778	0.00025	0.30025	3.331
28	1550.293	0.0006	5164.311	0.00019	0.30019	3.331
29	2015.381	0.0005	6714.604	0.00015	0.30015	3.332
30	2619.996	0.0004	8729.985	0.00011	0.30011	3.332
35	9727.8060	0.0001	32422.868	0.00003	0.30003	3.333

Table A-7. 40% Interest factors.

Period n	Single-payment compound-amount F/P Future value of \$1 $(1 + i)^n$	Single-payment present-worth P/F Present value of \$1 $\dfrac{1}{(1 + i)^n}$	Uniform series compound-amount F/A Future value of uniform series of \$1 $\dfrac{(1 + i)^n - 1}{i}$	Sinking-fund payment A/F Uniform series whose future value is \$1 $\dfrac{i}{(1 + i)^n - 1}$	Capital recovery A/P Uniform series with present value of \$1 $\dfrac{i(1 + i)^n}{(1 + i)^n - 1}$	Uniform-series present-worth P/A Present value of uniform series of \$1 $\dfrac{(1 + i)^n - 1}{i(1 + i)^n}$
1	1.400	0.7143	1.000	1.00000	1.40000	0.714
2	1.960	0.5102	2.400	0.41667	0.81667	1.224
3	2.744	0.3644	4.360	0.22936	0.62936	1.589
4	3.842	0.2603	7.104	0.14077	0.54077	1.849
5	5.378	0.1859	10.946	0.09136	0.49136	2.035
6	7.530	0.1328	16.324	0.06126	0.46126	2.168
7	10.541	0.0949	23.853	0.04192	0.44192	2.263
8	14.758	0.0678	34.395	0.02907	0.42907	2.331
9	20.661	0.0484	49.153	0.02034	0.42034	2.379
10	28.925	0.0346	69.814	0.01432	0.41432	2.414
11	40.496	0.0247	98.739	0.01013	0.41013	2.438
12	56.694	0.0176	139.235	0.00718	0.40718	2.456
13	79.371	0.0126	195.929	0.00510	0.40510	2.469
14	111.120	0.0090	275.300	0.00363	0.40363	2.478
15	155.568	0.0064	386.420	0.00259	0.40259	2.484
16	217.795	0.0046	541.988	0.00185	0.40185	2.489
17	304.913	0.0033	759.784	0.00132	0.40132	2.492
18	426.879	0.0023	1064.697	0.00094	0.40094	2.494
19	597.630	0.0017	1491.576	0.00067	0.40067	2.496
20	836.683	0.0012	2089.206	0.00048	0.40048	2.497
21	1171.356	0.0009	2925.889	0.00034	0.40034	2.498
22	1639.898	0.0006	4097.245	0.00024	0.40024	2.498
23	2295.857	0.0004	5737.142	0.00017	0.40017	2.499
24	3214.200	0.0003	8032.999	0.00012	0.40012	2.499
25	4499.880	0.0002	11247.199	0.00009	0.40009	2.499
26	6299.831	0.0002	15747.079	0.00006	0.40006	2.500
27	8819.764	0.0001	22046.910	0.00005	0.40005	2.500

Table A-8. 50% Interest factors.

Period n	Single-payment compound-amount F/P	Single-payment present-worth P/F	Uniform series compound-amount F/A	Sinking-fund payment A/F	Capital recovery A/P	Uniform-series present-worth P/A
	Future value of $1	Present value of $1	Future value of uniform series of $1	Uniform series whose future value is $1	Uniform series with present value of $1	Present value of uniform series of $1
	$(1+i)^n$	$\dfrac{1}{(1+i)^n}$	$\dfrac{(1+i)^n-1}{i}$	$\dfrac{i}{(1+i)^n-1}$	$\dfrac{i(1+i)^n}{(1+i)^n-1}$	$\dfrac{(1+i)^n-1}{i(1+i)^n}$
1	1.500	0.6667	1.000	1.00000	1.50000	0.667
2	2.250	0.4444	2.500	0.40000	0.90000	1.111
3	3.375	0.2963	4.750	0.21053	0.71053	1.407
4	5.062	0.1975	8.125	0.12308	0.62308	1.605
5	7.594	0.1317	13.188	0.07583	0.57583	1.737
6	11.391	0.0878	20.781	0.04812	0.54812	1.824
7	17.086	0.0585	32.172	0.03108	0.53108	1.883
8	25.629	0.0390	49.258	0.02030	0.52030	1.922
9	38.443	0.0260	74.887	0.01335	0.51335	1.948
10	57.665	0.0173	113.330	0.00882	0.50882	1.965
11	86.498	0.0116	170.995	0.00585	0.50585	1.977
12	129.746	0.0077	257.493	0.00388	0.50388	1.985
13	194.620	0.0051	387.239	0.00258	0.50258	1.990
14	291.929	0.0034	581.859	0.00172	0.50172	1.993
15	437.894	0.0023	873.788	0.00114	0.50114	1.995
16	656.841	0.0015	1311.682	0.00076	0.50076	1.997
17	985.261	0.0010	1968.523	0.00051	0.50051	1.998
18	1477.892	0.0007	2953.784	0.00034	0.50034	1.999
19	2216.838	0.0005	4431.676	0.00023	0.50023	1.999
20	3325.257	0.0003	6648.513	0.00015	0.50015	1.999
21	4987.885	0.0002	9973.770	0.00010	0.50010	2.000
22	7481.828	0.0001	14961.655	0.00007	0.50007	2.000

Table A-9. Five-year escalation table.

Present Worth of a Series of Escalating Payments Compounded Annually
Discount-Escalation Factors for $n = 5$ Years

| Discount | Annual Escalation Rate | | | | | |
Rate	0.10	0.12	0.14	0.16	0.18	0.20
0.10	5.000000	5.279234	5.572605	5.880105	6.202627	6.540569
0.11	4.866862	5.136200	5.420152	5.717603	6.029313	6.355882
0.12	4.738562	5.000000	5.274242	5.561868	5.863289	6.179066
0.13	4.615647	4.869164	5.133876	5.412404	5.704137	6.009541
0.14	4.497670	4.742953	5.000000	5.269208	5.551563	5.847029
0.15	4.384494	4.622149	4.871228	5.131703	5.404955	5.691165
0.16	4.275647	4.505953	4.747390	5.000000	5.264441	5.541511
0.17	4.171042	4.394428	4.628438	4.873699	5.129353	5.397964
0.18	4.070432	4.287089	4.513947	4.751566	5.000000	5.259749
0.19	3.973684	4.183921	4.403996	4.634350	4.875619	5.126925
0.20	3.880510	4.084577	4.298207	4.521178	4.755725	5.000000
0.21	3.790801	3.989001	4.196400	4.413341	4.640260	4.877689
0.22	3.704368	3.896891	4.098287	4.308947	4.529298	4.759649
0.23	3.621094	3.808179	4.003835	4.208479	4.422339	4.645864
0.24	3.540773	3.722628	3.912807	4.111612	4.319417	4.536517
0.25	3.463301	3.640161	3.825008	4.018249	4.220158	4.431144
0.26	3.388553	3.560586	3.740376	3.928286	4.124553	4.329514
0.27	3.316408	3.483803	3.658706	3.841442	4.032275	4.231583
0.28	3.246718	3.409649	3.579870	3.757639	3.943295	4.137057
0.29	3.179393	3.338051	3.503722	3.676771	3.857370	4.045902
0.30	3.114338	3.268861	3.430201	3.598653	3.774459	3.957921
0.31	3.051452	3.201978	3.359143	3.523171	3.694328	3.872901
0.32	2.990618	3.137327	3.290436	3.450224	3.616936	3.790808
0.33	2.939764	3.074780	3.224015	3.379722	3.542100	3.711472
0.34	2.874812	3.014281	3.159770	3.311524	3.469775	3.634758

Table A-10. Ten-year escalation table.

Present Worth of a Series of Escalating Payments Compounded Annually
Discount-Escalation Factors for $n = 10$ Years

Discount Rate	Annual Escalation Rate					
	0.10	0.12	0.14	0.16	0.18	0.20
0.10	10.000000	11.056250	12.234870	13.548650	15.013550	16.646080
0.11	9.518405	10.508020	11.613440	12.844310	14.215140	15.741560
0.12	9.068870	10.000000	11.036530	12.190470	13.474590	14.903510
0.13	8.650280	9.526666	10.498990	11.582430	12.786980	14.125780
0.14	8.259741	9.084209	10.000000	11.017130	12.147890	13.403480
0.15	7.895187	8.672058	9.534301	10.490510	11.552670	12.731900
0.16	7.554141	8.286779	9.099380	10.000000	10.998720	12.106600
0.17	7.234974	7.926784	8.693151	9.542653	10.481740	11.524400
0.18	6.935890	7.589595	8.312960	9.113885	10.000000	10.980620
0.19	6.655455	7.273785	7.957330	8.713262	9.549790	10.472990
0.20	6.392080	6.977461	7.624072	8.338518	9.128122	10.000000
0.21	6.144593	6.699373	7.311519	7.987156	8.733109	9.557141
0.22	5.911755	6.437922	7.017915	7.657542	8.363208	9.141752
0.23	5.692557	6.192047	6.742093	7.348193	8.015993	8.752133
0.24	5.485921	5.960481	6.482632	7.057347	7.690163	8.387045
0.25	5.290990	5.742294	6.238276	6.783767	7.383800	8.044173
0.26	5.106956	5.536463	6.008083	6.526298	7.095769	7.721807
0.27	4.933045	5.342146	5.790929	6.283557	6.824442	7.418647
0.28	4.768518	5.158489	5.585917	6.054608	6.568835	7.133100
0.29	4.612762	4.984826	5.392166	5.838531	6.327682	6.864109
0.30	4.465205	4.820429	5.209000	5.634354	6.100129	6.610435
0.31	4.325286	4.664669	5.035615	5.441257	5.885058	6.370867
0.32	4.192478	4.517015	4.871346	5.258512	5.681746	6.144601
0.33	4.066339	4.376884	4.715648	5.085461	5.489304	5.930659
0.34	3.946452	4.243845	4.567942	4.921409	5.307107	5.728189

Table A-11. Fifteen-year escalation table.

Present Worth of a Series of Escalating Payments Compounded Annually
Discount-Escalation Factors for $n = 15$ years

Discount Rate	Annual Escalation Rate					
	0.10	0.12	0.14	0.16	0.18	0.20
0.10	15.000000	17.377880	20.199780	23.549540	27.529640	32.259620
0.11	13.964150	16.126230	18.690120	21.727370	25.328490	29.601330
0.12	13.026090	15.000000	17.332040	20.090360	23.355070	27.221890
0.13	12.177030	13.981710	16.105770	18.616160	21.581750	25.087260
0.14	11.406510	13.057790	15.000000	17.287320	19.985530	23.169060
0.15	10.706220	12.220570	13.998120	16.086500	18.545150	21.442230
0.16	10.068030	11.459170	13.088900	15.000000	17.244580	19.884420
0.17	9.485654	10.766180	12.262790	14.015480	16.066830	18.477610
0.18	8.953083	10.133630	11.510270	13.118840	15.000000	17.203010
0.19	8.465335	9.555676	10.824310	12.303300	14.030830	16.047480
0.20	8.017635	9.026333	10.197550	11.560150	13.148090	15.000000
0.21	7.606115	8.540965	9.623969	10.881130	12.343120	14.046400
0.22	7.227109	8.094845	9.097863	10.259820	11.608480	13.176250
0.23	6.877548	7.684317	8.614813	9.690559	10.936240	12.381480
0.24	6.554501	7.305762	8.170423	9.167798	10.320590	11.655310
0.25	6.255518	6.956243	7.760848	8.687104	9.755424	10.990130
0.26	5.978393	6.632936	7.382943	8.244519	9.236152	10.379760
0.27	5.721101	6.333429	7.033547	7.836080	8.757889	9.819020
0.28	5.481814	6.055485	6.710042	7.458700	8.316982	9.302823
0.29	5.258970	5.797236	6.410005	7.109541	7.909701	8.827153
0.30	5.051153	5.556882	6.131433	6.785917	7.533113	8.388091
0.31	4.857052	5.332839	5.872303	6.485500	7.184156	7.982019
0.32	4.675478	5.123753	5.630905	6.206250	6.860492	7.606122
0.33	4.505413	4.928297	5.405771	5.946343	6.559743	7.257569
0.34	4.345926	4.745399	5.195502	5.704048	6.280019	6.933897

Table A-12. Twenty-year escalation table.

Present Worth of a Series of Escalating Payments Compounded Annually
Discount-Escalation Factors for $n = 20$ Years

Discount Rate	Annual Escalation Rate					
	0.10	0.12	0.14	0.16	0.18	0.20
0.10	20.000000	24.295450	29.722090	36.592170	45.308970	56.383330
0.11	18.213210	22.002090	26.776150	32.799710	40.417480	50.067940
0.12	16.642370	20.000000	24.210030	29.505400	36.181240	44.614710
0.13	15.259850	18.243100	21.964990	26.634490	32.502270	39.891400
0.14	14.038630	16.694830	20.000000	24.127100	29.298170	35.789680
0.15	12.957040	15.329770	18.271200	21.929940	26.498510	32.218060
0.16	11.995640	14.121040	16.746150	20.000000	24.047720	29.098950
0.17	11.138940	13.048560	15.397670	18.300390	21.894660	26.369210
0.18	10.373120	12.093400	14.201180	16.795710	20.000000	23.970940
0.19	9.686791	11.240870	13.137510	15.463070	18.326720	21.860120
0.20	9.069737	10.477430	12.188860	14.279470	16.844020	20.000000
0.21	8.513605	9.792256	11.340570	13.224610	15.527270	18.353210
0.22	8.010912	9.175267	10.579620	12.282120	14.355520	16.890730
0.23	7.555427	8.618459	9.895583	11.438060	13.309280	15.589300
0.24	7.141531	8.114476	9.278916	10.679810	12.373300	14.429370
0.25	6.764528	7.657278	8.721467	9.997057	11.533310	13.392180
0.26	6.420316	7.241402	8.216490	9.380883	10.778020	12.462340
0.27	6.105252	6.862203	7.757722	8.823063	10.096710	11.626890
0.28	5.816151	6.515563	7.339966	8.316995	9.480940	10.874120
0.29	5.550301	6.198027	6.958601	7.856833	8.922847	10.194520
0.30	5.305312	5.906440	6.609778	7.437339	8.416060	9.579437
0.31	5.079039	5.638064	6.289875	7.054007	7.954518	9.021190
0.32	4.869585	5.390575	5.995840	6.702967	7.533406	8.513612
0.33	4.675331	5.161809	5.725066	6.380829	7.148198	8.050965
0.34	4.494838	4.949990	5.475180	6.084525	6.795200	7.628322

Table A-13. Saturated steam

		Sat. Water	Sat. Steam	Temperature (°F)													
s. (psi) Temp.)				200	250	300	350	400	450	500	600	700	800	900	1000	1100	1200
1 (1.74)	Sh			98.26	148.26	198.26	248.26	298.26	348.26	398.26	498.26	598.26	698.26	798.26	898.26	998.26	1098.26
	v	0.01614	333.6	392.5	422.4	452.3	482.1	511.9	541.7	571.5	631.1	690.7	750.3	809.8	869.4	929.0	988.6
	h	69.73	1105.8	1150.2	1172.9	1195.7	1218.7	1241.8	1265.1	1288.6	1336.1	1384.5	1433.7	1483.8	1534.9	1586.8	1639.7
	s	0.1326	1.9781	2.0509	2.0841	2.1152	2.1445	2.1722	2.1985	2.2237	2.2708	2.3144	2.3551	2.3934	2.4296	2.4640	2.4969
5 (2.24)	Sh			37.76	87.76	137.76	187.76	237.76	287.76	337.76	437.76	537.76	637.76	737.76	837.76	937.76	1037.76
	v	0.01641	73.53	78.14	84.21	90.24	96.25	102.24	108.23	114.21	126.15	138.08	150.01	161.94	173.86	185.78	197.70
	h	130.20	1131.1	1148.6	1171.7	1194.8	1218.0	1241.3	1264.7	1288.2	1335.9	1384.3	1433.6	1483.7	1534.7	1586.7	1639.6
	s	0.2349	1.8443	1.8716	1.9054	1.9369	1.9664	1.9943	2.0208	2.0460	2.0932	2.1369	2.1776	2.2159	2.2521	2.2866	2.3194
10 (3.21)	Sh			6.79	56.79	106.79	156.79	206.79	256.79	306.79	406.79	506.79	606.79	706.79	806.79	906.79	1006.79
	v	0.01659	38.42	38.84	41.93	44.98	48.02	51.03	54.04	57.04	63.03	69.00	74.98	80.94	86.91	92.87	98.84
	h	161.26	1143.3	1146.6	1170.2	1193.7	1217.1	1240.6	1264.1	1287.8	1335.5	1384.0	1433.4	1483.5	1534.6	1586.6	1639.5
	s	0.2836	1.7879	1.7928	1.8273	1.8593	1.8892	1.9173	1.9439	1.9692	2.0166	2.0603	2.1011	2.1394	2.1757	2.2101	2.2430
.696 (2.00)	Sh				38.00	88.00	138.00	188.00	238.00	288.00	388.00	488.00	588.00	688.00	788.00	888.00	988.00
	v	.0167	26.799		28.42	30.52	32.60	34.67	36.72	38.77	42.86	46.93	51.00	55.06	59.13	63.19	67.25
	h	180.17	1150.5		1168.8	1192.6	1216.3	1239.9	1263.6	1287.4	1335.2	1383.8	1433.2	1483.4	1534.5	1586.5	1639.4
	s	.3121	1.7568		1.7833	1.8158	1.8459	1.8743	1.9010	1.9265	1.9739	2.0177	2.0585	2.0969	2.1332	2.1676	2.2005
15 (3.03)	Sh				36.97	86.97	136.97	186.97	236.97	286.97	386.97	486.97	586.97	686.97	786.97	886.97	986.97
	v	0.01673	26.290		27.837	29.899	31.939	33.963	35.977	37.985	41.986	45.978	49.964	53.946	57.926	61.905	65.882
	h	181.21	1150.9		1168.7	1192.5	1216.2	1239.9	1263.6	1287.3	1335.2	1383.8	1433.2	1483.4	1534.5	1586.5	1639.4
	s	0.3137	1.7552		1.7809	1.8134	1.8437	1.8720	1.8988	1.9242	1.9717	2.0155	2.0563	2.0946	2.1309	2.1653	2.1982
20 (7.96)	Sh				22.04	72.04	122.04	172.04	222.04	272.04	372.04	472.04	572.04	672.04	772.04	872.04	972.04
	v	0.01683	20.087		20.788	22.356	23.900	25.428	26.946	28.457	31.466	34.465	37.458	40.447	43.435	46.420	49.405
	h	196.27	1156.3		1167.1	1191.4	1215.4	1239.2	1263.0	1286.9	1334.9	1383.5	1432.9	1483.2	1534.3	1586.3	1639.3
	s	0.3358	1.7320		1.7475	1.7805	1.8111	1.8397	1.8666	1.8921	1.9397	1.9836	2.0244	2.0628	2.0991	2.1336	2.1982
25 (0.07)	Sh				9.93	59.93	109.93	159.93	209.93	259.93	359.93	459.93	559.93	659.93	759.93	859.93	959.93
	v	0.01693	16.301		16.558	17.829	19.076	20.307	21.527	22.740	25.153	27.557	29.954	32.348	34.740	37.130	39.518
	h	208.52	1160.6		1165.6	1190.2	1214.5	1238.5	1262.5	1286.4	1334.6	1383.3	1432.7	1483.0	1534.2	1586.2	1639.2
	s	0.3535	1.7141		1.7212	1.7547	1.7856	1.8145	1.8415	1.8672	1.9149	1.9588	1.9997	2.0381	2.0744	2.1089	2.1418
30 (0.34)	Sh					49.66	99.66	149.66	199.66	249.66	349.66	449.66	549.66	649.66	749.66	849.66	949.66
	v	0.01701	13.744			14.810	15.859	16.892	17.914	18.929	20.945	22.951	24.952	26.949	28.943	30.936	32.927
	h	218.93	1164.1			1189.0	1213.6	1237.8	1261.9	1286.0	1334.2	1383.0	1432.5	1482.8	1534.0	1586.1	1639.0
	s	0.3682	1.6995			1.7334	1.7647	1.7937	1.8210	1.8467	1.8946	1.9386	1.9795	2.0179	2.0543	2.0888	2.1217

Table A-13. Saturated steam (Continued).

Abs. Press. (psi) (Sat. Temp.)		Sat. Water	Sat. Steam	200	250	300	350	400	450	500	600	700	800	900	1000	1100	1200
35 (259.29)	Sh					40.71	90.71	140.71	190.71	240.71	340.71	440.71	540.71	640.71	740.71	840.71	940.71
	v	0.01708	11.896			12.654	13.562	14.453	15.334	16.207	17.939	19.662	21.379	23.092	24.803	26.512	28.220
	h	228.03	1167.1			1187.8	1212.7	1237.1	1261.3	1285.5	1333.9	1382.8	1432.3	1482.7	1533.9	1586.0	1638.9
	s	0.3809	1.6872			1.7152	1.7468	1.7761	1.8035	1.8294	1.8774	1.9214	1.9624	2.0009	2.0372	2.0717	2.1046
40 (267.25)	Sh					32.75	82.75	132.75	182.75	232.75	332.75	432.75	532.75	632.75	732.75	832.75	932.75
	v	0.01715	10.497			11.036	11.838	12.624	13.398	14.165	15.685	17.195	18.699	20.199	21.697	23.194	24.689
	h	236.14	1169.8			1186.6	1211.7	1236.4	1260.8	1285.0	1333.6	1382.5	1432.1	1482.5	1533.7	1585.8	1638.8
	s	0.3921	1.6765			1.6992	1.7312	1.7608	1.7883	1.8143	1.8624	1.9065	1.9476	1.9860	2.0224	2.0569	2.0899
45 (274.44)	Sh					25.56	75.56	125.56	175.56	225.56	325.56	425.56	525.56	625.56	725.56	825.56	925.56
	v	0.01721	9.399			9.777	10.497	11.201	11.892	12.577	13.932	15.276	16.614	17.950	19.282	20.613	21.943
	h	243.49	1172.1			1185.4	1210.4	1235.7	1260.2	1284.6	1333.3	1382.3	1431.9	1482.3	1533.6	1585.7	1638.7
	s	0.4021	1.6671			1.6849	1.7173	1.7471	1.7748	1.8010	1.8492	1.8934	1.9345	1.9730	2.0093	2.0439	2.0768
50 (281.02)	Sh					18.98	68.98	118.98	168.98	218.98	318.98	418.98	518.98	618.98	718.98	818.98	918.98
	v	0.01727	8.514			8.769	9.424	10.062	10.688	11.306	12.529	13.741	14.947	16.150	17.350	18.549	19.746
	h	250.21	1174.1			1184.1	1209.9	1234.9	1259.6	1284.1	1332.9	1382.0	1431.7	1482.2	1533.4	1585.6	1638.6
	s	0.4112	1.6586			1.6720	1.7048	1.7349	1.7628	1.7890	1.8374	1.8816	1.9227	1.9613	1.9977	2.0322	2.0652
55 (287.07)	Sh					12.93	62.93	112.93	162.93	212.93	312.93	412.93	512.93	612.93	712.93	812.93	912.93
	v	0.01733	7.787			7.945	8.546	9.130	9.702	10.267	11.381	12.485	13.583	14.677	15.769	16.859	17.948
	h	256.43	1175.9			1182.9	1208.9	1234.2	1259.1	1283.6	1332.6	1381.8	1431.5	1482.0	1533.3	1585.5	1638.5
	s	0.4196	1.6509			1.6601	1.6933	1.7237	1.7518	1.7781	1.8266	1.8710	1.9121	1.9507	1.987	2.022	2.055
60 (292.71)	Sh					7.29	57.29	107.29	157.29	207.29	307.29	407.29	507.29	607.29	707.29	807.29	907.29
	v	0.01738	7.174			7.257	7.815	8.354	8.881	9.400	10.425	11.438	12.446	13.450	14.452	15.452	16.450
	h	262.21	1177.6			1181.6	1208.0	1233.5	1258.5	1283.2	1332.3	1381.5	1431.3	1481.8	1533.2	1585.3	1638.4
	s	0.4273	1.6440			1.6492	1.6934	1.7134	1.7417	1.7681	1.8168	1.8612	1.9024	1.9410	1.9774	2.0120	2.0450
65 (297.98)	Sh					2.02	52.02	102.02	152.02	202.02	302.02	402.02	502.02	602.02	702.02	802.02	902.02
	v	0.01743	6.653			6.675	7.195	7.697	8.186	8.667	9.615	10.552	11.484	12.412	13.337	14.261	15.183
	h	267.63	1179.1			1180.3	1207.0	1232.7	1257.9	1282.7	1331.9	1381.3	1431.1	1481.6	1533.0	1585.2	1638.3
	s	0.4344	1.6375			1.6390	1.6731	1.7040	1.7324	1.7590	1.8077	1.8522	1.8935	1.9321	1.9685	2.0031	2.0361
70 (302.93)	Sh						47.07	97.07	147.07	197.07	297.07	397.07	497.07	597.07	697.07	797.07	897.07
	v	0.01748	6.205				6.664	7.133	7.590	8.039	8.922	9.793	10.659	11.522	12.382	13.240	14.097
	h	272.74	1180.6				1206.2	1232.0	1257.3	1282.2	1331.6	1381.0	1430.9	1481.5	1532.9	1585.1	1638.2
	s	0.4411	1.6316				1.6640	1.6951	1.7237	1.7504	1.7993	1.8439	1.8852	1.9238	1.9603	1.9949	2.0279

Temperature (°F)

Abs. Press. (psi) (Sat. Temp.)		Sat. Water	Sat. Steam	Temperature (°F)													
				350	400	450	500	550	600	700	800	900	1000	1100	1200	1300	1400
80 (312.04)	Sh			37.96	87.96	137.96	187.96	237.96	287.96	387.96	487.96	587.96	687.96	787.96	887.96	987.96	1087.96
	v	0.01757	5.471	5.801	6.218	6.622	7.018	7.408	7.794	8.560	9.319	10.075	10.829	11.581	12.331	13.081	13.829
	h	282.15	1183.1	1204.0	1230.5	1256.1	1281.3	1306.2	1330.9	1380.5	1430.5	1481.1	1532.6	1584.9	1638.0	1692.0	1746.8
	s	0.4534	1.6208	1.6473	1.6790	1.7080	1.7349	1.7602	1.7842	1.8289	1.8702	1.9089	1.9454	1.9800	2.0131	2.0446	2.0750
85 (316.26)	Sh			33.74	83.74	133.74	183.74	233.74	283.74	383.74	483.74	583.74	683.74	783.74	883.74	983.74	1083.74
	v	0.01762	5.167	5.445	5.840	6.223	6.597	6.966	7.330	8.052	8.768	9.480	10.190	10.898	11.604	12.310	13.014
	h	286.52	1184.2	1203.0	1229.7	1255.5	1280.8	1305.8	1330.6	1380.2	1430.3	1481.0	1532.4	1584.7	1637.9	1691.9	1746.8
	s	0.4590	1.6159	1.6398	1.6716	1.7008	1.7279	1.7532	1.7772	1.8220	1.8634	1.9021	1.9386	1.9733	2.0063	2.0379	2.0682
90 (320.28)	Sh			29.72	79.72	129.72	179.72	229.72	279.72	379.72	479.72	579.72	679.72	779.72	879.72	979.72	1079.72
	v	0.01766	4.895	5.128	5.505	5.869	6.223	6.572	6.917	7.600	8.277	8.950	9.621	10.290	10.958	11.625	12.290
	h	290.69	1185.3	1202.0	1228.9	1254.9	1280.3	1305.4	1330.2	1380.0	1430.1	1480.8	1532.3	1584.6	1637.8	1691.8	1746.7
	s	0.4643	1.6113	1.6323	1.6646	1.6940	1.7212	1.7467	1.7707	1.8156	1.8570	1.8957	1.9323	1.9669	2.0000	2.0316	2.0619
95 (324.13)	Sh			25.87	75.87	125.87	175.87	225.87	275.87	375.87	475.87	575.87	675.87	775.87	875.87	975.87	1075.87
	v	0.01770	4.651	4.845	5.205	5.551	5.889	6.221	6.548	7.196	7.838	8.477	9.113	9.747	10.380	11.012	11.643
	h	294.70	1186.2	1200.9	1228.1	1254.5	1279.9	1305.1	1329.9	1379.7	1429.9	1480.6	1532.1	1584.5	1637.7	1691.7	1746.6
	s	0.4694	1.6069	1.6253	1.6580	1.6876	1.7149	1.7404	1.7645	1.8094	1.8509	1.8897	1.9262	1.9609	1.9940	2.0256	2.0559
100 (327.82)	Sh			22.18	72.18	122.18	172.18	222.18	272.18	372.18	472.18	572.18	672.18	772.18	872.18	972.18	1072.18
	v	0.01774	4.431	4.590	4.935	5.266	5.588	5.904	6.216	6.833	7.443	8.050	8.655	9.258	9.860	10.460	11.060
	h	298.54	1187.2	1199.9	1227.4	1253.7	1279.3	1304.6	1329.6	1379.5	1429.7	1480.4	1532.0	1584.4	1637.6	1691.6	1746.5
	s	0.4743	1.6027	1.6187	1.6516	1.6814	1.7088	1.7344	1.7586	1.8036	1.8451	1.8839	1.9205	1.9552	1.9883	2.0199	2.0502
105 (331.37)	Sh			18.63	68.63	118.63	168.63	218.63	268.63	368.63	468.63	568.63	668.63	768.63	868.63	968.63	1068.63
	v	0.01778	4.231	4.359	4.690	5.007	5.315	5.617	5.915	6.504	7.086	7.665	8.241	8.816	9.389	9.961	10.532
	h	302.24	1188.0	1198.8	1226.6	1253.1	1278.8	1304.2	1329.2	1379.2	1429.4	1480.3	1531.8	1584.2	1637.5	1691.5	1746.4
	s	0.4790	1.5988	1.6122	1.6455	1.6755	1.7031	1.7288	1.7530	1.7981	1.8396	1.8785	1.9151	1.9498	1.9828	2.0145	2.0448
110 (334.79)	Sh			15.21	65.21	115.21	165.21	215.21	265.21	365.21	465.21	565.21	665.21	765.21	865.21	965.21	1065.21
	v	0.01782	4.048	4.149	4.468	4.772	5.068	5.357	5.642	6.205	6.761	7.314	7.865	8.413	8.961	9.507	10.053
	h	305.80	1188.9	1197.7	1225.8	1252.5	1278.3	1303.8	1328.9	1379.0	1429.2	1480.1	1531.7	1584.1	1637.4	1691.4	1746.4
	s	0.4834	1.5950	1.6061	1.6396	1.6698	1.6975	1.7233	1.7476	1.7928	1.8344	1.8732	1.9099	1.9446	1.9777	2.0093	2.0397
115 (338.08)	Sh			11.92	61.92	111.92	161.92	211.92	261.92	361.92	461.92	561.92	661.92	761.92	861.92	961.92	1061.92
	v	0.01785	3.881	3.957	4.265	4.558	4.841	5.119	5.392	5.932	6.465	6.994	7.521	8.046	8.570	9.093	9.615
	h	309.25	1189.6	1196.7	1225.0	1251.8	1277.9	1303.3	1328.6	1378.7	1429.0	1479.9	1531.6	1584.0	1637.2	1691.4	1746.3
	s	0.4877	1.5913	1.6001	1.6340	1.6644	1.6922	1.7181	1.7425	1.7877	1.8294	1.8682	1.9049	1.9396	1.9727	2.0044	2.0347

Table A-13. Saturated steam (Continued).

Abs. Press. (psi) (Sat. Temp.)		Sat. Water	Sat. Steam	350	400	450	500	550	600	700	800	900	1000	1100	1200	1300	1400
120 (341.27)	Sh			8.73	58.73	108.73	158.73	208.73	258.73	358.73	458.73	558.73	658.73	758.73	858.73	958.73	1058.73
	v	0.01789	3.7275	3.7815	4.0786	4.3610	4.6341	4.9009	5.1637	5.6813	6.1928	6.7006	7.2060	7.7096	8.2119	8.7130	9.2134
	h	312.58	1190.4	1195.6	1224.1	1251.2	1277.4	1302.9	1328.2	1378.4	1428.8	1479.8	1531.4	1583.9	1637.1	1691.3	1746.2
	s	0.4919	1.5879	1.5943	1.6286	1.6592	1.6872	1.7132	1.7376	1.7829	1.8246	1.8635	1.9001	1.9349	1.9680	1.9996	2.0300
130 (347.33)	Sh			2.67	52.67	102.67	152.67	202.67	252.67	352.67	452.67	552.67	652.67	752.67	852.67	952.67	1052.67
	v	0.01796	3.4544	3.4699	3.7489	4.0129	4.2672	4.5151	4.7589	5.2384	5.7118	6.1814	6.6486	7.1140	7.5781	8.0411	8.5033
	h	318.95	1191.7	1193.4	1222.5	1249.9	1276.4	1302.1	1327.5	1377.9	1428.4	1479.4	1531.1	1583.6	1636.9	1691.1	1746.1
	s	0.4998	1.5813	1.5833	1.6182	1.6493	1.6775	1.7037	1.7283	1.7737	1.8155	1.8545	1.8911	1.9259	1.9591	1.9907	2.0211
140 (353.04)	Sh				46.96	96.96	146.96	196.96	246.96	346.96	446.96	546.96	646.96	746.96	846.96	946.96	1046.96
	v	0.01803	3.2190		3.4661	3.7143	3.9526	4.1844	4.4119	4.8588	5.2995	5.7364	6.1709	6.6036	7.0349	7.4652	7.8946
	h	324.96	1193.0		1220.8	1248.7	1275.3	1301.3	1326.8	1377.4	1428.0	1479.1	1530.8	1583.4	1636.7	1690.9	1745.9
	s	0.5071	1.5752		1.6085	1.6400	1.6686	1.6949	1.7196	1.7652	1.8071	1.8461	1.8828	1.9176	1.9508	1.9825	2.0129
150 (358.43)	Sh				41.57	91.57	141.57	191.57	241.57	341.57	441.57	541.57	641.57	741.57	841.57	941.57	1041.57
	v	0.01809	3.0139		3.2208	3.4555	3.6799	3.8978	4.1112	4.5298	4.9421	5.3507	5.7568	6.1612	6.5642	6.9661	7.3671
	h	330.65	1194.1		1219.1	1247.4	1274.3	1300.5	1326.1	1376.9	1427.6	1478.7	1530.5	1583.1	1636.5	1690.7	1745.7
	s	0.5141	1.5695		1.5993	1.6313	1.6602	1.6867	1.7115	1.7573	1.7992	1.8383	1.8751	1.9099	1.9431	1.9748	2.0052
160 (363.55)	Sh				36.45	86.45	136.45	186.45	236.45	336.45	436.45	536.45	636.45	736.45	836.45	936.45	1036.45
	v	0.01815	2.8336		3.0060	3.2288	3.4413	3.6469	3.8480	4.2420	4.6295	5.0132	5.3945	5.7741	6.1522	6.5293	6.9055
	h	336.07	1195.1		1217.4	1246.0	1273.3	1299.6	1325.4	1376.4	1427.2	1478.4	1530.3	1582.9	1636.3	1690.5	1745.6
	s	0.5206	1.5641		1.5906	1.6231	1.6522	1.6790	1.7039	1.7499	1.7919	1.8310	1.8678	1.9027	1.9359	1.9676	1.9980
170 (368.42)	Sh				31.58	81.58	131.58	181.58	231.58	331.58	431.58	531.58	631.58	731.58	831.58	931.58	1031.58
	v	0.01821	2.6738		2.8162	3.0288	3.2306	3.4255	3.6158	3.9879	4.3536	4.7155	5.0749	5.4325	5.7888	6.1440	6.4983
	h	341.24	1196.0		1215.6	1244.7	1272.2	1298.8	1324.7	1375.8	1426.8	1478.0	1530.0	1582.6	1636.1	1690.4	1745.4
	s	0.5269	1.5591		1.5823	1.6152	1.6447	1.6717	1.6968	1.7428	1.7850	1.8241	1.8610	1.8959	1.9291	1.9608	1.9913
180 (373.08)	Sh				26.92	76.92	126.92	176.92	226.92	326.92	426.92	526.92	626.92	726.92	826.92	926.92	1026.92
	v	0.01827	2.5312		2.6474	2.8508	3.0433	3.2286	3.4093	3.7621	4.1084	4.4508	4.7907	5.1289	5.4657	5.8014	6.1363
	h	346.19	1196.9		1213.8	1243.4	1271.2	1297.9	1324.0	1375.3	1426.3	1477.7	1529.7	1582.4	1635.9	1690.2	1745.3
	s	0.5328	1.5543		1.5743	1.6078	1.6376	1.6647	1.6900	1.7362	1.7784	1.8176	1.8545	1.8894	1.9227	1.9545	1.9849
190 (377.53)	Sh				22.47	72.47	122.47	172.47	222.47	322.47	422.47	522.47	622.47	722.47	822.47	922.47	1022.47
	v	0.01833	2.4030		2.4961	2.6915	2.8756	3.0525	3.2246	3.5601	3.8889	4.2140	4.5365	4.8572	5.1766	5.4949	5.8124
	h	350.94	1197.6		1212.0	1242.0	1270.1	1297.1	1323.3	1374.8	1425.9	1477.4	1529.4	1582.1	1635.7	1690.0	1745.1
	s	0.5384	1.5498		1.5667	1.6006	1.6307	1.6581	1.6835	1.7299	1.7722	1.8115	1.8484	1.8834	1.9166	1.9484	1.9789
200	Sh				18.20	68.20	118.20	168.20	218.20	318.20	418.20	518.20	618.20	718.20	818.20	918.20	1018.20

Abs. Press. (psi) (Sat. Temp.)		Sat. Water	Sat. Steam	400	450	500	550	600	700	800	900	1000	1100	1200	1300	1400	1500
									Temperature (°F)								
210 (385.91)	Sh			14.09	64.09	114.09	164.09	214.09	314.09	414.09	514.09	614.09	714.09	814.09	914.09	1014.09	1114.09
	v	0.01844	2.1822	2.2364	2.4181	2.5880	2.7504	2.9078	3.2137	3.5128	3.8080	4.1007	4.3915	4.6811	4.9695	5.2571	5.5440
	h	359.91	1199.0	1208.02	1239.2	1268.0	1295.3	1321.9	1373.7	1425.1	1476.7	1528.8	1581.6	1635.2	1689.6	1744.8	1800.8
	s	0.5490	1.5413	1.5522	1.5872	1.6180	1.6458	1.6715	1.7182	1.7607	1.8001	1.8371	1.8721	1.9054	1.9372	1.9677	1.9970
220 (389.88)	Sh			10.12	60.12	110.12	160.12	210.12	310.12	410.12	510.12	610.12	710.12	810.12	910.12	1010.12	1110.12
	v	0.01850	2.0863	2.1240	2.2999	2.4638	2.6199	2.7710	3.0642	3.3504	3.6327	3.9125	4.1905	4.4671	4.7426	5.0173	5.2913
	h	364.17	1199.6	1206.3	1237.8	1266.9	1294.5	1321.2	1373.2	1424.7	1476.3	1528.5	1581.4	1635.0	1689.4	1744.7	1800.6
	s	0.5540	1.5374	1.5453	1.5808	1.6120	1.6400	1.6658	1.7128	1.7553	1.7948	1.8318	1.8668	1.9002	1.9320	1.9625	1.9919
230 (393.70)	Sh			6.30	56.30	106.30	156.30	206.30	306.30	406.30	506.30	606.30	706.30	806.30	906.30	1006.30	1106.30
	v	0.01855	1.9985	2.0212	2.1919	2.3503	2.5008	2.6461	2.9276	3.2020	3.4726	3.7406	4.0068	4.2717	4.5355	4.7984	5.0606
	h	368.28	1200.1	1204.4	1236.3	1265.7	1293.6	1320.4	1372.7	1424.2	1476.0	1528.2	1581.1	1634.8	1689.3	1744.5	1800.5
	s	0.5588	1.5336	1.5385	1.5747	1.6062	1.6344	1.6604	1.7075	1.7502	1.7897	1.8268	1.8618	1.8952	1.9270	1.9576	1.9869
240 (397.39)	Sh			2.61	52.61	102.61	152.61	202.61	302.61	402.61	502.61	602.61	702.61	802.61	902.61	1002.61	1102.61
	v	0.01860	1.9177	1.9268	2.0928	2.2462	2.3915	2.5316	2.8024	3.0661	3.3259	3.5831	3.8385	4.0926	4.3456	4.5977	4.8492
	h	372.27	1200.6	1202.4	1234.9	1264.6	1292.7	1319.7	1372.1	1423.8	1475.6	1527.9	1580.9	1634.6	1689.1	1744.3	1800.4
	s	0.5634	1.5299	1.5320	1.5687	1.6006	1.6291	1.6552	1.7025	1.7452	1.7848	1.8219	1.8570	1.8904	1.9223	1.9528	1.9822
250 (400.97)	Sh				49.03	99.03	149.03	199.03	299.03	399.03	499.03	599.03	699.03	799.03	899.03	999.03	1099.03
	v	0.01865	1.8432		2.0016	2.1504	2.2909	2.4262	2.6872	2.9410	3.1909	3.4382	3.6837	3.9278	4.1709	4.4131	4.6546
	h	376.14	1201.1		1233.4	1263.5	1291.8	1319.0	1371.6	1423.4	1475.3	1527.6	1580.6	1634.4	1688.9	1744.2	1800.2
	s	0.5679	1.5264		1.5629	1.5951	1.6239	1.6502	1.6976	1.7405	1.7801	1.8173	1.8524	1.8858	1.9177	1.9482	1.9776
260 (404.44)	Sh				45.56	95.56	145.56	195.56	295.56	395.56	495.56	595.56	695.56	795.56	895.56	995.56	1095.56
	v	0.01870	1.7742		1.9173	2.0619	2.1981	2.3289	2.5808	2.8256	3.0663	3.3044	3.5408	3.7758	4.0097	4.2427	4.4750
	h	379.90	1201.5		1231.9	1262.4	1290.9	1318.2	1371.1	1423.0	1474.9	1527.3	1580.4	1634.2	1688.7	1744.0	1800.1
	s	0.5722	1.5230		1.5573	1.5899	1.6189	1.6453	1.6930	1.7359	1.7756	1.8128	1.8480	1.8814	1.9133	1.9439	1.9732
270 (407.80)	Sh				42.20	92.20	142.20	192.20	292.20	392.20	492.20	592.20	692.20	792.20	892.20	992.20	1092.20
	v	0.01875	1.7101		1.8391	1.9799	2.1121	2.2388	2.4824	2.7186	2.9509	3.1806	3.4084	3.6349	3.8603	4.0849	4.3087

Table A-13. Saturated steam (Continued).

Abs. Press. p (psi)	Temp. t (°F)	Specific Volume Sat. Liquid v_f	Specific Volume Evap v_{fg}	Specific Volume Sat. Vapor v_g	Enthalpy Sat. Liquid h_f	Enthalpy Evap h_{fg}	Enthalpy Sat. Vapor h_g	Entropy Sat. Liquid s_f	Entropy Evap s_{fg}	Entropy Sat. Vapor s_g	Temp. t (°F)
508.0	731.40	0.02062	0.60530	0.62592	497.5	703.7	1201.1	0.6987	0.7271	1.4258	508.0
512.0	757.72	0.02072	0.58218	0.60289	502.3	698.2	1200.5	0.7036	0.7185	1.4221	512.0
516.0	784.76	0.02081	0.55997	0.58079	507.1	692.7	1199.8	0.7085	0.7099	1.4183	516.0
520.0	812.53	0.02091	0.53864	0.55956	512.0	687.0	1199.0	0.7133	0.7013	1.4146	520.0
524.0	841.04	0.02102	0.51814	0.53916	516.9	681.3	1198.2	0.7182	0.6926	1.4108	524.0
528.0	870.31	0.02112	0.49843	0.51955	521.8	675.5	1197.3	0.7231	0.6839	1.4070	528.0
532.0	900.34	0.02123	0.47947	0.50070	526.8	669.6	1196.4	0.7280	0.6752	1.4032	532.0
536.0	931.17	0.02134	0.46123	0.48257	531.7	663.6	1195.4	0.7329	0.6665	1.3993	536.0
540.0	962.79	0.02146	0.44367	0.46513	536.8	657.5	1194.3	0.7378	0.6577	1.3954	540.0
544.0	995.22	0.02157	0.42677	0.44834	541.8	651.3	1193.1	0.7427	0.6489	1.3915	544.0
548.0	1028.49	0.02169	0.41048	0.43217	546.9	645.0	1191.9	0.7476	0.6400	1.3876	548.0
552.0	1062.59	0.02182	0.39479	0.41660	552.0	638.5	1190.6	0.7525	0.6311	1.3837	552.0
556.0	1097.55	0.02194	0.37966	0.40160	557.2	632.0	1189.2	0.7575	0.6222	1.3797	556.0
560.0	1133.38	0.02207	0.36507	0.38714	562.4	625.3	1187.7	0.7625	0.6132	1.3757	560.0
564.0	1170.10	0.02221	0.35099	0.37320	567.6	618.5	1186.1	0.7674	0.6041	1.3716	564.0
568.0	1207.72	0.02235	0.33741	0.35975	572.9	611.5	1184.5	0.7725	0.5950	1.3675	568.0
572.0	1246.26	0.02249	0.32429	0.34678	578.3	604.5	1182.7	0.7775	0.5859	1.3634	572.0
576.0	1285.74	0.02264	0.31162	0.33426	583.7	597.2	1180.9	0.7825	0.5766	1.3592	576.0
580.0	1326.17	0.02279	0.29937	0.32216	589.1	589.9	1179.0	0.7876	0.5673	1.3550	580.0
584.0	1367.7	0.02295	0.28753	0.31048	594.6	582.4	1176.9	0.7927	0.5580	1.3507	584.0
588.0	1410.0	0.02311	0.27608	0.29919	600.1	574.7	1174.8	0.7978	0.5485	1.3464	588.0
592.0	1453.3	0.02328	0.26499	0.28827	605.7	566.8	1172.6	0.8030	0.5390	1.3420	592.0
596.0	1497.8	0.02345	0.25425	0.27770	611.4	558.8	1170.2	0.8082	0.5293	1.3375	596.0
600.0	1543.2	0.02364	0.24384	0.26747	617.1	550.6	1167.7	0.8134	0.5196	1.3330	600.0
604.0	1589.7	0.02382	0.23374	0.25757	622.9	542.2	1165.1	0.8187	0.5097	1.3284	604.0
608.0	1637.3	0.02402	0.22394	0.24796	628.8	533.6	1162.4	0.8240	0.4997	1.3238	608.0
612.0	1686.1	0.02422	0.21442	0.23865	634.8	524.7	1159.5	0.8294	0.4896	1.3190	612.0
616.0	1735.9	0.02444	0.20516	0.22960	640.8	515.6	1156.4	0.8348	0.4794	1.3141	616.0

620.0	1786.9	0.02466	0.19615	0.22081	646.9	506.3	1153.2	0.8403	0.4689	1.3092	620.0
624.0	1839.0	0.02489	0.18737	0.21226	653.1	496.6	1149.8	0.8458	0.4583	1.3041	624.0
628.0	1892.4	0.02514	0.17880	0.20394	659.5	486.7	1146.1	0.8514	0.4474	1.2988	628.0
632.0	1947.0	0.02539	0.17044	0.19583	665.9	476.4	1142.2	0.8571	0.4364	1.2934	632.0
636.0	2002.8	0.02566	0.16226	0.18792	672.4	465.7	1138.1	0.8628	0.4251	1.2879	636.0
640.0	2059.9	0.02595	0.15427	0.18021	679.1	454.6	1133.7	0.8686	0.4134	1.2821	640.0
644.0	2118.3	0.02625	0.14644	0.17269	685.9	443.1	1129.0	0.8746	0.4015	1.2761	644.0
648.0	2178.1	0.02657	0.13876	0.16534	692.9	431.1	1124.0	0.8806	0.3893	1.2699	648.0
652.0	2239.2	0.02691	0.13124	0.15816	700.0	418.7	1118.7	0.8868	0.3767	1.2634	652.0
656.0	2301.7	0.02728	0.12387	0.15115	707.4	405.7	1113.1	0.8931	0.3637	1.2567	656.0
660.0	2365.7	0.02768	0.11663	0.14431	714.9	392.1	1107.0	0.8995	0.3502	1.2498	660.0
664.0	2431.1	0.02811	0.10947	0.13757	722.9	377.7	1100.6	0.9064	0.3361	1.2425	664.0
668.0	2498.1	0.02858	0.10229	0.13087	731.5	362.1	1093.5	0.9137	0.3210	1.2347	668.0
672.0	2566.6	0.02911	0.09514	0.12424	740.2	345.7	1085.9	0.9212	0.3054	1.2266	672.0
676.0	2636.8	0.02970	0.08799	0.11769	749.2	328.5	1077.6	0.9287	0.2892	1.2179	676.0
680.0	2708.6	0.03037	0.08080	0.11117	758.5	310.1	1068.5	0.9365	0.2720	1.2086	680.0
684.0	2782.1	0.03114	0.07349	0.10463	768.2	290.2	1058.4	0.9447	0.2537	1.1984	684.0
688.0	2857.4	0.03204	0.06595	0.09799	778.8	268.2	1047.0	0.9535	0.2337	1.1872	688.0
692.0	2934.5	0.03313	0.05797	0.09110	790.5	243.1	1033.6	0.9634	0.2110	1.1744	692.0
696.0	3013.4	0.03455	0.04916	0.08371	804.4	212.8	1017.2	0.9749	0.1841	1.1591	696.0
700.0	3094.3	0.03662	0.03857	0.07519	822.4	172.7	995.2	0.9901	0.1490	1.1390	700.0
702.0	3135.5	0.03824	0.03173	0.06997	835.0	144.7	979.7	1.0006	0.1246	1.1252	702.0
704.0	3177.2	0.04108	0.02192	0.06300	854.2	102.0	956.2	1.0169	0.0876	1.1046	704.0
705.0	3198.3	0.04427	0.01304	0.05730	873.0	61.4	934.4	1.0329	0.0527	1.0856	705.0
705.47[b]	3208.2	0.05078	0.00000	0.05078	906.0	0.0	906.0	1.0612	0.0000	1.0612	705.47[b]

Table A-13. Saturated steam (Continued).

Pressure Table

Abs. Press. p (psi)	Temp. t (°F)	Specific Volume			Enthalpy			Entropy			Temp. t (°F)
		Sat. Liquid v_f	Evap v_{fx}	Sat. Vapor v_x	Sat. Liquid h_f	Evap h_{fx}	Sat. Vapor h_x	Sat. Liquid s_f	Evap s_{fx}	Sat. Vapor s_f	
0.08865	32.018	0.016022	3302.4	3302.4	0.0003	1075.5	1075.5	0.0000	2.1872	2.1872	0.08865
0.25	59.323	0.016032	1235.5	1235.5	27.382	1060.1	1087.4	0.0542	2.0425	2.0967	0.25
0.50	79.586	0.016071	641.5	641.5	47.623	1048.6	1096.3	0.0925	1.9446	2.0370	0.50
1.0	101.74	0.016136	333.59	333.60	69.73	1036.1	1105.8	0.1326	1.8455	1.9781	1.0
5.0	162.24	0.016407	73.515	73.532	130.20	1000.9	1131.1	0.2349	1.6094	1.8443	5.0
10.0	193.21	0.016592	38.404	38.420	161.26	982.1	1143.3	0.2836	1.5043	1.7879	10.0
14.696	212.00	0.016719	26.782	26.799	180.17	970.3	1150.5	0.3121	1.4447	1.7568	14.696
15.0	213.03	0.016726	26.274	26.290	181.21	969.7	1150.9	0.3137	1.4415	1.7552	15.0
20.0	227.96	0.016834	20.070	20.087	196.27	960.1	1156.3	0.3358	1.3962	1.7320	20.0
30.0	250.34	0.017009	13.7266	13.7436	218.9	945.2	1164.1	0.3682	1.3313	1.6995	30.0
40.0	267.25	0.017151	10.4794	10.4965	236.1	933.6	1169.8	0.3921	1.2844	1.6765	40.0
50.0	281.02	0.017274	8.4967	8.5140	250.2	923.9	1174.1	0.4112	1.2474	1.6586	50.0
60.0	292.71	0.017383	7.1562	7.1736	262.2	915.4	1177.6	0.4273	1.2167	1.6440	60.0
70.0	302.93	0.017482	6.1875	6.2050	272.7	907.8	1180.6	0.4411	1.1905	1.6316	70.0
80.0	312.04	0.017573	5.4536	5.4711	282.1	900.9	1183.1	0.4534	1.1675	1.6208	80.0
90.0	320.28	0.017659	4.8779	4.8953	290.7	894.6	1185.3	0.4643	1.1470	1.6113	90.0
100.0	327.82	0.017740	4.4133	4.4310	298.5	888.6	1187.2	0.4743	1.1284	1.6027	100.0
110.0	334.79	0.01782	4.0306	4.0484	305.8	883.1	1188.9	0.4834	1.1115	1.5950	110.0
120.0	341.27	0.01789	3.7097	3.7275	312.6	877.8	1190.4	0.4919	1.0960	1.5879	120.0
130.0	347.33	0.01796	3.4364	3.4544	319.0	872.8	1191.7	0.4998	1.0815	1.5813	130.0
140.0	353.04	0.01803	3.2010	3.2190	325.0	868.0	1193.0	0.5071	1.0681	1.5752	140.0
150.0	358.43	0.01809	2.9958	3.0139	330.6	863.4	1194.1	0.5141	1.0554	1.5695	150.0
160.0	363.55	0.01815	2.8155	2.8336	336.1	859.0	1195.1	0.5206	1.0435	1.5641	160.0
170.0	368.42	0.01821	2.6556	2.6738	341.2	854.8	1196.0	0.5269	1.0322	1.5591	170.0
180.0	373.08	0.01827	2.5129	2.5312	346.2	850.7	1196.9	0.5328	1.0215	1.5543	180.0
190.0	377.53	0.01833	2.3847	2.4030	350.9	846.7	1197.6	0.5384	1.0113	1.5498	190.0

200.0	381.80	0.01839	2.2689	2.2873	355.5	842.8	1198.3	0.5438	1.0016	1.5454	200.0
210.0	385.91	0.01844	2.16373	2.18217	359.9	839.1	1199.0	0.5490	0.9923	1.5413	210.0
220.0	389.88	0.01850	2.06779	2.08629	364.2	835.4	1199.6	0.5540	0.9834	1.5374	220.0
230.0	393.70	0.01855	1.97991	1.99846	368.3	831.8	1200.1	0.5588	0.9748	1.5336	230.0
240.0	397.39	0.01860	1.89909	1.91769	372.3	828.4	1200.6	0.5634	0.9665	1.5299	240.0
250.0	400.97	0.01865	1.82452	1.84317	376.1	825.0	1201.1	0.5679	0.9585	1.5264	250.0
260.0	404.44	0.01870	1.75548	1.77418	379.9	821.6	1201.5	0.5722	0.9508	1.5230	260.0
270.0	407.80	0.01875	1.69137	1.71013	383.6	818.3	1201.9	0.5764	0.9433	1.5197	270.0
280.0	411.07	0.01880	1.63169	1.65049	387.1	815.1	1202.3	0.5805	0.9361	1.5166	280.0
290.0	414.25	0.01885	1.57597	1.59482	390.6	812.0	1202.6	0.5844	0.9291	1.5135	290.0
300.0	417.35	0.01889	1.52384	1.54274	394.0	808.9	1202.9	0.5882	0.9223	1.5105	300.0
350.0	431.73	0.01912	1.30642	1.32554	409.8	794.2	1204.0	0.6059	0.8909	1.4968	350.0
400.0	444.60	0.01934	1.14162	1.16095	424.2	780.4	1204.6	0.6217	0.8630	1.4847	400.0
450.0	456.28	0.01954	1.01224	1.03179	437.3	767.5	1204.8	0.6360	0.8378	1.4738	450.0
500.0	467.01	0.01975	0.90787	0.92762	449.5	755.1	1204.7	0.6490	0.8148	1.4639	500.0
550.0	476.94	0.01994	0.82183	0.84177	460.9	743.3	1204.3	0.6611	0.7936	1.4547	550.0
600.0	486.20	0.02013	0.74962	0.76975	471.7	732.0	1203.7	0.6723	0.7738	1.4461	600.0
650.0	494.89	0.02032	0.68811	0.70843	481.9	720.9	1202.8	0.6828	0.7552	1.4381	650.0
700.0	503.08	0.02050	0.63505	0.65556	491.6	710.2	1201.8	0.6928	0.7377	1.4304	700.0
750.0	510.84	0.02069	0.58880	0.60949	500.9	699.8	1200.7	0.7022	0.7210	1.4232	750.0
800.0	518.21	0.02087	0.54809	0.56896	509.8	689.6	1199.4	0.7111	0.7051	1.4163	800.0
850.0	525.24	0.02105	0.51197	0.53302	518.4	679.5	1198.0	0.7197	0.6899	1.4096	850.0
900.0	531.95	0.02123	0.47968	0.50091	526.7	669.7	1196.4	0.7279	0.6753	1.4032	900.0
950.0	538.39	0.02141	0.45064	0.47205	534.7	660.0	1194.7	0.7358	0.6612	1.3970	950.0
1000.0	544.58	0.02159	0.42436	0.44596	542.6	650.4	1192.9	0.7434	0.6476	1.3910	1000.0
1050.0	550.53	0.02177	0.40047	0.42224	550.1	640.9	1191.0	0.7507	0.6344	1.3851	1050.0
1100.0	556.28	0.02195	0.37863	0.40058	557.5	631.5	1189.1	0.7578	0.6216	1.3794	1100.0
1150.0	561.82	0.02214	0.35859	0.38073	564.8	622.2	1187.0	0.7647	0.6091	1.3738	1150.0
1200.0	567.19	0.02232	0.34013	0.36245	571.9	613.0	1184.8	0.7714	0.5969	1.3683	1200.0

Table A-13. Saturated Steam (Continued).

| Temp. t (°F) | Abs. Press. p (psi) | Specific Volume | | | Enthalpy | | | Entropy | | | Sat. Temp. t (°F) |
		Sat. Liquid v_f	Evap v_{fg}	Sat. Vapor v_g	Sat. Liquid h_f	Evap h_{fg}	Sat. Vapor h_g	Sat. Liquid s_f	Evap s_{fg}	Sat. Vapor s_g	
					Temperature Table						
1250.0	572.38	0.02250	0.32306	0.34556	578.8	603.8	1182.6	0.7780	0.5850	1.3630	1250.0
1300.0	577.42	0.02269	0.30722	0.32991	585.6	594.6	1180.2	0.7843	0.5733	1.3577	1300.0
1350.0	582.32	0.02288	0.29250	0.31537	592.3	585.4	1177.8	0.7906	0.5620	1.3525	1350.0
1400.0	587.07	0.02307	0.27871	0.30178	598.8	576.5	1175.3	0.7966	0.5507	1.3474	1400.0
1450.0	591.70	0.02327	0.26584	0.28911	605.3	567.4	1172.8	0.8026	0.5397	1.3423	1450.0
1500.0	596.20	0.02346	0.25372	0.27719	611.7	558.4	1170.1	0.8085	0.5288	1.3373	1500.0
1550.0	600.59	0.02366	0.24235	0.26601	618.0	549.4	1167.4	0.8142	0.5182	1.3324	1550.0
1600.0	604.87	0.02387	0.23159	0.25545	624.2	540.3	1164.5	0.8199	0.5076	1.3274	1600.0
1650.0	609.05	0.02407	0.22143	0.24551	630.4	531.3	1161.6	0.8254	0.4971	1.3225	1650.0
1700.0	613.13	0.02428	0.21178	0.23607	636.5	522.2	1158.6	0.8309	0.4867	1.3176	1700.0
1750.0	617.12	0.02450	0.20263	0.22713	642.5	513.1	1155.6	0.8363	0.4765	1.3128	1750.0
1800.0	621.02	0.02472	0.19390	0.21861	648.5	503.8	1152.3	0.8417	0.4662	1.3079	1800.0
1850.0	624.83	0.02495	0.18558	0.21052	654.5	494.6	1149.0	0.8470	0.4561	1.3030	1850.0
1900.0	628.56	0.02517	0.17761	0.20278	660.4	485.2	1145.6	0.8522	0.4459	1.2981	1900.0
1950.0	632.22	0.02541	0.16999	0.19540	666.3	475.8	1142.0	0.8574	0.4358	1.2931	1950.0
2000.0	635.80	0.02565	0.16266	0.18831	672.1	466.2	1138.3	0.8625	0.4256	1.2881	2000.0
2100.0	642.76	0.02615	0.14885	0.17501	683.8	446.7	1130.5	0.8727	0.4053	1.2780	2100.0
2200.0	649.45	0.02669	0.13603	0.16272	695.5	426.7	1122.2	0.8828	0.3848	1.2676	2200.0
2300.0	655.89	0.02727	0.12406	0.15133	707.2	406.0	1113.2	0.8929	0.3640	1.2569	2300.0
2400.0	662.11	0.02790	0.11287	0.14076	719.0	384.8	1103.7	0.9031	0.3430	1.2460	2400.0
2500.0	668.11	0.02859	0.10209	0.13068	731.7	361.6	1093.3	0.9139	0.3206	1.2345	2500.0
2600.0	673.91	0.02938	0.09172	0.12110	744.5	337.6	1082.0	0.9247	0.2977	1.2225	2600.0
2700.0	679.53	0.03029	0.08165	0.11194	757.3	312.3	1069.7	0.9356	0.2741	1.2097	2700.0
2800.0	684.96	0.03134	0.07171	0.10305	770.7	285.1	1055.8	0.9468	0.2491	1.1958	2800.0
2900.0	690.22	0.03262	0.06158	0.09420	785.1	254.7	1039.8	0.9588	0.2215	1.1803	2900.0
3000.0	695.33	0.03428	0.05073	0.08500	801.8	218.4	1020.3	0.9728	0.1891	1.1619	3000.0
3100.0	700.28	0.03681	0.03771	0.07452	824.0	169.3	993.3	0.9914	0.1460	1.1373	3100.0
3200.0	705.08	0.04472	0.01191	0.05663	875.5	56.1	931.6	1.0351	0.0482	1.0832	3200.0
3208.2'	705.47	0.05078	0.00000	0.05078	906.0	0.0	906.0	1.0612	0.0000	1.0612	3208.2'

' Critical pressure.
Source: Copyright 1967 ASME (Abridged): reprinted by permission.

Table A-14. Superheated Steam*.

Abs. Press. (psi) (Sat. Temp.)		Sat. Water	Sat. Steam	Temperature (°F)													
				200	250	300	350	400	450	500	600	700	800	900	1000	1100	1200
1 (101.74)	Sh			98.26	148.26	198.26	248.26	298.26	348.26	398.26	498.26	598.26	698.26	798.26	898.26	998.26	1098.26
	v	0.01614	333.6	392.5	422.4	452.3	482.1	511.9	541.7	571.5	631.1	690.7	750.3	809.8	869.4	929.0	988.6
	h	69.73	1105.8	1150.2	1172.9	1195.7	1218.7	1241.8	1265.1	1288.6	1336.1	1384.5	1433.7	1483.8	1534.9	1586.8	1639.7
	s	0.1326	1.9781	2.0509	2.0841	2.1152	2.1445	2.1722	2.1985	2.2237	2.2708	2.3144	2.3551	2.3934	2.4296	2.4640	2.4969
5 (162.24)	Sh			37.76	87.76	137.76	187.76	237.76	287.76	337.76	437.76	537.76	637.76	737.76	837.76	937.76	1037.76
	v	0.01641	73.53	78.14	84.21	90.24	96.25	102.24	108.23	114.21	126.15	138.08	150.01	161.94	173.86	185.78	197.70
	h	130.20	1131.1	1148.6	1171.7	1194.8	1218.0	1241.3	1264.7	1288.2	1335.9	1384.3	1433.6	1483.7	1534.7	1586.7	1639.6
	s	0.2349	1.8443	1.8716	1.9054	1.9369	1.9664	1.9943	2.0208	2.0460	2.0932	2.1369	2.1776	2.2159	2.2521	2.2866	2.3194
10 (193.21)	Sh			6.79	56.79	106.79	156.79	206.79	256.79	306.79	406.79	506.79	606.79	706.79	806.79	906.79	1006.79
	v	0.01659	38.42	38.84	41.93	44.98	48.02	51.03	54.04	57.04	63.03	69.00	74.98	80.94	86.91	92.87	98.84
	h	161.26	1143.3	1146.6	1170.2	1193.7	1217.1	1240.6	1264.1	1287.8	1335.5	1384.0	1433.4	1483.5	1534.6	1586.6	1639.5
	s	0.2836	1.7879	1.7928	1.8273	1.8593	1.8892	1.9173	1.9439	1.9692	2.0166	2.0603	2.1011	2.1394	2.1757	2.2101	2.2430
14.696 (212.00)	Sh				38.00	88.00	138.00	188.00	238.00	288.00	388.00	488.00	588.00	688.00	788.00	888.00	988.00
	v	.0167	26.799		28.42	30.52	32.60	34.67	36.72	38.77	42.86	46.93	51.00	55.06	59.13	63.19	67.25
	h	180.17	1150.5		1168.8	1192.6	1216.3	1239.9	1263.6	1287.4	1335.2	1383.8	1433.2	1483.4	1534.5	1586.5	1639.4
	s	.3121	1.7568		1.7833	1.8158	1.8459	1.8743	1.9010	1.9265	1.9739	2.0177	2.0585	2.0969	2.1332	2.1676	2.2005
15 (213.03)	Sh				36.97	86.97	136.97	186.97	236.97	286.97	386.97	486.97	586.97	686.97	786.97	886.97	986.97
	v	0.01673	26.290		27.837	29.899	31.939	33.963	35.977	37.985	41.986	45.978	49.964	53.946	57.926	61.905	65.882
	h	181.21	1150.9		1168.7	1192.5	1216.2	1239.9	1263.6	1287.3	1335.2	1383.8	1433.2	1483.4	1534.5	1586.5	1639.4
	s	0.3137	1.7552		1.7809	1.8134	1.8437	1.8720	1.8988	1.9242	1.9717	2.0155	2.0563	2.0946	2.1309	2.1653	2.1982
20 (227.96)	Sh				22.04	72.04	122.04	172.04	222.04	272.04	372.04	472.04	572.04	672.04	772.04	872.04	972.04
	v	0.01683	20.087		20.788	22.356	23.900	25.428	26.946	28.457	31.466	34.465	37.458	40.447	43.435	46.420	49.405
	h	196.27	1156.3		1167.1	1191.4	1215.4	1239.2	1263.0	1286.9	1334.9	1383.5	1432.9	1483.2	1534.3	1586.3	1639.3
	s	0.3358	1.7320		1.7475	1.7805	1.8111	1.8397	1.8666	1.8921	1.9397	1.9836	2.0244	2.0628	2.0991	2.1336	2.1700
25 (240.07)	Sh				9.93	59.93	109.93	159.93	209.93	259.93	359.93	459.93	559.93	659.93	759.93	859.93	959.93
	v	0.01693	16.301		16.558	17.829	19.076	20.307	21.527	22.740	25.153	27.557	29.954	32.348	34.740	37.130	39.518
	h	208.52	1160.6		1165.6	1190.2	1214.5	1238.5	1262.5	1286.4	1334.6	1383.3	1432.7	1483.0	1534.2	1586.2	1639.2
	s	0.3535	1.7141		1.7212	1.7547	1.7856	1.8145	1.8415	1.8672	1.9149	1.9588	1.9997	2.0381	2.0744	2.1089	2.1418
30 (250.34)	Sh					49.66	99.66	149.66	199.66	249.66	349.66	449.66	549.66	649.66	749.66	849.66	949.66
	v	0.01701	13.744			14.810	15.859	16.892	17.914	18.929	20.945	22.951	24.952	26.949	28.943	30.936	32.927
	h	218.93	1164.1			1189.0	1213.6	1237.8	1261.9	1286.0	1334.2	1383.0	1432.5	1482.8	1534.0	1586.1	1639.0
	s	0.3682	1.6995			1.7334	1.7647	1.7937	1.8210	1.8467	1.8946	1.9386	1.9795	2.0179	2.0543	2.0888	2.1217
35 (259.29)	Sh					40.71	90.71	140.71	190.71	240.71	340.71	440.71	540.71	640.71	740.71	840.71	940.71
	v	0.01708	11.896			12.654	13.562	14.453	15.334	16.207	17.939	19.662	21.379	23.092	24.803	26.512	28.220
	h	228.03	1167.1			1187.8	1212.7	1237.1	1261.3	1285.5	1333.9	1382.8	1432.3	1482.7	1533.9	1586.0	1638.9
	s	0.3809	1.6872			1.7152	1.7468	1.7761	1.8035	1.8294	1.8774	1.9214	1.9624	2.0009	2.0372	2.0717	2.1046

Table A-14. Superheated Steam* (Continued).

Abs. Press. (psi) (Sat. Temp.)		Sat. Water	Sat. Steam	200	250	300	350	400	450	500	600	700	800	900	1000	1100	1200
35 (259.29)	Sh					40.71	90.71	140.71	190.71	240.71	340.71	440.71	540.71	640.71	740.71	840.71	940.71
	v	0.01708	11.896			12.654	13.562	14.453	15.334	16.207	17.939	19.662	21.379	23.092	24.803	26.512	28.220
	h	228.03	1167.1			1187.8	1212.7	1237.1	1261.3	1285.5	1333.9	1382.8	1432.3	1482.7	1533.9	1586.0	1638.9
	s	0.3809	1.6872			1.7152	1.7468	1.7761	1.8035	1.8294	1.8774	1.9214	1.9624	2.0009	2.0372	2.0717	2.1046
40 (267.25)	Sh					32.75	82.75	132.75	182.75	232.75	332.75	432.75	532.75	632.75	732.75	832.75	932.75
	v	0.01715	10.497			11.036	11.838	12.624	13.398	14.165	15.685	17.195	18.699	20.199	21.697	23.194	24.689
	h	236.14	1169.8			1186.6	1211.7	1236.4	1260.8	1285.0	1333.6	1382.5	1432.1	1482.5	1533.7	1585.8	1638.8
	s	0.3921	1.6765			1.6992	1.7312	1.7608	1.7883	1.8143	1.8624	1.9065	1.9476	1.9860	2.0224	2.0569	2.0899
45 (274.44)	Sh					25.56	75.56	125.56	175.56	225.56	325.56	425.56	525.56	625.56	725.56	825.56	925.56
	v	0.01721	9.399			9.777	10.497	11.201	11.892	12.577	13.932	15.276	16.614	17.950	19.282	20.613	21.943
	h	243.49	1172.1			1185.4	1210.4	1235.7	1260.2	1284.6	1333.3	1382.3	1431.9	1482.3	1533.6	1585.7	1638.7
	s	0.4021	1.6671			1.6849	1.7173	1.7471	1.7748	1.8010	1.8492	1.8934	1.9345	1.9730	2.0093	2.0439	2.0768
50 (281.02)	Sh					18.98	68.98	118.98	168.98	218.98	318.98	418.98	518.98	618.98	718.98	818.98	918.98
	v	0.01727	8.514			8.769	9.424	10.062	10.688	11.306	12.529	13.741	14.947	16.150	17.350	18.549	19.746
	h	250.21	1174.1			1184.1	1209.9	1234.9	1259.6	1284.1	1332.9	1382.0	1431.7	1482.2	1533.4	1585.6	1638.6
	s	0.4112	1.6586			1.6720	1.7048	1.7349	1.7628	1.7890	1.8374	1.8816	1.9227	1.9613	1.9977	2.0322	2.0652
55 (287.07)	Sh					12.93	62.93	112.93	162.93	212.93	312.93	412.93	512.93	612.93	712.93	812.93	912.93
	v	0.01733	7.787			7.945	8.546	9.130	9.702	10.267	11.381	12.485	13.583	14.677	15.769	16.859	17.948
	h	256.43	1175.9			1182.9	1208.9	1234.2	1259.1	1283.6	1332.6	1381.8	1431.5	1482.0	1533.3	1585.5	1638.5
	s	0.4196	1.6509			1.6601	1.6933	1.7237	1.7518	1.7781	1.8266	1.8710	1.9121	1.9507	1.987	2.022	2.055
60 (292.71)	Sh					7.29	57.29	107.29	157.29	207.29	307.29	407.29	507.29	607.29	707.29	807.29	907.29
	v	0.01738	7.174			7.257	7.815	8.354	8.881	9.400	10.425	11.438	12.446	13.450	14.452	15.452	16.450
	h	262.21	1177.6			1181.6	1208.0	1233.5	1258.5	1283.2	1332.3	1381.5	1431.3	1481.8	1533.2	1585.3	1638.4
	s	0.4273	1.6440			1.6492	1.6934	1.7134	1.7417	1.7681	1.8168	1.8612	1.9024	1.9410	1.9774	2.0120	2.0450
65 (297.98)	Sh					2.02	52.02	102.02	152.02	202.02	302.02	402.02	502.02	602.02	702.02	802.02	902.02
	v	0.01743	6.653			6.675	7.195	7.697	8.186	8.667	9.615	10.552	11.484	12.412	13.337	14.261	15.183
	h	267.63	1179.1			1180.3	1207.0	1232.7	1257.9	1282.7	1331.9	1381.3	1431.1	1481.6	1533.0	1585.2	1638.3
	s	0.4344	1.6375			1.6390	1.6731	1.7040	1.7324	1.7590	1.8077	1.8522	1.8935	1.9321	1.9685	2.0031	2.0361
70 (302.93)	Sh						47.07	97.07	147.07	197.07	297.07	397.07	497.07	597.07	697.07	797.07	897.07
	v	0.01748	6.205				6.664	7.133	7.590	8.039	8.922	9.793	10.659	11.522	12.382	13.240	14.097
	h	272.74	1180.6				1206.0	1232.0	1257.3	1282.2	1331.6	1381.0	1430.9	1481.5	1532.9	1585.1	1638.2
	s	0.4411	1.6316				1.6640	1.6951	1.7237	1.7504	1.7993	1.8439	1.8852	1.9238	1.9603	1.9949	2.0279
75 (307.61)	Sh						42.39	92.39	142.39	192.39	292.39	392.39	492.39	592.39	692.39	792.39	892.39
	v	0.01753	5.814				6.204	6.645	7.074	7.494	8.320	9.135	9.945	10.750	11.553	12.355	13.155
	h	277.56	1181.9				1205.0	1231.2	1256.7	1281.7	1331.3	1380.7	1430.7	1481.3	1532.7	1585.0	1638.1
	s	0.4474	1.6260				1.6554	1.6868	1.7156	1.7424	1.7915	1.8361	1.8774	1.9161	1.9526	1.9872	2.0202

Abs. Press. (psi) (Sat. Temp.)		Sat. Water	Sat. Steam	|	350	400	450	500	550	600	700	800	900	1000	1100	1200	1300	1400
				Temperature (°F)														
80 (312.04)	Sh				37.96	87.96	137.96	187.96	237.96	287.96	387.96	487.96	587.96	687.96	787.96	887.96	987.96	1087.96
	v	0.01757	5.471		5.801	6.218	6.622	7.018	7.408	7.794	8.560	9.319	10.075	10.829	11.581	12.331	13.081	13.829
	h	282.15	1183.1		1204.0	1230.5	1256.1	1281.3	1306.2	1330.9	1380.5	1430.5	1481.1	1532.6	1584.9	1638.0	1692.0	1746.8
	s	0.4534	1.6208		1.6473	1.6790	1.7080	1.7349	1.7602	1.7842	1.8289	1.8702	1.9089	1.9454	1.9800	2.0131	2.0446	2.0750
85 (316.26)	Sh				33.74	83.74	133.74	183.74	233.74	283.74	383.74	483.74	583.74	683.74	783.74	883.74	983.74	1083.74
	v	0.01762	5.167		5.445	5.840	6.223	6.597	6.966	7.330	8.052	8.768	9.480	10.190	10.898	11.604	12.310	13.014
	h	286.52	1184.2		1203.0	1229.7	1255.5	1280.8	1305.8	1330.6	1380.2	1430.3	1481.0	1532.4	1584.7	1637.9	1691.9	1746.8
	s	0.4590	1.6159		1.6396	1.6716	1.7008	1.7279	1.7532	1.7772	1.8220	1.8634	1.9021	1.9386	1.9733	2.0063	2.0379	2.0682
90 (320.28)	Sh				29.72	79.72	129.72	179.72	229.72	279.72	379.72	479.72	579.72	679.72	779.72	879.72	979.72	1079.72
	v	0.01766	4.895		5.128	5.505	5.869	6.223	6.572	6.917	7.600	8.277	8.950	9.621	10.290	10.958	11.625	12.290
	h	290.69	1185.3		1202.0	1228.9	1254.9	1280.3	1305.4	1330.2	1380.0	1430.1	1480.8	1532.3	1584.6	1637.8	1691.8	1746.7
	s	0.4643	1.6113		1.6323	1.6646	1.6940	1.7212	1.7467	1.7707	1.8156	1.8570	1.8957	1.9323	1.9669	2.0000	2.0316	2.0619
95 (324.13)	Sh				25.87	75.87	125.87	175.87	225.87	275.87	375.87	475.87	575.87	675.87	775.87	875.87	975.87	1075.87
	v	0.01770	4.651		4.845	5.205	5.551	5.889	6.221	6.548	7.196	7.838	8.477	9.113	9.747	10.380	11.012	11.643
	h	294.70	1186.2		1200.9	1228.1	1254.3	1279.8	1305.0	1329.9	1379.7	1429.9	1480.6	1532.1	1584.5	1637.7	1691.7	1746.6
	s	0.4694	1.6069		1.6253	1.6580	1.6876	1.7149	1.7404	1.7645	1.8094	1.8509	1.8897	1.9262	1.9609	1.9940	2.0256	2.0559
100 (327.82)	Sh				22.18	72.18	122.18	172.18	222.18	272.18	372.18	472.18	572.18	672.18	772.18	872.18	972.18	1072.18
	v	0.01774	4.431		4.590	4.935	5.266	5.588	5.904	6.216	6.833	7.443	8.050	8.655	9.258	9.860	10.460	11.060
	h	298.54	1187.2		1199.9	1227.4	1253.7	1279.3	1304.6	1329.6	1379.5	1429.7	1480.4	1532.0	1584.4	1637.6	1691.6	1746.5
	s	0.4743	1.6027		1.6187	1.6516	1.6814	1.7088	1.7344	1.7586	1.8036	1.8451	1.8839	1.9205	1.9552	1.9883	2.0199	2.0502
105 (331.37)	Sh				18.63	68.63	118.63	168.63	218.63	268.63	368.63	468.63	568.63	668.63	768.63	868.63	968.63	1068.63
	v	0.01778	4.231		4.359	4.690	5.007	5.315	5.617	5.915	6.504	7.086	7.665	8.241	8.816	9.389	9.961	10.532
	h	302.24	1188.0		1198.8	1226.6	1253.1	1278.8	1304.2	1329.2	1379.2	1429.4	1480.3	1531.8	1584.2	1637.5	1691.5	1746.4
	s	0.4790	1.5988		1.6122	1.6455	1.6755	1.7031	1.7288	1.7530	1.7981	1.8396	1.8785	1.9151	1.9498	1.9828	2.0145	2.0448
110 (334.79)	Sh				15.21	65.21	115.21	165.21	215.21	265.21	365.21	465.21	565.21	665.21	765.21	865.21	965.21	1065.21
	v	0.01782	4.048		4.149	4.468	4.772	5.068	5.357	5.642	6.205	6.761	7.314	7.865	8.413	8.961	9.507	10.053
	h	305.80	1188.9		1197.7	1225.8	1252.5	1278.3	1303.8	1328.8	1379.0	1429.2	1480.1	1531.7	1584.1	1637.4	1691.4	1746.4
	s	0.4834	1.5950		1.6061	1.6396	1.6698	1.6975	1.7233	1.7476	1.7928	1.8344	1.8732	1.9099	1.9446	1.9777	2.0093	2.0397
115 (338.08)	Sh				11.92	61.92	111.92	161.92	211.92	261.92	361.92	461.92	561.92	661.92	761.92	861.92	961.92	1061.92
	v	0.01785	3.881		3.957	4.265	4.558	4.841	5.119	5.392	5.932	6.465	6.994	7.521	8.046	8.570	9.093	9.615
	h	309.25	1189.6		1196.7	1225.0	1251.8	1277.9	1303.3	1328.6	1378.7	1429.0	1479.9	1531.6	1584.0	1637.2	1691.4	1746.3
	s	0.4877	1.5913		1.6001	1.6340	1.6644	1.6922	1.7181	1.7425	1.7877	1.8294	1.8682	1.9049	1.9396	1.9727	2.0044	2.0347

Table A-14. Superheated Steam* (Continued).

Abs. Press. (psi) (Sat. Temp.)		Sat. Water	Sat. Steam	350	400	450	500	550	600	700	800	900	1000	1100	1200	1300	1400
										Temperature (°F)							
120 (341.27)	Sh			8.73	58.73	108.73	158.73	208.73	258.73	358.73	458.73	558.73	658.73	758.73	858.73	958.73	1058.73
	v	0.01789	3.7275	3.7815	4.0786	4.3610	4.6341	4.9009	5.1637	5.6813	6.1928	6.7006	7.2060	7.7096	8.2119	8.7130	9.2134
	h	312.58	1190.4	1195.6	1224.1	1251.2	1277.4	1302.9	1328.2	1378.4	1428.8	1479.8	1531.4	1583.9	1637.1	1691.3	1746.2
	s	0.4919	1.5879	1.5943	1.6286	1.6592	1.6872	1.7132	1.7376	1.7829	1.8246	1.8635	1.9001	1.9349	1.9680	1.9996	2.0300
130 (347.33)	Sh			2.67	52.67	102.67	152.67	202.67	252.67	352.67	452.67	552.67	652.67	752.67	852.67	952.67	1052.67
	v	0.01796	3.4544	3.4699	3.7489	4.0129	4.2672	4.5151	4.7589	5.2384	5.7118	6.1814	6.6486	7.1140	7.5781	8.0411	8.5033
	h	318.95	1191.7	1193.4	1222.5	1249.9	1276.4	1302.1	1327.5	1377.9	1428.4	1479.4	1531.1	1583.6	1636.9	1691.1	1746.1
	s	0.4998	1.5813	1.5833	1.6182	1.6493	1.6775	1.7037	1.7283	1.7737	1.8155	1.8545	1.8911	1.9259	1.9591	1.9907	2.0211
140 (353.04)	Sh				46.96	96.96	146.96	196.96	246.96	346.96	446.96	546.96	646.96	746.96	846.96	946.96	1046.96
	v	0.01803	3.2190		3.4661	3.7143	3.9526	4.1844	4.4119	4.8588	5.2995	5.7364	6.1709	6.6036	7.0349	7.4652	7.8946
	h	324.96	1193.0		1220.8	1248.7	1275.3	1301.3	1326.8	1377.4	1428.0	1479.1	1530.8	1583.4	1636.7	1690.9	1745.9
	s	0.5071	1.5752		1.6085	1.6400	1.6686	1.6949	1.7196	1.7652	1.8071	1.8461	1.8828	1.9176	1.9508	1.9825	2.0129
150 (358.43)	Sh				41.57	91.57	141.57	191.57	241.57	341.57	441.57	541.57	641.57	741.57	841.57	941.57	1041.57
	v	0.01809	3.0139		3.2208	3.4555	3.6799	3.8978	4.1112	4.5298	4.9421	5.3507	5.7568	6.1612	6.5642	6.9661	7.3671
	h	330.65	1194.1		1219.1	1247.4	1274.3	1300.5	1326.1	1376.9	1427.6	1478.7	1530.5	1583.1	1636.5	1690.7	1745.7
	s	0.5141	1.5695		1.5993	1.6313	1.6602	1.6867	1.7115	1.7573	1.7992	1.8383	1.8751	1.9099	1.9431	1.9748	2.0052
160 (363.55)	Sh				36.45	86.45	136.45	186.45	236.45	336.45	436.45	536.45	636.45	736.45	836.45	936.45	1036.45
	v	0.01815	2.8336		3.0060	3.2288	3.4413	3.6469	3.8480	4.2420	4.6295	5.0132	5.3945	5.7741	6.1522	6.5293	6.9055
	h	336.07	1195.1		1217.4	1246.0	1273.3	1299.6	1325.4	1376.4	1427.2	1478.4	1530.3	1582.9	1636.3	1690.5	1745.6
	s	0.5206	1.5641		1.5906	1.6231	1.6522	1.6790	1.7039	1.7499	1.7919	1.8310	1.8678	1.9027	1.9359	1.9676	1.9980
170 (368.42)	Sh				31.58	81.58	131.58	181.58	231.58	331.58	431.58	531.58	631.58	731.58	831.58	931.58	1031.58
	v	0.01821	2.6738		2.8162	3.0288	3.2306	3.4255	3.6158	3.9879	4.3536	4.7155	5.0749	5.4325	5.7888	6.1440	6.4983
	h	341.24	1196.0		1215.6	1244.7	1272.2	1298.8	1324.7	1375.8	1426.8	1478.0	1530.0	1582.6	1636.1	1690.4	1745.4
	s	0.5269	1.5591		1.5823	1.6152	1.6447	1.6717	1.6968	1.7428	1.7850	1.8241	1.8610	1.8959	1.9291	1.9608	1.9913
180 (373.08)	Sh				26.92	76.92	126.92	176.92	226.92	326.92	426.92	526.92	626.92	726.92	826.92	926.92	1026.92
	v	0.01827	2.5312		2.6474	2.8508	3.0433	3.2286	3.4093	3.7621	4.1084	4.4508	4.7907	5.1289	5.4657	5.8014	6.1363
	h	346.19	1196.9		1213.8	1243.4	1271.2	1297.9	1324.0	1375.3	1426.3	1477.7	1529.7	1582.4	1635.9	1690.2	1745.3
	s	0.5328	1.5543		1.5743	1.6078	1.6376	1.6647	1.6900	1.7362	1.7784	1.8176	1.8545	1.8894	1.9227	1.9545	1.9849
190 (377.53)	Sh				22.47	72.47	122.47	172.47	222.47	322.47	422.47	522.47	622.47	722.47	822.47	922.47	1022.47
	v	0.01833	2.4030		2.4961	2.6915	2.8756	3.0525	3.2246	3.5601	3.8889	4.2140	4.5365	4.8572	5.1766	5.4949	5.8124
	h	350.94	1197.6		1212.0	1242.0	1270.1	1297.1	1323.3	1374.8	1425.9	1477.4	1529.4	1582.1	1635.7	1690.0	1745.1
	s	0.5384	1.5498		1.5667	1.6006	1.6307	1.6581	1.6835	1.7299	1.7722	1.8115	1.8484	1.8834	1.9166	1.9484	1.9789
200 (381.80)	Sh				18.20	68.20	118.20	168.20	218.20	318.20	418.20	518.20	618.20	718.20	818.20	918.20	1018.20
	v	0.01839	2.2873		2.3598	2.5480	2.7247	2.8939	3.0583	3.3783	3.6915	4.0008	4.3077	4.6128	4.9165	5.2191	5.5209
	h	355.51	1198.3		1210.1	1240.6	1269.0	1296.2	1322.6	1374.3	1425.5	1477.0	1529.1	1581.9	1635.4	1689.8	1745.0
	s	0.5438	1.5454		1.5593	1.5938	1.6242	1.6518	1.6773	1.7239	1.7663	1.8057	1.8426	1.8776	1.9109	1.9427	1.9732

Abs. Press. (psi) (Sat. Temp.)		Sat. Water	Sat. Steam	Temperature (°F) 400	450	500	550	600	700	800	900	1000	1100	1200	1300	1400	1500
210 (385.91)	Sh			14.09	64.09	114.09	164.09	214.09	314.09	414.09	514.09	614.09	714.09	814.09	914.09	1014.09	1114.09
	v	0.01844	2.1822	2.2364	2.4181	2.5880	2.7504	2.9078	3.2137	3.5128	3.8080	4.1007	4.3915	4.6811	4.9695	5.2571	5.5440
	h	359.91	1199.0	1208.02	1239.2	1268.0	1295.3	1321.9	1373.7	1425.1	1476.7	1528.8	1581.6	1635.2	1689.6	1744.8	1800.8
	s	0.5490	1.5413	1.5522	1.5872	1.6180	1.6458	1.6715	1.7182	1.7607	1.8001	1.8371	1.8721	1.9054	1.9372	1.9677	1.9970
220 (389.88)	Sh			10.12	60.12	110.12	160.12	210.12	310.12	410.12	510.12	610.12	710.12	810.12	910.12	1010.12	1110.12
	v	0.01850	2.0863	2.1240	2.2999	2.4638	2.6199	2.7710	3.0642	3.3504	3.6327	3.9125	4.1905	4.4671	4.7426	5.0173	5.2913
	h	364.17	1199.6	1206.3	1237.8	1266.9	1294.5	1321.2	1373.2	1424.7	1476.3	1528.5	1581.4	1635.0	1689.4	1744.7	1800.6
	s	0.5540	1.5374	1.5453	1.5808	1.6120	1.6400	1.6658	1.7128	1.7553	1.7948	1.8318	1.8668	1.9002	1.9320	1.9625	1.9919
230 (393.70)	Sh			6.30	56.30	106.30	156.30	206.30	306.30	406.30	506.30	606.30	706.30	806.30	906.30	1006.30	1106.30
	v	0.01855	1.9985	2.0212	2.1919	2.3503	2.5008	2.6461	2.9276	3.2020	3.4726	3.7406	4.0068	4.2717	4.5355	4.7984	5.0606
	h	368.28	1200.1	1204.4	1236.3	1265.7	1293.6	1320.4	1372.7	1424.2	1476.0	1528.2	1581.1	1634.8	1689.3	1744.5	1800.5
	s	0.5588	1.5336	1.5385	1.5747	1.6062	1.6344	1.6604	1.7075	1.7502	1.7897	1.8268	1.8618	1.8952	1.9270	1.9576	1.9869
240 (397.39)	Sh			2.61	52.61	102.61	152.61	202.61	302.61	402.61	502.61	602.61	702.61	802.61	902.61	1002.61	1102.61
	v	0.01860	1.9177	1.9268	2.0928	2.2462	2.3915	2.5316	2.8024	3.0661	3.3259	3.5831	3.8385	4.0926	4.3456	4.5977	4.8492
	h	372.27	1200.6	1202.4	1234.9	1264.6	1292.7	1319.7	1372.1	1423.8	1475.6	1527.9	1580.9	1634.6	1689.1	1744.3	1800.4
	s	0.5634	1.5299	1.5320	1.5687	1.6006	1.6291	1.6552	1.7025	1.7452	1.7848	1.8219	1.8570	1.8904	1.9223	1.9528	1.9822
250 (400.97)	Sh				49.03	99.03	149.03	199.03	299.03	399.03	499.03	599.03	699.03	799.03	899.03	999.03	1099.03
	v	0.01865	1.8432		2.0016	2.1504	2.2909	2.4262	2.6872	2.9410	3.1909	3.4382	3.6837	3.9278	4.1709	4.4131	4.6546
	h	376.14	1201.1		1233.4	1263.5	1291.8	1319.0	1371.6	1423.4	1475.3	1527.6	1580.6	1634.4	1688.9	1744.2	1800.2
	s	0.5679	1.5264		1.5629	1.5951	1.6239	1.6502	1.6976	1.7405	1.7801	1.8173	1.8524	1.8858	1.9177	1.9482	1.9776
260 (404.44)	Sh				45.56	95.56	145.56	195.56	295.56	395.56	495.56	595.56	695.56	795.56	895.56	995.56	1095.56
	v	0.01870	1.7742		1.9173	2.0619	2.1981	2.3289	2.5808	2.8256	3.0663	3.3044	3.5408	3.7758	4.0097	4.2427	4.4750
	h	379.90	1201.5		1231.9	1262.4	1290.9	1318.2	1371.1	1423.0	1474.9	1527.3	1580.4	1634.2	1688.7	1744.0	1800.1
	s	0.5722	1.5230		1.5573	1.5899	1.6189	1.6453	1.6930	1.7359	1.7756	1.8128	1.8480	1.8814	1.9133	1.9439	1.9732
270 (407.80)	Sh				42.20	92.20	142.20	192.20	292.20	392.20	492.20	592.20	692.20	792.20	892.20	992.20	1092.20
	v	0.01875	1.7101		1.8391	1.9799	2.1121	2.2388	2.4824	2.7186	2.9509	3.1806	3.4084	3.6349	3.8603	4.0849	4.3087
	h	383.56	1201.9		1230.4	1261.2	1290.0	1317.5	1370.5	1422.6	1474.6	1527.1	1580.1	1634.0	1688.5	1743.9	1800.0
	s	0.5764	1.5197		1.5518	1.5848	1.6140	1.6406	1.6885	1.7315	1.7713	1.8085	1.8437	1.8771	1.9090	1.9396	1.9690
280 (411.07)	Sh				38.93	88.93	138.93	188.93	288.93	388.93	488.93	588.93	688.93	788.93	888.93	988.93	1088.93
	v	0.01880	1.6505		1.7665	1.9037	2.0322	2.1551	2.3909	2.6194	2.8437	3.0655	3.2855	3.5042	3.7217	3.9384	4.1543
	h	387.12	1202.3		1228.8	1260.0	1289.1	1316.8	1370.0	1422.1	1474.2	1526.8	1579.9	1633.8	1688.4	1743.7	1799.8
	s	0.5805	1.5166		1.5464	1.5798	1.6093	1.6361	1.6841	1.7273	1.7671	1.8043	1.8395	1.8730	1.9050	1.9356	1.9649

Table A-14. Superheated Steam* (Continued).

Abs. Press. (psi) (Sat. Temp.)		Sat. Water	Sat. Steam	400	450	500	550	600	700	800	900	1000	1100	1200	1300	1400	1500
580 (482.57)	Sh				17.43	67.43	117.43	167.43	217.43	317.43	417.43	517.43	617.43	717.43	817.43	917.43	1017.43
	v	0.02006	0.7971		0.8287	0.9100	0.9824	1.0492	1.1125	1.2324	1.3473	1.4593	1.5693	1.6780	1.7855	1.8921	1.9980
	h	467.47	1203.9		1219.1	1258.0	1292.1	1323.4	1353.0	1409.2	1463.7	1518.0	1572.4	1627.4	1682.9	1739.1	1795.9
	s	0.6679	1.4495		1.4654	1.5049	1.5380	1.5668	1.5929	1.6394	1.6811	1.7196	1.7556	1.7898	1.8223	1.8533	1.8831
600 (486.20)	Sh				13.80	63.80	113.80	163.80	213.80	313.80	413.80	513.80	613.80	713.80	813.80	913.80	1013.80
	v	0.02013	0.7697		0.7944	0.8746	0.9456	1.0109	1.0726	1.1892	1.3008	1.4093	1.5160	1.6211	1.7252	1.8284	1.9309
	h	471.70	1203.7		1215.9	1255.6	1290.3	1322.0	1351.8	1408.3	1463.0	1517.4	1571.9	1627.0	1682.6	1738.8	1795.6
	s	0.6723	1.4461		1.4590	1.4993	1.5329	1.5621	1.5884	1.6351	1.6769	1.7155	1.7517	1.7859	1.8184	1.8494	1.8792
650 (494.89)	Sh				5.11	55.11	105.11	155.11	205.11	305.11	405.11	505.11	605.11	705.11	805.11	905.11	1005.11
	v	0.02032	0.7084		0.7173	0.7954	0.8634	0.9254	0.9835	1.0929	1.1969	1.2979	1.3969	1.4944	1.5909	1.6864	1.7813
	h	481.89	1202.8		1207.6	1249.6	1285.7	1318.3	1348.7	1406.0	1461.2	1515.9	1570.7	1625.9	1681.6	1738.0	1794.9
	s	1.6828	1.4381		1.4430	1.4858	1.5207	1.5507	1.5775	1.6249	1.6671	1.7059	1.7422	1.7765	1.8092	1.8403	1.8701
700 (503.08)	Sh					46.92	96.92	146.92	196.92	296.92	396.92	496.92	596.92	696.92	796.92	896.92	996.92
	v	0.02050	0.6556			0.7271	0.7928	0.8520	0.9072	1.0102	1.1078	1.2023	1.2948	1.3858	1.4757	1.5647	1.6530
	h	491.60	1201.8			1243.4	1281.0	1314.6	1345.6	1403.7	1459.4	1514.4	1569.4	1624.8	1680.7	1737.2	1794.3
	s	1.6928	1.4304			1.4726	1.5090	1.5399	1.5673	1.6154	1.6580	1.6970	1.7335	1.7679	1.8006	1.8318	1.8617
750 (510.84)	Sh					39.16	89.16	139.16	189.16	289.16	389.16	489.16	589.16	689.16	789.16	889.16	989.16
	v	0.02069	0.6095			0.6676	0.7313	0.7882	0.8409	0.9386	1.0306	1.1195	1.2063	1.2916	1.3759	1.4592	1.5419
	h	500.89	1200.7			1236.9	1276.1	1310.7	1342.5	1401.5	1457.6	1512.9	1568.2	1623.8	1679.8	1736.4	1793.6
	s	0.7022	1.4232			1.4598	1.4977	1.5296	1.5577	1.6065	1.6494	1.6886	1.7252	1.7598	1.7926	1.8239	1.8538
800 (518.21)	Sh					31.79	81.79	131.79	181.79	281.79	381.79	481.79	581.79	681.79	781.79	881.79	981.79
	v	0.02087	0.5690			0.6151	0.6774	0.7323	0.7828	0.8759	0.9631	1.0470	1.1289	1.2093	1.2885	1.3669	1.4446
	h	509.81	1199.4			1230.1	1271.1	1306.8	1339.3	1399.1	1455.8	1511.4	1566.9	1622.7	1678.9	1735.7	1792.9
	s	0.7111	1.4163			1.4472	1.4869	1.5198	1.5484	1.5980	1.6413	1.6807	1.7175	1.7522	1.7851	1.8164	1.8464
850 (525.24)	Sh					24.76	74.76	124.76	174.76	274.76	374.76	474.76	574.76	674.76	774.76	874.76	974.76
	v	0.02105	0.5330			0.5683	0.6296	0.6829	0.7315	0.8205	0.9034	0.9830	1.0606	1.1366	1.2115	1.2855	1.3588
	h	518.40	1198.0			1223.0	1265.9	1302.8	1336.0	1396.8	1454.0	1510.0	1565.7	1621.6	1678.0	1734.9	1792.3
	s	0.7197	1.4096			1.4347	1.4763	1.5102	1.5396	1.5899	1.6336	1.6733	1.7102	1.7450	1.7780	1.8094	1.8395
900 (531.95)	Sh					18.05	68.05	118.05	168.05	268.05	368.05	468.05	568.05	668.05	768.05	868.05	968.05
	v	0.02123	0.5009			0.5263	0.5869	0.6388	0.6858	0.7713	0.8504	0.9262	0.9998	1.0720	1.1430	1.2131	1.2825
	h	526.70	1196.4			1215.5	1260.6	1298.6	1332.7	1394.4	1452.2	1508.5	1564.4	1620.6	1677.1	1734.1	1791.6
	s	0.7279	1.4032			1.4223	1.4659	1.5010	1.5311	1.5822	1.6263	1.6662	1.7033	1.7382	1.7713	1.8028	1.8329

Temperature (°F)

Abs. Press. (psi) (Sat. Temp.)		Sat. Water	Sat. Steam	Temperature (°F)													
				450	500	550	600	650	700	800	900	1000	1100	1200	1300	1400	1500
400 (444.60)	Sh			5.40	55.40	105.40	155.40	205.40	255.40	355.40	455.40	555.40	655.40	755.40	855.40	955.40	1055.40
	v	0.01934	1.1610	1.1738	1.2841	1.3836	1.4763	1.5646	1.6499	1.8151	1.9759	2.1139	2.2901	2.4450	2.5987	2.7515	2.9037
	h	424.17	1204.6	1208.8	1245.1	1277.5	1307.4	1335.9	1363.4	1417.0	1470.1	1523.3	1576.2	1631.2	1686.2	1741.9	1798.2
	s	0.6217	1.4847	1.4894	1.5282	1.5611	1.5901	1.6163	1.6406	1.6850	1.7255	1.7632	1.7988	1.8325	1.8647	1.8955	1.9250
420 (449.40)	Sh			.60	50.60	100.60	150.60	200.60	250.60	350.60	450.60	550.60	650.60	750.60	850.60	950.60	1050.60
	v	0.01942	1.1057	1.1071	1.2148	1.3113	1.4007	1.4856	1.5676	1.7258	1.8795	2.0304	2.1795	2.3273	2.4739	2.6196	2.7647
	h	429.56	1204.7	1205.2	1242.4	1275.4	1305.8	1334.5	1362.3	1416.2	1469.4	1522.7	1576.4	1630.8	1685.8	1741.6	1798.0
	s	0.6276	1.4802	1.4808	1.5206	1.5542	1.5835	1.6100	1.6345	1.6791	1.7197	1.7575	1.7932	1.8269	1.8591	1.8899	1.9195
440 (454.03)	Sh				45.97	95.97	145.97	195.97	245.97	345.97	445.97	545.97	645.97	745.97	845.97	945.97	1045.97
	v	0.01950	1.0554		1.1517	1.2454	1.3319	1.4138	1.4926	1.6445	1.7918	1.9363	2.0790	2.2203	2.3605	2.4998	2.6384
	h	434.77	1204.8		1239.7	1273.4	1304.2	1333.2	1361.1	1415.3	1468.7	1522.1	1575.9	1630.4	1685.5	1741.2	1797.7
	s	0.6332	1.4759		1.5132	1.5474	1.5772	1.6040	1.6286	1.6734	1.7142	1.7521	1.7878	1.8216	1.8538	1.8847	1.9143
460 (458.50)	Sh				41.50	91.50	141.50	191.50	241.50	341.50	441.50	541.50	641.50	741.50	841.50	941.50	1041.50
	v	0.01959	1.0092		1.0939	1.1852	1.2691	1.3482	1.4242	1.5703	1.7117	1.8504	1.9872	2.1226	2.2569	2.3903	2.5230
	h	439.83	1204.8		1236.9	1271.3	1302.5	1331.8	1360.0	1414.4	1468.0	1521.5	1575.4	1629.9	1685.1	1740.9	1797.4
	s	0.6387	1.4718		1.5060	1.5409	1.5711	1.5982	1.6230	1.6680	1.7089	1.7469	1.7826	1.8165	1.8488	1.8797	1.9093
480 (462.82)	Sh				37.18	87.18	137.18	187.18	237.18	337.18	437.18	537.18	637.18	737.18	837.18	937.18	1037.18
	v	0.01967	0.9668		1.0409	1.1300	1.2115	1.2881	1.3615	1.5023	1.6384	1.7716	1.9030	2.0330	2.1619	2.2900	2.4173
	h	444.75	1204.8		1234.1	1269.1	1300.8	1330.5	1358.8	1413.6	1467.3	1520.9	1574.9	1629.5	1684.7	1740.6	1797.2
	s	0.6439	1.4677		1.4990	1.5346	1.5652	1.5925	1.6176	1.6628	1.7038	1.7419	1.7777	1.8116	1.8439	1.8748	1.9045
500 (467.01)	Sh				32.99	82.99	132.99	182.99	232.99	332.99	432.99	532.99	632.99	732.99	832.99	932.99	1032.99
	v	0.01975	0.9276		0.9919	1.0791	1.1584	1.2327	1.3037	1.4397	1.5708	1.6992	1.8256	1.9507	2.0746	2.1977	2.3200
	h	449.52	1204.7		1231.2	1267.0	1299.1	1329.1	1357.7	1412.7	1466.6	1520.3	1574.4	1629.1	1684.4	1740.3	1796.9
	s	0.6490	1.4639		1.4921	1.5284	1.5595	1.5871	1.6123	1.6578	1.6990	1.7371	1.7730	1.8069	1.8393	1.8702	1.8998
520 (471.07)	Sh				28.93	78.93	128.93	178.93	228.93	328.93	428.93	528.93	628.93	728.93	828.93	928.93	1028.93
	v	0.01982	0.8914		0.9466	1.0321	1.1094	1.1816	1.2504	1.3819	1.5085	1.6323	1.7542	1.8746	1.9940	2.1125	2.2302
	h	454.18	1204.5		1228.3	1264.8	1297.4	1327.7	1356.5	1411.8	1465.9	1519.7	1573.9	1628.7	1684.0	1740.0	1796.7
	s	0.6540	1.4601		1.4853	1.5223	1.5539	1.5818	1.6072	1.6530	1.6943	1.7325	1.7684	1.8024	1.8348	1.8657	1.8954
540 (475.01)	Sh				24.99	74.99	124.99	174.99	224.99	324.99	424.99	524.99	624.99	724.99	824.99	924.99	1024.99
	v	0.01990	0.8577		0.9145	0.9884	1.0640	1.1342	1.2010	1.3284	1.4508	1.5704	1.6880	1.8042	1.9193	2.0336	2.1471
	h	458.71	1204.4		1225.3	1262.5	1295.7	1326.3	1355.3	1410.9	1465.1	1519.1	1573.4	1628.2	1683.6	1739.7	1796.4
	s	0.6587	1.4565		1.4786	1.5164	1.5485	1.5767	1.6023	1.6483	1.6897	1.7280	1.7640	1.7981	1.8305	1.8615	1.8911
560 (478.84)	Sh				21.16	71.16	121.16	171.16	221.16	321.16	421.16	521.16	621.16	721.16	821.16	921.16	1021.16
	v	0.01998	0.8264		0.8653	0.9479	1.0217	1.0902	1.1552	1.2787	1.3972	1.5129	1.6266	1.7388	1.8500	1.9603	2.0699
	h	463.14	1204.2		1222.2	1260.3	1293.9	1324.9	1354.2	1410.0	1464.4	1518.6	1572.9	1627.8	1683.3	1739.4	1796.1
	s	0.6634	1.4529		1.4720	1.5106	1.5431	1.5717	1.5975	1.6438	1.6853	1.7237	1.7598	1.7939	1.8263	1.8573	1.8870

Table A-14. Superheated Steam* (Continued).

Abs. Press. (psi) (Sat. Temp.)		Sat. Water	Sat. Steam	450	500	550	600	650	700	800	900	1000	1100	1200	1300	1400	1500
580 (482.57)	Sh				17.43	67.43	117.43	167.43	217.43	317.43	417.43	517.43	617.43	717.43	817.43	917.43	1017.43
	v	0.02006	0.7971		0.8287	0.9100	0.9824	1.0492	1.1125	1.2324	1.3473	1.4593	1.5693	1.6780	1.7855	1.8921	1.9980
	h	467.47	1203.9		1219.1	1258.0	1292.1	1323.4	1353.0	1409.2	1463.7	1518.0	1572.4	1627.4	1682.9	1739.1	1795.9
	s	0.6679	1.4495		1.4654	1.5049	1.5380	1.5668	1.5929	1.6394	1.6811	1.7196	1.7556	1.7898	1.8223	1.8533	1.8831
600 (486.20)	Sh				13.80	63.80	113.80	163.80	213.80	313.80	413.80	513.80	613.80	713.80	813.80	913.80	1013.80
	v	0.02013	0.7697		0.7944	0.8746	0.9456	1.0109	1.0726	1.1892	1.3008	1.4093	1.5160	1.6211	1.7252	1.8284	1.9309
	h	471.70	1203.7		1215.9	1255.6	1290.3	1322.0	1351.8	1408.3	1463.0	1517.4	1571.9	1627.0	1682.6	1738.8	1795.6
	s	0.6723	1.4461		1.4590	1.4993	1.5329	1.5621	1.5884	1.6351	1.6769	1.7155	1.7517	1.7859	1.8184	1.8494	1.8792
650 (494.89)	Sh				5.11	55.11	105.11	155.11	205.11	305.11	405.11	505.11	605.11	705.11	805.11	905.11	1005.11
	v	0.02032	0.7084		0.7173	0.7954	0.8634	0.9254	0.9835	1.0929	1.1969	1.2979	1.3969	1.4944	1.5909	1.6864	1.7813
	h	481.89	1202.8		1207.6	1249.6	1285.7	1318.3	1348.7	1406.0	1461.2	1515.9	1570.7	1625.9	1681.6	1738.0	1794.9
	s	0.6828	1.4381		1.4430	1.4858	1.5207	1.5507	1.5775	1.6249	1.6671	1.7059	1.7422	1.7765	1.8092	1.8403	1.8701
700 (503.08)	Sh					46.92	96.92	146.92	196.92	296.92	396.92	496.92	596.92	696.92	796.92	896.92	996.92
	v	0.02050	0.6556			0.7271	0.7928	0.8520	0.9072	1.0102	1.1078	1.2023	1.2948	1.3858	1.4757	1.5647	1.6530
	h	491.60	1201.8			1243.4	1281.0	1314.6	1345.6	1403.7	1459.4	1514.4	1569.4	1624.8	1680.7	1737.2	1794.3
	s	0.6928	1.4304			1.4726	1.5090	1.5399	1.5673	1.6154	1.6580	1.6970	1.7335	1.7679	1.8006	1.8318	1.8617
750 (510.84)	Sh					39.16	89.16	139.16	189.16	289.16	389.16	489.16	589.16	689.16	789.16	889.16	989.16
	v	0.02069	0.6095			0.6676	0.7313	0.7882	0.8409	0.9386	1.0306	1.1195	1.2063	1.2916	1.3759	1.4592	1.5419
	h	500.89	1200.7			1236.9	1276.1	1310.7	1342.5	1401.5	1457.6	1512.9	1568.2	1623.8	1679.8	1736.4	1793.6
	s	0.7022	1.4232			1.4598	1.4977	1.5296	1.5577	1.6065	1.6494	1.6886	1.7252	1.7598	1.7926	1.8239	1.8538
800 (518.21)	Sh					31.79	81.79	131.79	181.79	281.79	381.79	481.79	581.79	681.79	781.79	881.79	981.79
	v	0.02087	0.5690			0.6151	0.6774	0.7323	0.7828	0.8759	0.9631	1.0470	1.1289	1.2093	1.2885	1.3669	1.4446
	h	509.81	1199.4			1230.1	1271.1	1306.8	1339.3	1399.1	1455.8	1511.4	1566.9	1622.7	1678.9	1735.7	1792.9
	s	0.7111	1.4163			1.4472	1.4869	1.5198	1.5484	1.5980	1.6413	1.6807	1.7175	1.7522	1.7851	1.8164	1.8464
850 (525.24)	Sh					24.76	74.76	124.76	174.76	274.76	374.76	474.76	574.76	674.76	774.76	874.76	974.76
	v	0.02105	0.5330			0.5683	0.6296	0.6829	0.7315	0.8205	0.9034	0.9830	1.0606	1.1366	1.2115	1.2855	1.3588
	h	518.40	1198.0			1223.0	1265.9	1302.8	1336.0	1396.8	1454.0	1510.0	1565.7	1621.6	1678.0	1734.9	1792.3
	s	0.7197	1.4096			1.4347	1.4763	1.5102	1.5396	1.5899	1.6336	1.6733	1.7102	1.7450	1.7780	1.8094	1.8395
900 (531.95)	Sh					18.05	68.05	118.05	168.05	268.05	368.05	468.05	568.05	668.05	768.05	868.05	968.05
	v	0.02123	0.5009			0.5263	0.5869	0.6388	0.6858	0.7713	0.8504	0.9262	0.9998	1.0720	1.1430	1.2131	1.2825
	h	526.70	1196.4			1215.5	1260.6	1298.6	1332.7	1394.4	1452.2	1508.5	1564.4	1620.6	1677.1	1734.1	1791.6
	s	0.7279	1.4032			1.4223	1.4659	1.5010	1.5311	1.5822	1.6263	1.6662	1.7033	1.7382	1.7713	1.8028	1.8329

Temperature (°F)

		Sat. Water	Sat. Steam	*Temperature (°F)*													
Abs. Press. (psi) (Sat. Temp.)				550	600	650	700	750	800	850	900	1000	1100	1200	1300	1400	1500
950 (538.39)	Sh			11.61	61.61	111.61	161.61	211.61	261.61	311.61	361.61	461.61	561.61	661.61	761.61	861.61	961.61
	v	0.02141	0.4721	0.4883	0.5485	0.5993	0.6449	0.6871	0.7272	0.7656	0.8030	0.8753	0.9455	1.0142	1.0817	1.1484	1.2143
	h	534.74	1194.7	1207.6	1255.1	1294.4	1329.4	1361.5	1392.0	1421.5	1450.3	1507.0	1563.2	1619.5	1676.2	1733.3	1791.0
	s	0.7358	1.3970	1.4098	1.4557	1.4921	1.5228	1.5500	1.5748	1.5977	1.6193	1.6595	1.6967	1.7317	1.7649	1.7965	1.8267
1000 (544.58)	Sh			5.42	55.42	105.42	155.42	205.42	255.42	305.42	355.42	455.42	555.42	655.42	755.42	855.42	955.42
	v	0.02159	0.4460	0.4535	0.5137	0.5636	0.6080	0.6489	0.6875	0.7245	0.7603	0.8295	0.8966	0.9622	1.0266	1.0901	1.1529
	h	542.55	1192.9	1199.3	1249.3	1290.1	1325.9	1358.7	1389.6	1419.4	1448.5	1505.4	1561.9	1618.4	1675.3	1732.5	1790.3
	s	0.7434	1.3910	1.3973	1.4457	1.4833	1.5149	1.5426	1.5677	1.5908	1.6126	1.6530	1.6905	1.7256	1.7589	1.7905	1.8207
1050 (550.53)	Sh				49.47	99.47	149.47	199.47	249.47	299.47	349.47	449.47	549.47	649.47	749.47	849.47	949.47
	v	0.02177	0.4222		0.4821	0.5312	0.5745	0.6142	0.6515	0.6872	0.7216	0.7881	0.8524	0.9151	0.9767	1.0373	1.0973
	h	550.15	1191.0		1243.4	1285.7	1322.4	1355.8	1387.2	1417.3	1446.6	1503.9	1560.7	1617.4	1674.4	1731.8	1789.6
	s	0.7507	1.3851		1.4358	1.4748	1.5072	1.5354	1.5608	1.5842	1.6062	1.6469	1.6845	1.7197	1.7531	1.7848	1.8151
1100 (556.28)	Sh				43.72	93.72	143.72	193.72	243.72	293.72	343.72	443.72	543.72	643.72	743.72	843.72	943.72
	v	0.02195	0.4006		0.4531	0.5017	0.5440	0.5826	0.6188	0.6533	0.6865	0.7505	0.8121	0.8723	0.9313	0.9894	1.0468
	h	557.55	1189.1		1237.3	1281.2	1318.8	1352.9	1384.7	1415.2	1444.7	1502.4	1559.4	1616.3	1673.5	1731.0	1789.0
	s	0.7578	1.3794		1.4259	1.4664	1.4996	1.5284	1.5542	1.5779	1.6000	1.6410	1.6787	1.7141	1.7475	1.7793	1.8097
1150 (561.82)	Sh				39.18	89.18	139.18	189.18	239.18	289.18	339.18	439.18	539.18	639.18	739.18	839.18	939.18
	v	0.02214	0.3807		0.4263	0.4746	0.5162	0.5538	0.5889	0.6223	0.6544	0.7161	0.7754	0.8332	0.8899	0.9456	1.0007
	h	564.78	1187.0		1230.9	1276.6	1315.2	1349.9	1382.2	1413.0	1442.8	1500.9	1558.1	1615.2	1672.6	1730.2	1788.3
	s	0.7647	1.3738		1.4160	1.4582	1.4923	1.5216	1.5478	1.5717	1.5941	1.6353	1.6732	1.7087	1.7422	1.7741	1.8045
1200 (567.19)	Sh				32.81	82.81	132.81	182.81	232.81	282.81	332.81	432.81	532.81	632.81	732.81	832.81	932.81
	v	0.02232	0.3624		0.4016	0.4497	0.4905	0.5273	0.5615	0.5939	0.6250	0.6845	0.7418	0.7974	0.8519	0.9055	0.9584
	h	571.85	1184.8		1224.2	1271.8	1311.5	1346.9	1379.7	1410.8	1440.9	1499.4	1556.9	1614.2	1671.6	1729.4	1787.6
	s	0.7714	1.3683		1.4061	1.4501	1.4851	1.5150	1.5415	1.5658	1.5883	1.6298	1.6679	1.7035	1.7371	1.7691	1.7996
1300 (577.42)	Sh				22.58	72.58	122.58	172.58	222.58	272.58	322.58	422.58	522.58	622.58	722.58	822.58	922.58
	v	0.02269	0.3299		0.3570	0.4052	0.4451	0.4804	0.5129	0.5436	0.5729	0.6287	0.6822	0.7341	0.7847	0.8345	0.8836
	h	585.58	1180.2		1209.9	1261.9	1303.9	1340.8	1374.6	1406.4	1437.1	1496.3	1554.3	1612.0	1669.8	1727.9	1786.3
	s	0.7843	1.3577		1.3860	1.4340	1.4711	1.5022	1.5296	1.5544	1.5773	1.6194	1.6578	1.6937	1.7275	1.7596	1.7902
1400 (587.07)	Sh				12.93	62.93	112.93	162.93	212.93	262.93	312.93	412.93	512.93	612.93	712.93	812.93	912.93
	v	0.02307	0.3018		0.3176	0.3667	0.4059	0.4400	0.4712	0.5004	0.5282	0.5809	0.6311	0.6798	0.7272	0.7737	0.8195
	h	598.83	1175.3		1194.1	1251.4	1296.1	1334.5	1369.3	1402.0	1433.2	1493.2	1551.8	1609.9	1668.0	1726.3	1785.0
	s	0.7966	1.3474		1.3652	1.4181	1.4575	1.4900	1.5182	1.5436	1.5670	1.6096	1.6484	1.6845	1.7185	1.7508	1.7815
1500 (596.20)	Sh				3.80	53.80	103.80	153.80	203.80	253.80	303.80	403.80	503.80	603.80	703.80	803.80	903.80
	v	0.02346	0.2772		0.2820	0.3328	0.3717	0.4049	0.4350	0.4629	0.4894	0.5394	0.5869	0.6327	0.6773	0.7210	0.7639
	h	611.68	1170.1		1176.3	1240.2	1287.9	1328.0	1364.0	1397.4	1429.2	1490.1	1549.2	1607.7	1666.2	1724.8	1783.7
	s	0.8085	1.3373		1.3431	1.4022	1.4443	1.4782	1.5073	1.5333	1.5572	1.6004	1.6395	1.6759	1.7101	1.7425	1.7734

Table A-14. Superheated Steam* (Continued).

Abs. Press. (psi) (Sat. Temp.)		Sat. Water	Sat. Steam	\<550\>	\<600\>	650	700	750	800	850	900	1000	1100	1200	1300	1400	1500
1600 (604.87)	Sh					45.13	95.13	145.13	195.13	245.13	295.13	395.13	495.13	595.13	695.13	795.13	895.13
	v	0.02387	0.2555			0.3026	0.3415	0.3741	0.4032	0.4301	0.4555	0.5031	0.5482	0.5915	0.6336	0.6748	0.7153
	h	624.20	1164.5			1228.3	1279.4	1321.4	1358.5	1392.8	1425.2	1486.9	1546.6	1605.6	1664.3	1723.2	1782.3
	s	0.8199	1.3274			1.3861	1.4312	1.4667	1.4968	1.5235	1.5478	1.5916	1.6312	1.6678	1.7022	1.7347	1.7657
1700 (613.13)	Sh					36.87	86.87	136.87	186.87	236.87	286.87	386.87	486.87	586.87	686.87	786.87	886.87
	v	0.02428	0.2361			0.2754	0.3147	0.3468	0.3751	0.4011	0.4255	0.4711	0.5140	0.5552	0.5951	0.6341	0.6724
	h	636.45	1158.6			1215.3	1270.5	1314.5	1352.9	1388.1	1421.2	1483.8	1544.0	1603.4	1662.5	1721.7	1781.0
	s	0.8309	1.3176			1.3697	1.4183	1.4555	1.4867	1.5140	1.5388	1.5833	1.6232	1.6601	1.6947	1.7274	1.7585
1800 (621.02)	Sh					28.98	78.98	128.98	178.98	228.98	278.98	378.98	478.98	578.98	678.98	778.98	878.98
	v	0.02472	0.2186			0.2505	0.2906	0.3223	0.3500	0.3752	0.3988	0.4426	0.4836	0.5229	0.5609	0.5980	0.6343
	h	648.49	1152.3			1201.2	1261.1	1307.4	1347.2	1383.3	1417.1	1480.6	1541.4	1601.2	1660.7	1720.1	1779.7
	s	0.8417	1.3079			1.3526	1.4054	1.4446	1.4768	1.5049	1.5302	1.5753	1.6156	1.6528	1.6876	1.7204	1.7516
1900 (628.56)	Sh					21.44	71.44	121.44	171.44	221.44	271.44	371.44	471.44	571.44	671.44	771.44	871.44
	v	0.02517	0.2028			0.2274	0.2687	0.3004	0.3275	0.3521	0.3749	0.4171	0.4565	0.4940	0.5303	0.5656	0.6002
	h	660.36	1145.6			1185.7	1251.3	1300.2	1341.4	1378.4	1412.9	1477.4	1538.8	1599.1	1658.8	1718.6	1778.4
	s	0.8522	1.2981			1.3346	1.3925	1.4338	1.4672	1.4960	1.5219	1.5677	1.6084	1.6458	1.6808	1.7138	1.7451
2000 (635.80)	Sh					14.20	64.20	114.20	164.20	214.20	264.20	364.20	464.20	564.20	664.20	764.20	864.20
	v	0.02565	0.1883			0.2056	0.2488	0.2805	0.3072	0.3312	0.3534	0.3942	0.4320	0.4680	0.5027	0.5365	0.5695
	h	672.11	1138.3			1168.3	1240.9	1292.6	1335.4	1373.5	1408.7	1474.1	1536.2	1596.9	1657.0	1717.0	1771.1
	s	0.8625	1.2881			1.3154	1.3794	1.4231	1.4578	1.4874	1.5138	1.5603	1.6014	1.6391	1.6743	1.7075	1.7389
2100 (642.76)	Sh					7.24	57.24	107.24	157.24	207.24	257.24	357.24	457.24	557.24	657.24	757.24	857.24
	v	0.02615	0.1750			0.1847	0.2304	0.2624	0.2888	0.3123	0.3339	0.3734	0.4099	0.4445	0.4778	0.5101	0.5418
	h	683.79	1130.5			1148.5	1229.8	1284.9	1329.3	1368.4	1404.4	1470.9	1533.6	1594.7	1655.2	1715.4	1775.7
	s	0.8727	1.2780			1.2942	1.3661	1.4125	1.4486	1.4790	1.5060	1.5532	1.5948	1.6327	1.6681	1.7014	1.7330
2200 (649.45)	Sh					.55	50.55	100.55	150.55	200.55	250.55	350.55	450.55	550.55	650.55	750.55	850.55
	v	0.02669	0.1627			0.1636	0.2134	0.2458	0.2720	0.2950	0.3161	0.3545	0.3897	0.4231	0.4551	0.4862	0.5165
	h	695.46	1122.2			1123.9	1218.0	1276.8	1323.1	1363.3	1400.0	1467.6	1530.9	1592.5	1653.3	1713.9	1774.4
	s	0.8828	1.2676			1.2691	1.3523	1.4020	1.4395	1.4708	1.4984	1.5463	1.5883	1.6266	1.6622	1.6956	1.7273
2300 (655.89)	Sh						44.11	94.11	144.11	194.11	244.11	344.11	444.11	544.11	644.11	744.11	844.11
	v	0.02727	0.1513				0.1975	0.2305	0.2566	0.2793	0.2999	0.3372	0.3714	0.4035	0.4344	0.4643	0.4935
	h	707.18	1113.2				1205.3	1268.4	1316.7	1358.1	1395.7	1464.2	1528.3	1590.3	1651.5	1712.3	1773.1
	s	0.8929	1.2569				1.3381	1.3914	1.4305	1.4628	1.4910	1.5397	1.5821	1.6207	1.6565	1.6901	1.7219

*Temperature (°F)

Abs. Press. (psi) (Sat. Temp.)		Sat. Water	Sat. Steam	700	750	800	850	900	950	1000	1050	1100	1150	1200	1300	1400	1500
									Temperature (°F)								
2400 (662.11)	Sh			37.89	87.89	137.89	187.89	237.89	287.89	337.89	387.89	437.89	487.89	537.89	637.89	737.89	837.89
	v	0.02790	0.1408	0.1824	0.2164	0.2424	0.2648	0.2850	0.3037	0.3214	0.3382	0.3545	0.3703	0.3856	0.4155	0.4443	0.4724
	h	718.95	1103.7	1191.6	1259.7	1310.1	1352.8	1391.2	1426.9	1460.9	1493.7	1525.6	1557.0	1588.1	1649.6	1710.8	1771.8
	s	0.9031	1.2460	1.3232	1.3808	1.4217	1.4549	1.4837	1.5095	1.5332	1.5553	1.5761	1.5959	1.6149	1.6509	1.6847	1.7167
2500 (668.11)	Sh			31.89	81.89	131.89	181.89	231.89	281.89	331.89	381.89	431.89	481.89	531.89	631.89	731.89	831.89
	v	0.02859	0.1307	0.1681	0.2032	0.2293	0.2514	0.2712	0.2896	0.3068	0.3232	0.3390	0.3543	0.3692	0.3980	0.4259	0.4529
	h	731.71	1093.3	1176.7	1250.6	1303.4	1347.4	1386.7	1423.1	1457.5	1490.7	1522.9	1554.6	1585.9	1647.8	1709.2	1770.4
	s	0.9139	1.2345	1.3076	1.3701	1.4129	1.4472	1.4766	1.5029	1.5269	1.5492	1.5703	1.5903	1.6094	1.6456	1.6796	1.7116
2600 (673.91)	Sh			26.09	76.09	126.09	176.09	226.09	276.09	326.09	376.09	426.09	476.09	526.09	626.09	726.09	826.09
	v	0.02938	0.1211	0.1544	0.1909	0.2171	0.2390	0.2585	0.2765	0.2933	0.3093	0.3247	0.3395	0.3540	0.3819	0.4088	0.4350
	h	744.47	1082.0	1160.2	1241.1	1296.5	1341.9	1382.1	1419.2	1454.1	1487.7	1520.2	1552.2	1583.7	1646.0	1707.7	1769.1
	s	0.9247	1.2225	1.2908	1.3592	1.4042	1.4395	1.4696	1.4964	1.5208	1.5434	1.5646	1.5848	1.6040	1.6405	1.6746	1.7068
2700 (679.53)	Sh			20.47	70.47	120.47	170.47	220.47	270.47	320.47	370.47	420.47	470.47	520.47	620.47	720.47	820.47
	v	0.03029	0.1119	0.1411	0.1794	0.2058	0.2275	0.2468	0.2644	0.2809	0.2965	0.3114	0.3259	0.3399	0.3670	0.3931	0.4184
	h	757.34	1069.7	1142.0	1231.1	1289.5	1336.3	1377.5	1415.2	1450.7	1484.6	1517.5	1549.8	1581.5	1644.1	1706.1	1767.8
	s	0.9356	1.2097	1.2727	1.3481	1.3954	1.4319	1.4628	1.4900	1.5148	1.5376	1.5591	1.5794	1.5988	1.6355	1.6697	1.7021
2800 (684.96)	Sh			15.04	65.04	115.04	165.04	215.04	265.04	315.04	365.04	415.04	465.04	515.04	615.04	715.04	815.04
	v	0.03134	0.1030	0.1278	0.1685	0.1952	0.2168	0.2358	0.2531	0.2693	0.2845	0.2991	0.3132	0.3268	0.3532	0.3785	0.4030
	h	770.69	1055.8	1121.2	1220.6	1282.2	1330.7	1372.8	1411.2	1447.2	1481.6	1514.8	1547.3	1579.3	1642.2	1704.5	1766.5
	s	0.9468	1.1958	1.2527	1.3368	1.3867	1.4245	1.4561	1.4838	1.5089	1.5321	1.5537	1.5742	1.5938	1.6306	1.6651	1.6975
2900 (690.22)	Sh			9.78	59.78	109.78	159.78	209.78	259.78	309.78	359.78	409.78	459.78	509.78	609.78	709.78	809.78
	v	0.03262	0.0942	0.1138	0.1581	0.1853	0.2068	0.2256	0.2427	0.2585	0.2734	0.2877	0.3014	0.3147	0.3403	0.3649	0.3887
	h	785.13	1039.8	1095.3	1209.6	1274.7	1324.7	1368.0	1407.2	1443.7	1478.5	1512.1	1544.9	1577.0	1640.4	1703.0	1765.2
	s	0.9588	1.1803	1.2283	1.3251	1.3780	1.4171	1.4494	1.4777	1.5032	1.5266	1.5485	1.5692	1.5889	1.6259	1.6605	1.6931
3000 (695.33)	Sh			4.67	54.67	104.67	154.67	204.67	254.67	304.67	354.67	404.67	454.67	504.67	604.67	704.67	804.67
	v	0.03428	0.0850	0.0982	0.1483	0.1759	0.1975	0.2161	0.2329	0.2484	0.2630	0.2770	0.2904	0.3033	0.3282	0.3522	0.3753
	h	801.84	1020.3	1060.5	1197.9	1267.0	1319.0	1363.2	1403.1	1440.2	1475.4	1509.4	1542.4	1574.8	1638.5	1701.4	1763.8
	s	0.9728	1.1619	1.1966	1.3131	1.3692	1.4097	1.4429	1.4717	1.4976	1.5213	1.5434	1.5642	1.5841	1.6214	1.6561	1.6888
3100 (700.28)	Sh				49.72	99.72	149.72	199.72	249.72	299.72	349.72	399.72	449.72	499.72	599.72	699.72	799.72
	v	0.03681	0.0745		0.1389	0.1671	0.1887	0.2071	0.2237	0.2390	0.2533	0.2670	0.2800	0.2927	0.3170	0.3403	0.3628
	h	823.97	993.3		1185.4	1259.1	1313.0	1358.4	1399.0	1436.7	1472.3	1506.6	1539.9	1572.6	1636.7	1699.8	1762.5
	s	0.9914	1.1373		1.3007	1.3604	1.4024	1.4364	1.4658	1.4920	1.5161	1.5384	1.5594	1.5794	1.6169	1.6518	1.6847
3200 (705.08)	Sh				44.92	94.92	144.92	194.92	244.92	294.92	344.92	394.92	444.92	494.92	594.92	694.92	794.92
	v	0.04472	0.0566		0.1300	0.1588	0.1804	0.1987	0.2151	0.2301	0.2442	0.2576	0.2704	0.2827	0.3065	0.3291	0.3510
	h	875.54	931.6		1172.3	1250.9	1306.9	1353.4	1394.9	1433.1	1469.2	1503.8	1537.4	1570.3	1634.8	1698.3	1761.2
	s	1.0351	1.0832		1.2877	1.3515	1.3951	1.4300	1.4600	1.4866	1.5110	1.5335	1.5547	1.5749	1.6126	1.6477	1.6806

Table A-14. Superheated Steam* (Continued).

Temperature (°F)

Columns for Sat. Water, Sat. Steam, and 700°F are blank throughout this section; values are tabulated from 750°F onward. "Sh" denotes the superheat rows.

Abs. Press. (psi) (Sat. Temp.)		750	800	850	900	950	1000	1050	1100	1150	1200	1300	1400	1500
---	v	0.1213	0.1510	0.1727	0.1908	0.2070	0.2218	0.2357	0.2488	0.2613	0.2734	0.2966	0.3187	0.3400
	h	1158.2	1242.5	1300.7	1348.4	1390.7	1429.5	1466.1	1501.0	1534.9	1568.1	1632.9	1696.7	1759.9
	s	1.2742	1.3425	1.3879	1.4237	1.4542	1.4813	1.5059	1.5287	1.5501	1.5704	1.6084	1.6436	1.6767
3400 Sh	v	0.1129	0.1435	0.1653	0.1834	0.1994	0.2140	0.2276	0.2405	0.2528	0.2646	0.2872	0.3088	0.3296
	h	1143.2	1233.7	1294.3	1343.4	1386.4	1425.9	1462.9	1498.3	1532.4	1565.8	1631.1	1695.1	1758.2
	s	1.2600	1.3334	1.3807	1.4174	1.4486	1.4761	1.5010	1.5240	1.5456	1.5660	1.6042	1.6396	1.6728
3500 Sh	v	0.1048	0.1364	0.1583	0.1764	0.1922	0.2066	0.2200	0.2326	0.2447	0.2563	0.2784	0.2995	0.3198
	h	1127.1	1224.6	1287.8	1338.2	1382.2	1422.2	1459.7	1495.5	1529.9	1563.6	1629.2	1693.6	1757.2
	s	1.2450	1.3242	1.3734	1.4112	1.4430	1.4709	1.4962	1.5194	1.5412	1.5618	1.6002	1.6358	1.6691
3600 Sh	v	0.0966	0.1296	0.1517	0.1697	0.1854	0.1996	0.2128	0.2252	0.2371	0.2485	0.2702	0.2908	0.3106
	h.	1108.6	1215.3	1281.2	1333.0	1377.9	1418.6	1456.5	1492.6	1527.4	1561.3	1627.3	1692.0	1755.9
	s	1.2281	1.3148	1.3662	1.4050	1.4374	1.4658	1.4914	1.5149	1.5369	1.5576	1.5962	1.6320	1.6654
3800 Sh	v	0.0799	0.1169	0.1395	0.1574	0.1729	0.1868	0.1996	0.2116	0.2231	0.2340	0.2549	0.2746	0.2936
	h	1064.2	1195.5	1267.6	1322.4	1369.1	1411.2	1450.1	1487.0	1522.4	1556.8	1623.6	1688.9	1753.2
	s	1.1888	1.2955	1.3517	1.3928	1.4265	1.4558	1.4821	1.5061	1.5284	1.5495	1.5886	1.6247	1.6584
4000 Sh	v	0.0631	0.1052	0.1284	0.1463	0.1616	0.1752	0.1877	0.1994	0.2105	0.2210	0.2411	0.2601	0.2783
	h	1007.4	1174.3	1253.4	1311.6	1360.2	1403.6	1443.6	1481.3	1517.3	1552.2	1619.8	1685.7	1750.6
	s	1.1396	1.2754	1.3371	1.3807	1.4158	1.4461	1.4730	1.4976	1.5203	1.5417	1.5812	1.6177	1.6516
4200 Sh	v	0.0498	0.0945	0.1183	0.1362	0.1513	0.1647	0.1769	0.1883	0.1991	0.2093	0.2287	0.2470	0.2645
	h	950.1	1151.6	1238.6	1300.4	1351.2	1396.0	1437.1	1475.5	1512.2	1547.6	1616.1	1682.6	1748.0
	s	1.0905	1.2544	1.3223	1.3686	1.4053	1.4366	1.4642	1.4893	1.5124	1.5341	1.5742	1.6109	1.6452
4400 Sh	v	0.0421	0.0846	0.1090	0.1270	0.1420	0.1552	0.1671	0.1782	0.1887	0.1986	0.2174	0.2351	0.2519
	h	909.5	1127.3	1223.3	1289.0	1342.0	1388.3	1430.4	1469.7	1507.1	1543.0	1612.3	1679.4	1745.3
	s	1.0556	1.2325	1.3073	1.3566	1.3949	1.4272	1.4556	1.4812	1.5048	1.5268	1.5673	1.6044	1.6389

Abs. Press. (psi) (Sat. Temp.)		Sat. Water	Sat. Steam	Temperature (°F)													
				750	800	850	900	950	1000	1050	1100	1150	1200	1250	1300	1400	1500
4600	Sh																
	v			0.0380	0.0751	0.1005	0.1186	0.1335	0.1465	0.1582	0.1691	0.1792	0.1889	0.1982	0.2071	0.2242	0.2404
	h			883.8	1100.0	1207.3	1277.2	1332.6	1380.5	1423.7	1463.9	1501.9	1538.4	1573.8	1608.5	1676.3	1742.7
	s			1.0331	1.2084	1.2922	1.3446	1.3847	1.4181	1.4472	1.4734	1.4974	1.5197	1.5407	1.5607	1.5982	1.6330
4800	Sh																
	v			0.0355	0.0665	0.0927	0.1109	0.1257	0.1385	0.1500	0.1606	0.1706	0.1800	0.1890	0.1977	0.2142	0.2299
	h			866.9	1071.2	1190.7	1265.2	1323.1	1372.6	1417.0	1458.0	1496.7	1533.8	1569.7	1604.7	1673.1	1740.0
	s			1.0180	1.1835	1.2768	1.3327	1.3745	1.4090	1.4390	1.4657	1.4901	1.5128	1.5341	1.5543	1.5921	1.6272
5000	Sh																
	v			0.0338	0.591	0.0855	0.1038	0.1185	0.1312	0.1425	0.1529	0.1626	0.1718	0.1806	0.1890	0.2050	0.2203
	h			854.9	1042.9	1173.6	1252.9	1313.5	1364.6	1410.2	1452.1	1491.5	1529.1	1565.5	1600.9	1670.0	1737.4
	s			1.0070	1.1593	1.2612	1.3207	1.3645	1.4001	1.4309	1.4582	1.4831	1.5061	1.5277	1.5481	1.5863	1.6216
5200	Sh																
	v			0.0326	0.0531	0.0789	0.0973	0.1119	0.1244	0.1356	0.1458	0.1553	0.1642	0.1728	0.1810	0.1966	0.2114
	h			845.8	1016.9	1156.0	1240.4	1303.7	1356.6	1403.4	1446.2	1486.3	1524.5	1561.3	1597.2	1666.8	1734.7
	s			0.9985	1.1370	1.2455	1.3088	1.3545	1.3914	1.4229	1.4509	1.4762	1.4995	1.5214	1.5420	1.5806	1.6161
5400	Sh																
	v			0.0317	0.0483	0.0728	0.0912	0.1058	0.1182	0.1292	0.1392	0.1485	0.1572	0.1656	0.1736	0.1888	0.2031
	h			838.5	994.3	1138.1	1227.7	1293.7	1348.4	1396.5	1440.3	1481.1	1519.8	1557.1	1593.4	1663.7	1732.1
	s			0.9915	1.1175	1.2296	1.2969	1.3446	1.3827	1.4151	1.4437	1.4694	1.4931	1.5153	1.5362	1.5750	1.6109
5600	Sh																
	v			0.0309	0.0447	0.0672	0.0856	0.1001	0.1124	0.1232	0.1331	0.1422	0.1508	0.1589	0.1667	0.1815	0.1954
	h			832.4	975.0	1119.9	1214.8	1283.7	1340.2	1389.6	1434.3	1475.9	1515.2	1552.9	1589.6	1660.5	1729.5
	s			0.9855	1.1008	1.2137	1.2850	1.3348	1.3742	1.4075	1.4366	1.4628	1.4869	1.5093	1.5304	1.5697	1.6058
5800	Sh																
	v			0.0303	0.0419	0.0622	0.0805	0.0949	0.1070	0.1177	0.1274	0.1363	0.1447	0.1527	0.1603	0.1747	0.1883
	h			827.3	958.8	1101.8	1201.8	1273.6	1332.0	1382.6	1428.3	1470.6	1510.5	1548.7	1585.8	1657.4	1726.8
	s			0.9803	1.0867	1.1981	1.2732	1.3250	1.3658	1.3999	1.4297	1.4564	1.4808	1.5035	1.5248	1.5644	1.6008
6000	Sh																
	v			0.0298	0.0397	0.0579	0.0757	0.0900	0.1020	0.1126	0.1221	0.1309	0.1391	0.1469	0.1544	0.1684	0.1817
	h			822.9	945.1	1084.6	1188.8	1263.4	1323.6	1375.7	1422.3	1465.4	1505.9	1544.6	1582.0	1654.2	1724.2
	s			0.9758	1.0746	1.1833	1.2615	1.3154	1.3574	1.3925	1.4229	1.4500	1.4748	1.4978	1.5194	1.5593	1.5960
6500	Sh																
	v			0.0287	0.0358	0.0495	0.0655	0.0793	0.0909	0.1012	0.1104	0.1188	0.1266	0.1340	0.1411	0.1544	0.1669
	h			813.9	919.5	1046.7	1156.3	1237.8	1302.7	1358.1	1407.3	1452.2	1494.2	1534.1	1572.5	1646.4	1717.6
	s			0.9661	1.0515	1.1506	1.2328	1.2917	1.3370	1.3743	1.4064	1.4347	1.4604	1.4841	1.5062	1.5471	1.5844

Table A-14. Superheated Steam* (Continued).

Abs. Press. (psi) (Sat. Temp.)		Sat. Water	Sat. Steam	750	800	850	900	950	1000	1050	1100	1150	1200	1250	1300	1400	1500
7000	Sh																
	v			0.0279	0.0334	0.0438	0.0573	0.0704	0.0816	0.0915	0.1004	0.1085	0.1160	0.1231	0.1298	0.1424	0.1542
	h			806.9	901.8	1016.5	1124.9	1212.6	1281.7	1340.5	1392.2	1439.1	1482.6	1523.7	1563.1	1638.6	1711.1
	s			0.9582	1.0350	1.1243	1.2055	1.2689	1.3171	1.3567	1.3904	1.4200	1.4466	1.4710	1.4938	1.5355	1.5735
7500	Sh																
	v			0.0272	0.0318	0.0399	0.0512	0.0631	0.0737	0.0833	0.0918	0.0996	0.1068	0.1136	0.1200	0.1321	0.1433
	h			801.3	889.0	992.9	1097.7	1188.3	1261.0	1322.9	1377.2	1426.0	1471.0	1513.3	1553.7	1630.8	1704.6
	s			0.9514	1.0224	1.1033	1.1818	1.2473	1.2980	1.3397	1.3751	1.4059	1.4335	1.4586	1.4819	1.5245	1.5632
8000	Sh																
	v			0.0267	0.0306	0.0371	0.0465	0.0571	0.0671	0.0762	0.0845	0.0920	0.0989	0.1054	0.1115	0.1230	0.1338
	h			796.6	879.1	974.4	1074.3	1165.4	1241.0	1305.5	1362.2	1413.0	1459.6	1503.1	1544.5	1623.1	1698.1
	s			0.9455	1.0122	1.0864	1.1613	1.2271	1.2798	1.3233	1.3603	1.3924	1.4208	1.4467	1.4705	1.5140	1.5533
8500	Sh																
	v			0.0262	0.0296	0.0350	0.0429	0.0522	0.0615	0.0701	0.0780	0.0853	0.0919	0.0982	0.1041	0.1151	0.1254
	h			792.7	871.2	959.8	1054.5	1144.0	1221.9	1288.5	1347.5	1400.2	1448.2	1492.9	1535.3	1615.4	1691.7
	s			0.9402	1.0037	1.0727	1.1437	1.2084	1.2627	1.3076	1.3460	1.3793	1.4087	1.4352	1.4597	1.5040	1.5439
9000	Sh																
	v			0.0258	0.0288	0.0335	0.0402	0.0483	0.0568	0.0649	0.0724	0.0794	0.0858	0.0918	0.0975	0.1081	0.1179
	h			789.3	864.7	948.0	1037.6	1125.4	1204.1	1272.1	1333.0	1387.5	1437.1	1482.9	1526.3	1607.9	1685.3
	s			0.9354	0.9964	1.0613	1.1285	1.1918	1.2468	1.2926	1.3323	1.3667	1.3970	1.4243	1.4492	1.4944	1.5349
9500	Sh																
	v			0.0254	0.0282	0.0322	0.0380	0.0451	0.0528	0.0603	0.0675	0.0742	0.0804	0.0862	0.0917	0.1019	0.1113
	h			786.4	859.2	938.3	1023.4	1108.9	1187.7	1256.6	1318.9	1375.1	1426.1	1473.1	1517.3	1600.4	1679.0
	s			0.9310	0.9900	1.0516	1.1153	1.1771	1.2320	1.2785	1.3191	1.3546	1.3858	1.4137	1.4392	1.4851	1.5263
10000	Sh																
	v			0.0251	0.0276	0.0312	0.0362	0.0425	0.0495	0.0565	0.0633	0.0697	0.0757	0.0812	0.0865	0.0963	0.1054
	h			783.8	854.5	930.2	1011.3	1094.2	1172.6	1242.0	1305.3	1362.9	1415.3	1463.4	1508.6	1593.1	1672.8
	s			0.9270	0.9842	1.0432	1.1039	1.1638	1.2185	1.2652	1.3065	1.3429	1.3749	1.4035	1.4295	1.4763	1.5180
10500	Sh																
	v			0.0248	0.0271	0.0303	0.0347	0.0404	0.0467	0.0532	0.0595	0.0656	0.0714	0.0768	0.0818	0.0913	0.1001
	h			781.5	850.5	923.4	1001.0	1081.3	1158.9	1228.4	1292.4	1351.1	1404.7	1453.9	1500.0	1585.8	1666.7
	s			0.9232	0.9790	1.0358	1.0939	1.1519	1.2060	1.2529	1.2946	1.3371	1.3644	1.3937	1.4202	1.4677	1.5100

Temperature (°F)

Table A-14. Superheated Steam* (Concluded).

Abs. Press. (psi) (Sat. Temp.)		Sat. Water	Sat. Steam	750	800	850	900	950	1000	1050	1100	1150	1200	1250	1300	1400	1500
11000	v			0.0245	0.0267	0.0296	0.0335	0.0386	0.0443	0.0503	0.0562	0.0620	0.0676	0.0727	0.0776	0.0868	0.0952
	h			779.5	846.9	917.5	992.1	1069.9	1146.3	1215.9	1280.2	1339.7	1394.4	1444.6	1491.5	1578.7	1660.6
	s			0.9196	0.9742	1.0292	1.0851	1.1412	1.1945	1.2414	1.2833	1.3209	1.3544	1.3842	1.4112	1.4595	1.5023
11500	v			0.0243	0.0263	0.0290	0.0325	0.0370	0.0423	0.0478	0.0534	0.0588	0.0641	0.0691	0.0739	0.0827	0.0909
	h			777.7	843.8	912.4	984.5	1059.8	1134.9	1204.3	1268.7	1328.8	1384.4	1435.5	1483.2	1571.8	1654.7
	s			0.9163	0.9698	1.0232	1.0772	1.1316	1.1840	1.2308	1.2727	1.3107	1.3446	1.3750	1.4025	1.4515	1.4949
12000	v			0.0241	0.0260	0.0284	0.0317	0.0357	0.0405	0.0456	0.0508	0.0560	0.0610	0.0659	0.0704	0.0790	0.0869
	h			776.1	841.0	907.9	977.8	1050.9	1124.5	1193.7	1258.0	1318.5	1374.7	1426.6	1475.1	1564.9	1648.8
	s			0.9131	0.9657	1.0177	1.0701	1.1229	1.1742	1.2209	1.2627	1.3010	1.3353	1.3662	1.3941	1.4438	1.4877
12500	v			0.0238	0.0256	0.0279	0.0309	0.0346	0.0390	0.0437	0.0486	0.0535	0.0583	0.0629	0.0673	0.0756	0.0832
	h			774.7	838.6	903.9	971.9	1043.1	1115.2	1184.1	1247.9	1308.8	1365.4	1418.0	1467.2	1558.2	1643.1
	s			0.9101	0.9618	1.0127	1.0637	1.1151	1.1653	1.2117	1.2534	1.2918	1.3264	1.3576	1.3860	1.4363	1.4808
13000	v			0.0236	0.0253	0.0275	0.0302	0.0336	0.0376	0.0420	0.0466	0.0512	0.0558	0.0602	0.0645	0.0725	0.0799
	h			773.5	836.3	900.4	966.8	1036.2	1106.7	1174.8	1238.5	1299.6	1356.5	1409.6	1459.4	1551.6	1637.4
	s			0.9073	0.9582	1.0080	1.0578	1.1079	1.1571	1.2030	1.2445	1.2831	1.3179	1.3494	1.3781	1.4291	1.4741
13500	v			0.0235	0.0251	0.0271	0.0297	0.0328	0.0364	0.0405	0.0448	0.0492	0.0535	0.0577	0.0619	0.0696	0.0768
	h			772.3	834.4	897.2	962.2	1030.0	1099.1	1166.3	1229.7	1291.0	1348.1	1401.5	1451.8	1545.2	1631.9
	s			0.9045	0.9548	1.0037	1.0524	1.1014	1.1495	1.1948	1.2361	1.2749	1.3098	1.3415	1.3705	1.4221	1.4675
14000	v			0.0233	0.0248	0.0267	0.0291	0.0320	0.0354	0.0392	0.0432	0.0474	0.0515	0.0555	0.0595	0.0670	0.0740
	h			771.3	832.6	894.3	958.0	1024.5	1092.3	1158.5	1221.4	1283.0	1340.2	1393.8	1444.4	1538.8	1626.5
	s			0.9019	0.9515	0.9996	1.0473	1.0953	1.1426	1.1872	1.2282	1.2671	1.3021	1.3339	1.3631	1.4153	1.4612
14500	v			0.0231	0.0246	0.0264	0.0287	0.0314	0.0345	0.0380	0.0418	0.0458	0.0496	0.0534	0.0573	0.0646	0.0714
	h			770.4	831.0	891.7	954.3	1019.6	1086.2	1151.4	1213.8	1275.4	1332.9	1386.4	1437.3	1532.6	1621.1
	s			0.8994	0.9484	0.9957	1.0426	1.0897	1.1362	1.1801	1.2208	1.2597	1.2949	1.3266	1.3560	1.4087	1.4551
15000	v			0.0230	0.0244	0.0261	0.0282	0.0308	0.0337	0.0369	0.0405	0.0443	0.0479	0.0516	0.0552	0.0624	0.0690
	h			769.6	829.5	889.3	950.9	1015.1	1080.6	1144.9	1206.8	1268.1	1326.0	1379.4	1430.3	1526.4	1615.9
	s			0.8970	0.9455	0.9920	1.0382	1.0846	1.1302	1.1735	1.2139	1.2525	1.2880	1.3197	1.3491	1.4022	1.4491
15500	v			0.0228	0.0242	0.0258	0.0278	0.0302	0.0329	0.0360	0.0393	0.0429	0.0464	0.0499	0.0534	0.0603	0.0668
	h			768.9	828.2	887.2	947.8	1011.1	1075.7	1139.0	1200.3	1261.1	1319.6	1372.8	1423.6	1520.4	1610.8
	s			0.8946	0.9427	0.9886	1.0340	1.0797	1.1247	1.1674	1.2073	1.2457	1.2815	1.3131	1.3424	1.3959	1.4433

*Sh-superheat, °F; v specific volume, ft³/lb; h enthalpy, Btu/lb; s entropy. Btu/°F · lb.

Source: Copyright 1967 ASME (Abridged): reprinted by permission.

References

1. Verderber, R.R. and O. Morse. Lawrence Berkeley Laboratory, Applied Science Division. "Preformance of Electronic Ballasts and Other New Lighting Equipment (phase II: The 34-Watt F40 Rapid Start T-12 Fluorescent Lamp)." University of California, February, 1988.
2. Morse, O. and R. Verderber. Lawrence Berkeley Laboratory, Applied Science Division. "Cost Effective Lighting." University of California, July, 1987.
3. Hollister, D.D. Lawrence Berkeley Laboratory, Applied Science Division. "Overview of Advances in Light Sources." University of California, June, 1986.
4. Verderber, R. and O. Morse. Lawrence Berkeley Laboratory, Applied Science Division. "Performance of Electronic Ballasts and Other New Lighting Equipment: Final Report." University of California, October, 1985.
5. Lovins, Amory B. "The State of the Art: Space Cooling." Rocky Mountain Institute, Old Snowmass, CO, November, 1986.
6. U.S. Department of Energy, Assistant Secretary, Conservation and Renewable Energy, Office of Buildings and Community Systems. "Energy Conservation Goals for Buildings." Washington, DC, May, 1988.
7. Rubinstein, F., T. Clark, M. Simonovitch, and R. Verderber. Lawrence Berkeley Laboratory, Applied Science Division. "The Effect of Lighting System Components on Lighting Quality, Energy Use, and Life-Cycle Cost." University of California, July, 1986.
8. Manufacturing Energy Consumption Survey: Consumption of Energy 1985 Energy Information Administration, November 18, 1988.
9. Heery Energy Consultants, Inc. "Georgia Cogeneration Handbook." The Governor's Office of Energy Resources, August, 1988.
10. Thermodynamics, New York, F. Sears, Reading, MA, Addison-Wesley Publishing Company, Inc., 1952.
11. Climatic Atlas of the U.S., reprinted under the title: Weather Atlas of the United States, Gale Research Co., 1975.
12. Thumann, Albert, "Plant Engineers and Managers Guide to Energy Conservation - 3rd Edition." Fairmont Press, 1987.
13. Thumann, Albert. "Handbook of Energy Audits - 2nd Edition." Fairmont Press, 1983.
14. Kreider and McNeil. "Waste Heat Management Guidebook." NBS Handbook 121.
15. U.S. Department of Energy, Office of the Assistant Secretary for Conservation and Solar, Office of Building and Community Systems, Oak

Ridge National Laboratory, Oak Ridge, TN, University of Massachusetts Cooperative Extensive Service Energy Education Center, Amherst, MA, The Solar Energy Research Institute, Golden, CO. "Residential Energy Audit Manual." Fairmont Press, 1981.

16. American Society of Heating, Refrigerating and Air-Conditioning Engineers, Inc. "ASHRAE Handbook, 1982 Fundamentals." Chapter 2 - Heat Transfer; Chapter II - Applied Heat Pump Systems; Chapter 23 - Design Heat Transmission Coefficients; Chapter 25 - Heating Load; Chapter 26 - Air Conditioning Cooling Load. Atlanta, GA, 1982.

17. American Society of Heating, Refrigerating and Air-Conditioning Engineers, Inc. "ASHRAE Handbook, 1982 Applications." Chapter 57 - Solar Energy Utilization for Heating and Cooling. Atlanta, GA, 1982.

18. American Society of Heating, Refrigerating and Air-Conditioning Engineers, Inc. "ASHRAE Handbook & Product Directory, 1982 Systems." Chapter 7 - Heat Recovery Systems; Chapter 43 - Energy Estimating Methods. Atlanta, GA, 1982.

19. Trane Air Conditioning Manual. The Trane Company, La Crosse WI.

Index